# Estuarine Acidification

Abhijit Mitra · Sufia Zaman

# Estuarine Acidification

## Exploring the Situation of Mangrove Dominated Indian Sundarban Estuaries

Springer

Abhijit Mitra
Department of Marine Science
University of Calcutta
Kolkata, India

Sufia Zaman
Department of Oceanography
Techno India University
West Bengal, Kolkata, India

ISBN 978-3-030-84791-3 ISBN 978-3-030-84792-0 (eBook)
https://doi.org/10.1007/978-3-030-84792-0



This Springer imprint is published by the registered company Springer Nature Switzerland AG
The registered company address is: Gewerbestrasse 11, 6330 Cham, Switzerland

# Acknowledgements

We are thankful to our scholar pool who raised the following questions that triggered us to conceptualize the theme of estuarine acidification (EA):

1. How can we beat EA and have a well-balanced aquatic ecosystem?
2. Why is EA an adverse by-product of rapid industrial development?
3. How are the population explosion and subsequent urban development related to EA?
4. What is the actual price of regulating service of mangroves in context to carbon scrubbing? And will the expansion of blue carbon be able to reduce the pace of EA?
5. Are there any strategies to manage EA at a local scale?
6. What will be the cost of EA management? And who will pay those bills?

These questions basically pushed us to dive deep in the subject and collect relevant data from various sources.

We are thankful to Dr. Pardis Fazli of Department of Biological and Agricultural Engineering, University Putra, Selangor, Malaysia, for her effort in representing our data in graphical forms.

Finally, Dr. Abhijit Mitra expresses his gratefulness to his wife Shampa, daughter Ankita and late mother Manjulika whose inspirations and encouragements act as a booster to complete the manuscript. The statement of late Dhanesh Chandra Mitra, father of Dr. Abhijit Mitra, to create a strong footprint in the life still boosts the author with extra heavenly energy. The sudden demise of mother of Dr. Mitra during the COVID phase retarded the pace significantly, but with the active support of his well-wisher and Chancellor of Techno India University, West Bengal, Dr. Goutam Roy Chowdhury, the author could finally serve the ground-zero data of Sundarban on the reader's plate.

Dr. Sufia Zaman expresses her deepest gratitude to her mother Mrs. Ayesha Zaman for her unconditional love, and practical day-to-day support from, father, Mr. Salim-uz-Zaman who gave her immense moral support. Dr. Zaman also acknowledges the support of her beloved husband Dr. Sahid Imam Mallick and her baby girl Shanza Mallick. Dr. Zaman wishes to accord her deep sense of gratitude to her family members including her uncle (Mr. Pradip Kumar Mitra) and aunt (Late

Mrs. Kanika Mitra), younger sister (Ms. Sharmilee Zaman), brother-in-law (Kazi Dr. Sazzad Manir), her nephew (baby Diyan Kazi), her in-laws and her beloved grandmother (Late Mrs. Shibani Dhar) for their encouragement and inspiration throughout the strenuous period of manuscript preparation. Dr. Zaman still recalls her granny late Shibani Dhar, who shaped her life and always wished to see her as an author of a subject that can create mark on the society.

Finally, our goal is to analyze the potential of different verticals of blue carbon (like mangroves, saltmarsh grass, sea grass, etc.) in retarding the pace of EA. Our aim is to bring the importance of EA in the climate change domain and draw the attention of busy decision makers, political leaders establishing national priorities, funding officers allocating research projects or young couples wishing to take bath in the coastal water in a sunny day. We hope this book will be a useful source of information for people of all ranks of the society.

# Contents

# About the Authors

**Dr. Abhijit Mitra** Associate Professor and former Head, Department of Marine Science, University of Calcutta (India), has been active in the sphere of oceanography since 1985. He obtained his Ph.D. as NET qualified scholar in 1994. Since then, he joined Calcutta Port Trust and WWF (World Wide Fund for Nature-India), in various capacities to carry out research programs on environmental science, biodiversity conservation, climate change and carbon sequestration. Presently, he is serving as Director, Research of Techno India University, West Bengal. He has to his credit about 625 scientific publications in various national and international journals and 54 books of postgraduate standards. He is presently Member of several committees like PACON International, IUCN, SIOS, Mangrove Society of India, The West Bengal National University of Juridical Sciences, etc., and has successfully completed about 17 projects on biodiversity loss in fishery sector, coastal pollution, aquaculture, alternative livelihood, climate change and carbon sequestration. He also visited as Faculty Member and Invited Speaker in several universities of Singapore, Kenya, Oman and USA. In 2008, he was invited as Visiting Fellow at University of Massachusetts at Dartmouth, USA, to deliver a series of lecture on climate change. He also successfully guided 41 Ph.D. students. Presently, his research areas include environmental science, mangrove ecology, sustainable aquaculture, alternative livelihood, blue economy, climate change and carbon sequestration.

**Dr. Sufia Zaman** presently serving as Head, Department of Oceanography in Techno India University, West Bengal, started her career in the field of Marine Science since 2001. She worked in the rigorous region of Indian Sundarbans and has wide range of experience in exploring the floral and faunal diversity of Sundarbans. She has published five books on carbon sequestration, 265 scientific papers and contributed chapters in several books on biodiversity, environmental science, aquaculture and livelihood development. She is presently Member of Fisheries Society of India. She is also running projects on carbon sequestration by mangroves of Indian Sundarbans. She is Recipient of DST Women Scientist and Jawaharlal Memorial Doctoral fellowship awards. Her areas of research include aquaculture, fish nutrition, phytoplankton diversity, climate change mangrove ecology and alternative livelihood. She is also the first researcher in the maritime state of West Bengal (India), who initiated trial experiments on iron fertilization and subsequent enhancement of primary (phytoplankton) and secondary (fish) productions in the brackish water ponds of Indian Sundarbans with the financial assistance of Department of Science and Technology, Government of India. She is also providing consultancy on green technology to several industries, NGOs and corporate sectors.

# Chapter 1
# Estuarine Acidification: An Overview

**Abstract** Rapid industrialization, unplanned urbanization, deforestation for aquaculture and tourism in coastal areas and change in land use pattern have raised carbon dioxide level in the atmosphere, which on mixing with the coastal and estuarine waters shift the pH value left to the normal value of the aquatic phase. It is now well established that the average pH of the world ocean surface waters has already fallen by about 0.1 units from an average value of about 8.21 to 8.10 since the beginning of the industrial revolution. Many Indian estuaries today exhibit an aquatic pH of 7.9–8.0 compared to the value around 8.3 in early eighties. This has altered the biodiversity spectrum of the estuarine ecosystem particularly the shelled organisms in terms of their community structure, morphology and behavior, which have been displayed in the chapter with specific case studies from Sundarban estuaries. The decadal data bank on aquatic pH in major estuaries of Indian sub-continent is a hallmark of the chapter.

## 1.1 Estuary: Definition and Types

### *1.1.1 Definition*

Estuaries are the zones of intersection between the saline water (of oceans, seas and bays) and fresh water of the rivers. An estuary is a highly dynamic system that exhibits changes in hydrological variables with tides. It is the reservoir of several living and non-living resources that run the wheel of human civilization.

Several definitions of estuary have been forwarded by researchers as highlighted here.

Cameron and Pritchard (1963) defined estuary as "a semi-enclosed coastal body of water which has a free connection with the open sea and within which seawater is measurably diluted with fresh water derived from land drainage".

Pritchard (1967) emphasized on the tidal actions in the estuarine system and stated the ecotone as "a semi-enclosed coastal body of water, which has a free connection with the open sea; it is thus strongly affected by tidal action and within it seawater is mixed with fresh water from land drainage".

 
A. Mitra and S. Zaman, *Estuarine Acidification*,
https://doi.org/10.1007/978-3-030-84792-0_1

**Table 1.1** Physico-chemical characteristics of surface water at high tide condition during 26th May, 2019

| Station no. | Water temperature (°C) | Salinity (psu) | pH | Alkalinity (mg/l) | DO (mg/l) |
|---|---|---|---|---|---|
| 1 | 35.9 | 2.27 | 7.86 | 148 | 5.40 |
| 2 | 35.9 | 4.05 | 7.74 | 159 | 5.17 |
| 3 | 36.1 | 5.78 | 7.72 | 163 | 5.80 |
| 4 | 35.9 | 12.39 | 8.02 | 189 | 5.20 |
| 5 | 35.8 | 11.44 | 8.12 | 175 | 4.78 |
| 6 | 35.8 | 12.99 | 8.10 | 169 | 4.82 |
| 7 | 36.0 | 14.77 | 8.15 | 194 | 4.79 |
| 8 | 35.9 | 14.95 | 8.13 | 189 | 4.66 |
| 9 | 35.9 | 17.08 | 8.17 | 203 | 4.72 |
| 10 | 35.8 | 18.75 | 8.18 | 220 | 4.81 |
| 11 | 36.2 | 19.99 | 8.22 | 229 | 4.91 |
| 12 | 35.9 | 25.96 | 8.28 | 278 | 4.85 |

Perillo (1995) defined estuary as "a semi-enclosed coastal body of water that extends to the effective limit of tidal influence, within which seawater entering from one or more free connections with the open sea or any other saline coastal bodies of water is significantly diluted with fresh water derived from land drainage and can sustain euryhaline biological species, either a part or whole of their life cycle".

The tidal actions in estuaries regulate the physico-chemical variables (preferably the salinity) including the species of heavy metals as all these are functions of salinity. It is observed that exactly in the same location (that can be fixed by GPS), an environmental variable alters significantly with the change in tidal phase.

A study conducted by the authors during May 2019 highlights the spatio-tidal variations of hydrological parameters (considering 12 stations) in the Indian Sundarban Hooghly estuary (a continuation of Ganga–Bhagirathi River system) as depicted in Tables 1.1 and 1.2.

A comparison between Tables 1.1 and 1.2 exhibit significant tidal variations in relation to the hydrological parameters and therefore it is strongly suggested to consider the tidal effects while monitoring the water quality of an estuary.

### *1.1.2 Types of Estuaries*

In standard text books on oceanography or marine science, four basic types of estuaries are usually described (Table 1.3).

**Table 1.2** Physico-chemical characteristics of surface water at low tide condition during 26th May, 2019

| Station no. | Water temperature (°C) | Salinity (psu) | pH | Alkalinity (mg/l) | DO (mg/l) |
|---|---|---|---|---|---|
| 1 | 35.6 | 1.22 | 7.83 | 154 | 6.10 |
| 2 | 35.7 | 2.20 | 7.59 | 167 | 5.94 |
| 3 | 35.7 | 3.58 | 7.78 | 181 | 5.92 |
| 4 | 35.7 | 9.89 | 7.99 | 195 | 5.95 |
| 5 | 35.6 | 9.47 | 8.08 | 178 | 5.50 |
| 6 | 35.5 | 10.45 | 8.07 | 179 | 5.70 |
| 7 | 35.8 | 11.48 | 8.11 | 208 | 5.98 |
| 8 | 35.8 | 12.79 | 8.10 | 199 | 5.80 |
| 9 | 35.5 | 14.90 | 8.16 | 206 | 5.24 |
| 10 | 35.6 | 15.74 | 8.16 | 230 | 5.75 |
| 11 | 35.5 | 17.68 | 8.20 | 240 | 5.77 |
| 12 | 35.6 | 21.42 | 8.25 | 273 | 5.58 |

Pritchard (1967) classified estuaries on the basis of the four important features. They are (a) Salinity (b) Geomorphology (c) Water circulation and stratification and (d) Systems energetics.

On the basis of salinity, estuaries can be grouped into three major categories (Table 1.4).

The dynamic nature of the estuaries is witnessed by the significant spatio-temporal variation of hydrological parameters in the Hooghly-Matla estuarine complex of Indian Sundarbans. The estuarine complex in this mangrove dominated deltaic complex can be sub-divided into three sectors namely western, central and eastern sectors (Fig. 1.1) each with different signatures of salinity (Fig. 1.2a, b, c). More than two decades of data (1984–2013) were compiled from the archives of the Department of Marine Science, University of Calcutta for this study. A number of studies on different aspects of the Sundarban complex have been published over the years, which include description of the data (and methods) at different times over the past three decades (Mitra et al. 1992; Chakraborty and Choudhury 1985; Mitra et al. 1987; Mitra and Choudhury 1994; Saha et al. 1999; Banerjee et al. 2002; Banerjee et al. 2003; Mondal et al. 2006; Mitra et al. 2009; Banerjee et al. 2013; Sengupta et al. 2013; Mitra and Zaman 2016). Real time data (through field sampling by the authors) were also collected simultaneously since 1998 from 18 sampling stations (Table 1.5) in the lower Gangetic region during high tide condition to assure quality and continuity to the data bank. For each observational station, at least 30 samples were collected within 500 m of each other and the mean value of 30 observations was considered for statistical interpretations.

In the western sector, the salinity decrease ranged from 0.58 psu/year (at Jambu Island) to 1.46 psu/ year (at Harinbari) (Fig. 1.2a). Considering all the six stations

**Table 1.3** Types of estuaries and its characteristics

| Name | Figure | Characteristic feature |
|---|---|---|
| Salt wedge estuary | 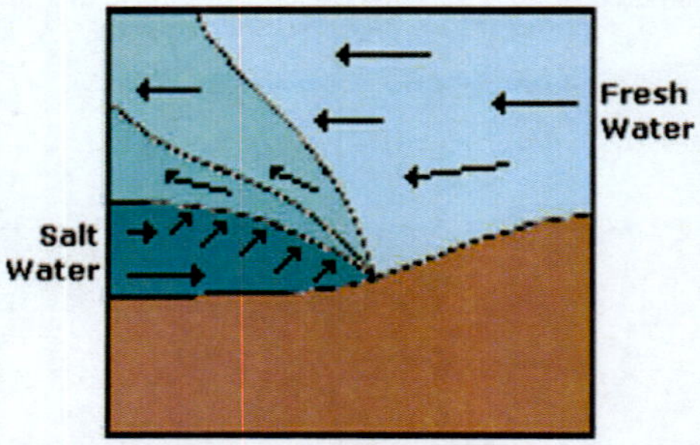  | The high flow rate of the river holds back the lesser flow of salt water. The salt water is drawn upward into the fast-moving river flow |
| Well-mixed estuary | 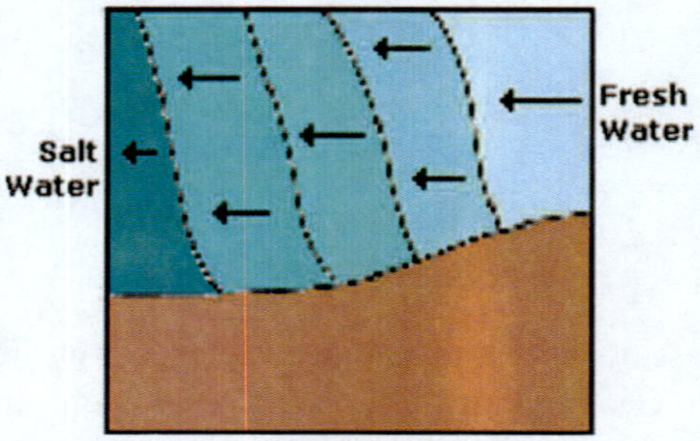  | Strong tidal currents distribute and mix the seawater throughout the shallow estuary. The net flow is weak and seaward at all depths |
| Partially-mixed estuary | 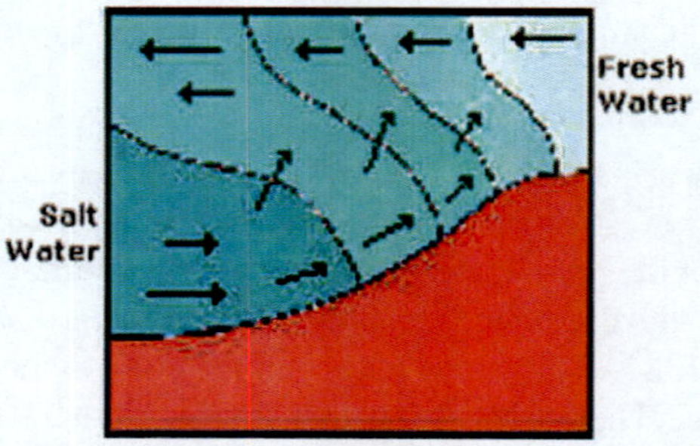  | Seawater enters below the mixed water that is flowing seaward at the surface. Seaward surface net flow is larger than river flow alone |
| Fjord-type estuary | 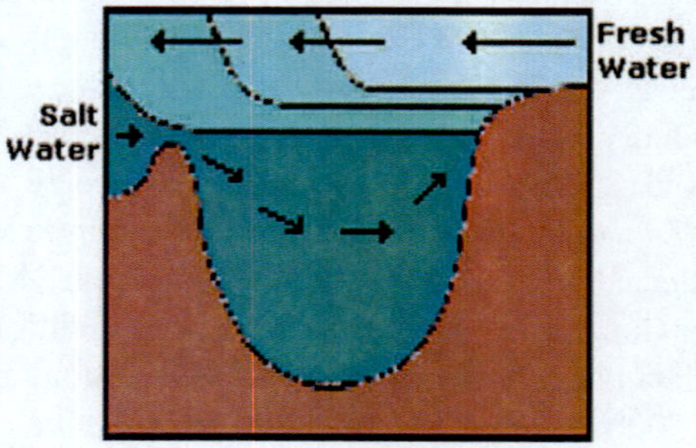  | River water flows seaward over the surface of the deeper seawater and gains salt slowly. The deeper layers may become stagnant due to the slow inflow rate of salt water |

**Table 1.4** Categories of estuary on the basis of salinity

| Category | Description | Example |
|---|---|---|
| Oligohaline | In this type of estuary, the fresh water mixes with the saline water in such proportion that the water ultimately becomes uniformly saline. Such estuaries are common where small rivers meet the seas | Haldi river of West Bengal (India) |
| Mesohaline | In this type of estuary the aquatic phase has medium salinity due to proportional mixing of the fresh and saline water | Mahanadi river of Odisha (in the East coast of India) |
| Polyhaline | Such types of estuaries exhibit distinct spatial variation of salinity owing to the variation in tidal water intrusion into the river mouth. The magnitude/volume of head water discharge also regulates the salinity of estuary | Estuaries of Sundarbans |

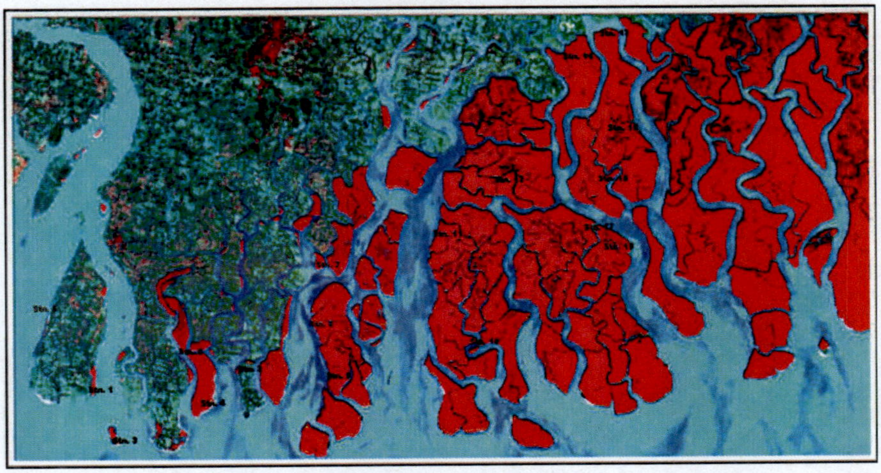

**Fig. 1.1** Location of sector-wise sampling stations in Indian Sundarbans; the red colour indicates the mangrove vegetation (*Source* Mitra and Zaman 2015)

in the western sector, the decadal decrease of salinity is 7.50 psu per decade. In the western sector, the salinity intrusion is confined to 70 km from the mouth even during the dry season. The tidal variation at the mouth is from 6.1 m at springs to 0.22 m at neaps. The fresh water discharge ranges from a peak value of 4250 $m^3\ s^{-1}$ to almost zero in the dry season. The average values of fresh water discharge are 3000 $m^3\ s^{-1}$ during SW monsoon season (June–September) and 1000 $m^3\ s^{-1}$ during a dry season (November–May). Normally, the fresh water discharges are regulated from Farakka barrage to maintain water levels at Calcutta (Biswas 1985; Sadhuram et al. 2005).

The central sector presents a completely reverse picture in terms of aquatic salinity. Irrespective of stations, salinity has increased (Fig. 1.2b between the range 1.05

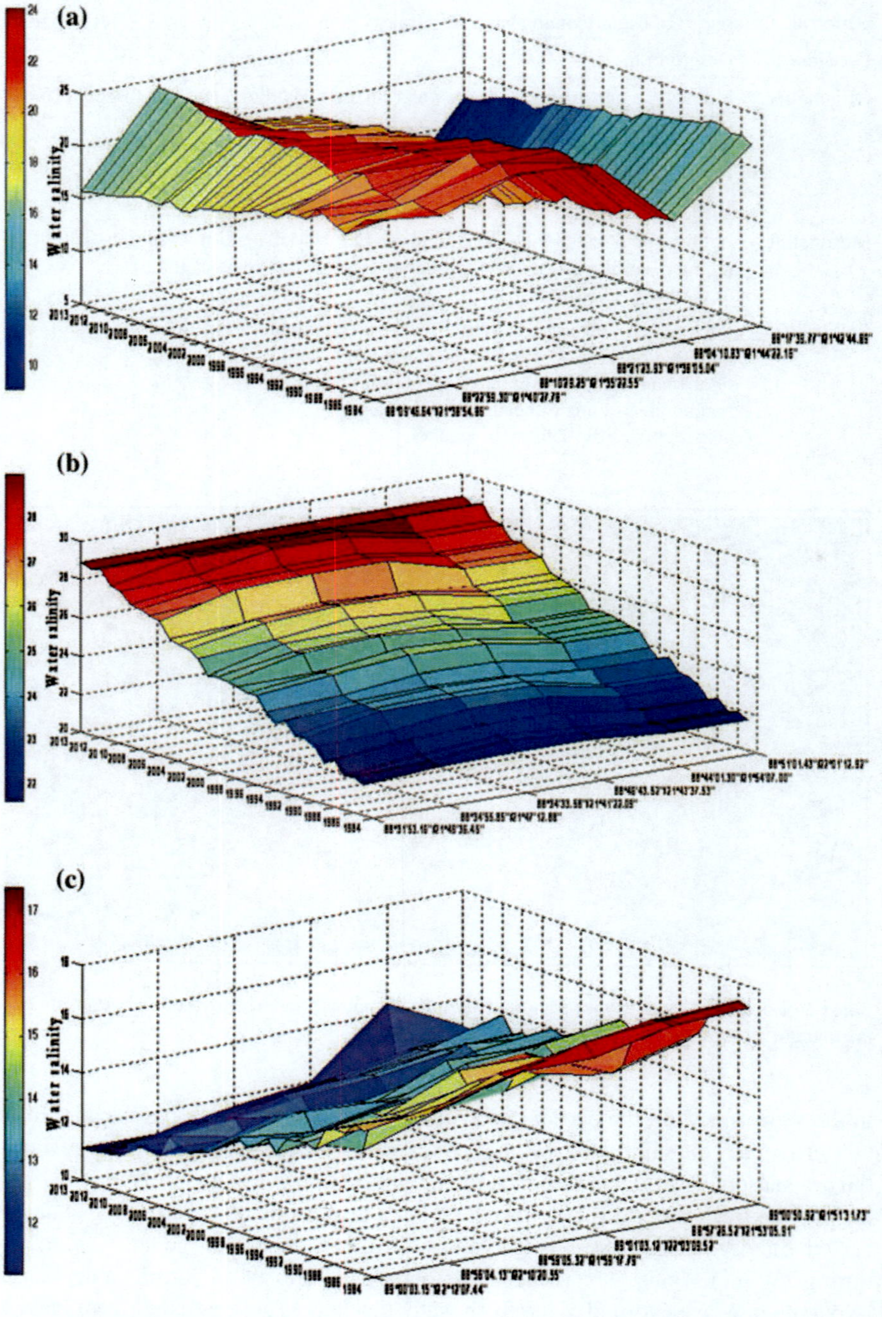


**Fig. 1.2** **a** Surface water salinity in six stations of western Indian Sundarbans. **b** Surface water salinity in six stations of central Indian Sundarbans. **c** Surface water salinity in six stations of eastern Indian Sundarbans (*Source* Mitra and Zaman 2015)

**Table 1.5** Sampling stations in the western, central and eastern sectors of Indian Sundarbans in the lower Gangetic delta region (*Source* Mitra and Zaman 2015)

| Sectors | | Sampling station | Latitude | Longitude |
|---|---|---|---|---|
| Western sector | Stn. 1 | Chemaguri ($W_1$) | 21° 38′ 25.86" N | 88° 08′ 53.55" E |
| | Stn. 2 | Saptamukhi ($W_2$) | 21° 40′ 02.33" N | 88° 23′ 27.18" E |
| | Stn. 3 | Jambu Island ($W_3$) | 21° 35′ 42.03" N | 88° 10′ 22.76" E |
| | Stn. 4 | Lothian ($W_4$) | 21° 38′ 21.20" N | 88° 20′ 29.32" E |
| | Stn. 5 | Harinbari ($W_5$) | 21° 44′ 22.55" N | 88° 04′ 32.97" E |
| | Stn. 6 | Prentice Island ($W_6$) | 21° 42′ 47.88" N | 88° 17′ 55.05" E |
| Central sector | Stn. 7 | Thakuran Char ($C_1$) | 21° 49′ 53.17" N | 88° 31′ 25.57" E |
| | Stn. 8 | Dhulibasani ($C_2$) | 21° 47′ 06.62" N | 88° 33′ 48.20" E |
| | Stn. 9 | Chulkathi ($C_3$) | 21° 41′ 53.62" N | 88° 34′ 10.31" E |
| | Stn. 10 | Goashaba ($C_4$) | 21° 43′ 50.64" N | 88° 46′ 41.44" E |
| | Stn. 11 | Matla ($C_5$) | 21° 53′ 15.30" N | 88° 44′ 08.74" E |
| | Stn. 12 | Pirkhali ($C_6$) | 22° 06′ 00.97" N | 88° 51′ 06.04" E |
| Eastern sector | Stn. 13 | Arbesi ($E_1$) | 22° 11′ 43.14" N | 89° 01′ 09.04" E |
| | Stn. 14 | Jhilla ($E_2$) | 22° 09′ 51.53" N | 88° 57′ 57.07" E |
| | Stn. 15 | Harinbhanga ($E_3$) | 21° 57′ 17.85" N | 88° 59′ 33.24" E |
| | Stn. 16 | Khatuajhuri ($E_4$) | 22° 03′ 06.55" N | 89° 01′ 05.33" E |
| | Stn. 17 | Chamta ($E_5$) | 21° 53′ 18.56" N | 88° 57′ 11.40" E |
| | Stn. 18 | Chandkhali ($E_6$) | 21° 51′ 13.59" N | 89° 00′ 44.68" E |

psu/year (in Chulkathi) to 1.12 psu/year (in Matla and Pirkhali). The average decadal increase of salinity is 13.04 psu per decade.

In the eastern sector, salinity has decreased (Fig. 1.2c) which ranges from 0.54 psu/year (in Chamta) to 0.98 psu/year (in Jhilla). The average decadal decrease of the salinity in this sector is 10.28 psu.

On the basis of geomorphology, estuaries can be grouped into four major categories (Table 1.6).

An estuary can also be classified on the basis of water circulation and stratification as highlighted in Table 1.7.

Estuaries may also be classified as tide dominated and wave dominated (Fig. 1.3) estuary (positive and negative) on the basis of several important hydrological processes (Table 1.8).

Estuaries can be categorized on the basis of systems energetics (Table 1.9).

Impounded estuarine system causes adverse impact on the natural ecosystem services of estuaries as the water flow and tidal actions are obstructed due to which the services offered by estuaries like bioremediation, regulation of biogeochemical cycles, breeding cycle of fishes etc. are significantly impacted. In countries like India and Bangladesh, huge numbers of impounded waterbodies have been created to promote shrimp or tiger prawn (*Penaeus monodon*) culture in the coastal areas,

**Table 1.6** Classification of estuaries on the basis of geomorphology (*Source* Mitra and Zaman 2015)

| Type of estuary | Description | Example |
|---|---|---|
| Drowned river valleys | In this type, the coastlines have relatively low and extensively wide coastal plains | Chesapeake Bay on the mid-Atlantic coast of the United States |
| Fjord type estuaries | The deep U-shaped indentures due glacial erosion with a shallow sill at their mouth formed by terminal glacial deposits | Norway, British Columbia and Alaska |
| Bar-built estuaries | These are shallow basins often partly exposed at low tide enclosed by a chain of offshore bars or barrier islands, broken at intervals by inlets. These bars are sometimes deposited offshore or are remnants of former coastal dunes | North Carolina and Georgia |
| Estuaries produced by tectonic processes | Geological faulting or local subsidence results in the formation of coastal indented structures | San Francisco Bay |

**Table 1.7** Classification of estuaries on the basis of water circulation and stratification (*Source* Mitra and Zaman 2015)

| Type of estuary | Description | Example |
|---|---|---|
| Highly stratified or 'salt-wedge' estuary | When the fresh water discharge from the rivers dominate over the tidal action, the fresh water tends to overflow the heavier salt water which forms a "wedge" extending along the bottom for a considerable distance upstream. This flow of fresh water is again governed by the Coriolis force, which forces the fresh water to move strongly along the right shore (if the observer faces the sea in the Northern Hemisphere and vice versa in the Southern Hemisphere). Such "stratified" or "bi-layered estuary will exhibit a salinity profile with a "halocline" or zone of sharp change in salinity from top to bottom | Mississippi River |
| Partially mixed or moderately stratified estuary | When the fresh water and the tidal water mix in equal proportion due to turbulence caused by periodicity of waves, such type of estuary is formed. Due to this, the energy is dissipated in vertical mixing thus creating a complex pattern of layers and water masses | Chesapeake Bay |
| Completely mixed or vertically homogenous estuary | When the tidal action is more than the fresh water discharge, the water tends to mix well from top to bottom and the salinity is relatively high. When there is wide variation in salinity and temperature, then horizontal estuaries are formed | Bar-built estuaries |

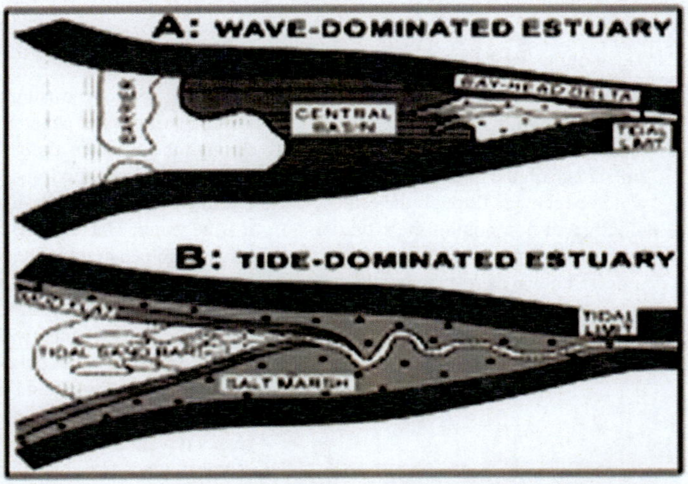


**Fig. 1.3** Wave and tide dominated estuary

due to which the ecology and environment of the surrounding areas have been greatly damaged. The rate of reclamation of marshes and particularly mangrove swamps has accelerated in recent years in some parts of the tropics due to the rapid expansion of pond farming of shrimps for export (Fig. 1.4). About 50% of the mangrove forests in

**Table 1.8** Comparative study of tide and wave dominated estuaries (*Source* Mitra and Zaman 2015)

| Point | Tide dominated estuary | Wave dominated estuary |
|---|---|---|
| Catchment's input | Freshwater entering from the catchments is relatively lower. In negative estuaries, the net inflow of marine water exceeds the outflow of catchment—derived (fresh) water. In such cases, the hypersaline water is usually exported to the ocean | Freshwater enters from the catchments. Although the volume of freshwater input varies spatially and temporally (depending on local catchments and climate conditions), it is often relatively high in positive estuaries |
| Freshwater input | The volume of freshwater entering the estuary is too low to cause significant level of stratification. High tidal ranges may tend to accelerate mixing of any freshwater inputs, and marine water | Water circulation in wave-dominated estuaries generally ranges from well mixed to salinity-stratified, depending on the degree of wave mixing, volume of freshwater input, and climate. 'Positive' wave-dominated estuaries have lower salinity water towards their head. The volume of freshwater causes stratification (or layering) in the water column, which varies with seasonal flow. Buoyant low-salinity fresh water floats above the denser, high-salinity ocean water |

(continued)

**Table 1.8** (continued)

| Point | Tide dominated estuary | Wave dominated estuary |
|---|---|---|
| Salinity | High rates of evaporation cause increases in salinity within the estuary. The resulting high-density hypersaline water sinks beneath the buoyant marine water which penetrates through the estuary mouth, and flows out of the estuarine entrance into the coastal ocean through a process known as reverse stratification. A large degree of mixing occurs between the two layers | A 'salt-wedge' or intrusion of denser saline marine water penetrates through the entrance along the bed of the estuary. Some mixing occurs at the interface between the fresh and marine water. The distance that the salt-wedge penetrates is dependent on tidal range and the amount of fluvial flow received by the estuary. During high fluvial flow events (which may be seasonal), fresh floodwater may push the salt water beyond the mouth. However, the large volume of central basins typical of wave-dominated estuaries tends to reduce this effect |
| Evaporation | Evaporation is the dominant process in negative tide-dominated estuaries due to arid climatic conditions, and the extensive area of shallow intertidal environments. Aridity and the degree of evaporation may vary seasonally; however by definition evaporation in 'negative' estuaries is much larger than freshwater input. Consequently, negative estuaries tend to have longer residence times than positive estuaries | While significant evaporation can occur in wave-dominated estuaries characterized by positive circulation, evaporation (by definition) does not exceed the amount of freshwater input |
| Water exchange | Exchange of sea water and estuarine water occurs through the wide entrance of the estuary. Flood and ebb tides may follow different routes into and out of the estuary, and the tidal prism tends to be large. In negative estuaries, the net inflow of marine water exceeds the outflow of freshwater derived from the catchments. In such cases, the hypersaline water is usually exported to the ocean | Exchange of ocean water and estuarine water occurs through the entrance of the estuary, although the magnitude of exchange depends on the size and length of the entrance channel. In positive wave-dominated estuaries, the outflow of freshwater exceeds the inflow of marine water. During dry conditions, the entrance of the estuary may be intermittently closed |

the Philippines have been converted into brackishwater fish ponds (Saclauso 1989). The area converted in Thailand is estimated to be about 27% and in Ecuador about 13–14%. Such large scale conversions have aroused considerable environmental concern among the public and development agencies.

Tabuchi (2003) estimated that on a global scale, the area under mangroves is shrinking by 100,000 hectares annually due to clear cutting of timber and conversion into aquaculture projects.

**Table 1.9** Classification of estuaries on the basis of systems energetics (*Source* Mitra and Zaman 2015)

| Type | Salient features | Biotic community | Example |
|---|---|---|---|
| Physically stressed estuarine system | (i) Strong tidal currents<br>(ii) Waves with high energy (breaking waves)<br>(iii) Extreme temperature variation<br>(iv) Extreme salinity variation<br>(v) Low DO during night | Species diversity is extremely poor and only limited number of species survive (opportunistic species) with high adaptive potential | Rocky beach |
| Natural polar ecosystem | (i) Extremely low temperature<br>(ii) Light is limiting due to short summer season<br>(iii) Poor productivity | Species diversity of phytoplankton is extremely low and food chain is relatively short; due to extreme coldness, the phytoplankton remain in encysted condition | Arctic and Antarctic coasts |
| Natural temperate coastal ecosystem | (i) Moderate waves, currents and tides<br>(ii) Environmental stress is relatively low<br>(iii) Nutrient level is considerably high<br>(iv) Anthropogenic activities are high due to discharge of wastes from industries, dredging activities etc. | Considerable species richness with presence of salt marshes, seaweeds, kelps, oyster beds and mud flats; dense population of clams and sea worms are observed | Long Island sound, USA |
| Natural tropical coastal ecosystem | (i) Minimum environmental stress<br>(ii) Relatively less wave and tidal actions<br>(iii) Rich in nutrients<br>(iv) Productivity is high<br>(v) Aquaculture is practiced in this ecosystem | Species diversity is extremely high because of congenial environment. Mangroves, sea grass, meadows, saltmarsh grass, coral reefs, seaweed communities and oyster beds are common in this ecosystem | Estuaries of Sundarbans mangrove ecosystem in India and Bangladesh |
| Impounded Estuarine System | (i) Estuaries are impounded at the cost of mangroves and salt marshes (Fig. 1.3)<br>(ii) Estuarine water is stocked and impounded for aquaculture related activities<br>(iii) Highly polluted with organic load and nutrients from shrimp farms<br>(iv) Absence of natural capacity of waste treatment | | Estuaries of Sundarban mangroves (few pockets), Philippines etc. |

**Fig. 1.4** Shrimp culture ponds in mangrove patches: A common scene in tropical mangrove ecosystem; Photograph taken on 04.12.2016 by Ms. Ankita Mitra (Environmentalist and a researcher)

## 1.2 Ecosystem Services of Estuaries

Estuaries are noted for taxonomic diversity of species, biological productivity and unique aesthetic value. The water, sediment and biotic communities of the estuaries offer several ecosystem services that provide both direct and indirect benefits to run the wheel of human civilization. Based on the living and non-living resources of the estuaries the concept of 'Blue Economy' (BE) has emerged as a new paradigm for sustainable development of marine and estuarine resources. Few common ecosystem services on which the pillars of blue economy stand are highlighted here in points:

1. Estuaries are the breeding, nursery and survival zones of a wide spectrum of finfish and shellfish species, which are commercially important and are also included in the export basket of the nation. The estuaries of Sundarbans, for example, are the survival ground for several species of finfish, shrimps and crabs that add foreign currency to the national economy.
2. Estuaries sustain mangroves, seagrasses, saltmarsh grasses and seaweeds, which have the potential to bioaccumulate conservative wastes like heavy metals in their body tissues. Reports of heavy metal accumulation by macrophytes of estuaries are plenty (Mitra et al. 1994; Trivedi et al. 1994), which show

the adaptive capability of estuarine flora to thrive in contaminated media (both water and sediment). Such biological treatments of conservative pollutants almost at zero cost have immense economic and ecological importance.

3. Estuaries are the sites for recycling of nutrients where microbes are the key players. These nutrients (preferably nitrate and phosphate) trigger the growth of phytoplankton and thereby sustain the food chain in the aquatic system.
4. Estuaries offer sustainable fisheries with high commercial value to the local population. In this context the Indian shad, *Tenualosa ilisha* needs to be mentioned as the species has demand both in the internal market and export basket.
5. The mangroves of estuaries provide timber, fuel, wax, honey and fodder to the island dwellers and local people of the fringe villages (Table 1.10). This provisional service of the estuarine vegetation plays a great role in the upliftment of local economy.
7. Estuaries are the home ground for a variety of migratory birds and endangered species that can serve as the foundation of eco-tourism in real sense. The intertidal mudflats, the aquatic system and floral communities of the estuaries attract the tourist with gorgeous and eye catching array of herons, pelicans, storks, eagles, ospreys, plovers, gulls, sandpipers and kingfishers. About 95 varieties of migratory birds come all the way in the estuaries of India from the Caspian Sea in Russia and the northern parts of the Himalayan ranges. The Sundarban estuaries are the home ground of Royal Bengal Tiger (*Panthera tigris tigris*), which can swim across the estuary over a distance of 15 km. The basking crocodiles on the intertidal mudflats of Sundarbans are special attractions for the tourists (Fig. 1.5).

**Table 1.10** Traditional uses of mangrove species

| Mangrove species | Use |
|---|---|
| *Aegiceras corniculatum* | Bark used as fish poison; also contains tannin |
| *Avicennia alba* | Used for fodder and fuel |
| *Avicennia officinalis* | Used for firewood; bark contains tannin |
| *Bruguiera caryophylloides* | Wood used for firewood and timber; bark used as tannin |
| *Bruguiera gymnorhiza* | Wood used for house posts; excellent fuel; bark contains tannin |
| *B. parviflora* | Timber used as fuel; leaves and bark contain tannin |
| *B. sexangula* | Timber used in house building |
| *Ceriops tagal* | Used for keels of boats and house posts. Provides good fuel charcoal. Bark rich in tannin used for dyeing fishing nets |
| *Rhizophora mucronata* | Bark used for tannin and cattle fodder |
| *Sonneratia apetala* | Wood used in house building, packing cases and yields excellent fuel |
| *Xylocarpus granatum* | Yields gum, resin which is used in local medicine; bark contains tannin |

**Fig. 1.5** Crocodile on the estuarine mudflat of Indian Sundarbans

8. The aquatic system of the estuary is extensively used for aquaculture, although in many pockets of lower Gangetic region the adverse impacts of shrimp culture have been documented with facts and figures (Mitra 1998). Aquaculture, preferably shrimp culture has upgraded the local economy in most cases, but in many places mangrove patches and saltmarsh grass beds have been destroyed in an unplanned way to construct shrimp ponds.
9. Estuaries are treated as the dumping sites of various categories of wastes. Deterioration of the water quality and commercially important fauna has been reported from many estuaries of the world. The present authors carried out a survey on the microbial load (Fecal coliform and Total coliform count) in cultured *Penaeus monodon* collected from Henry's Island and Sagar Island of western Indian Sundarbans during 2016. The count was considerably high in the shrimp samples, water and sediment (Table 1.11), which is an indication of deterioration of estuarine environment due to release of untreated wastes from coastal industries, tourism units and fish landing stations existing in the region.
10. The biotic communities of the estuarine system (particularly the mangroves, saltmarsh grass, seaweeds etc.) and the underlying soil act as the reservoir of carbon. This ecosystem service is of great relevance in context to carbon dioxide rise and subsequent climate change. A study conducted by

**Table 1.11** A comparative table showing the microbial load in the ambient media and shrimp muscle sampled from the two study sites in western Indian Sundarbans during the month of May, 2016

| Microbial load | Henry's Island | | Sagar Island | |
|---|---|---|---|---|
| | Total coliform | Fecal coliform | Total coliform | Fecal coliform |
| (a) Water (MPN/100 ml) | 22 | 8 | 4 | 3 |
| (b) Sediment (MPN/ gm) | $25 \times 10^2$ | $18 \times 10^2$ | $3 \times 10^2$ | $2 \times 10^2$ |
| (c) *P. monodon* (MPN/gm) | $6.5 \times 10^3$ | $4.1 \times 10^3$ | $2.9 \times 10^3$ | $1.6 \times 10^3$ |

the present authors (Mitra and Zaman 2016) in 24 stations of Indian Sundarbans (Fig. 1.5) exhibits considerable percentage of soil organic carbon in the intertidal mudflats ranging from 0.66 to 1.41% (Figs. 1.6 and 1.7).

11. India has taken great initiative in the sector of 'Blue Economy' through rational utilization of natural resources of oceans, seas, bays and estuaries. Researchers at Techno India University, West Bengal have developed few innovative food products from the floral species of Sundarban mangrove ecosystem. The salt-marsh grass found abundantly in the intertidal mudflats of estuaries is used for fish feed preparation. Research findings also reveal that fish feed incorporated with estuarine seaweed (*Enteromorpha intestinalis*) has resulted in improved performance, better feed efficiency, better pellet stability and improved animal (fish) product quality, which may be attributed to the enormous genetic potential of these lower group of plants with a genome that is more than twice the size of yeast. This quality has brought the seaweeds in serving many industries such as food, animal feed, cosmetics, pharmaceuticals and biofuels.
12. Flora and fauna of the estuaries are today used as sources of bioactive substances, which have immense importance in the pharmaceutical industries. Examples of **CAL** (Carcinoscorpious Amoebocyte Lysate) and **TAL** (Tachypleus Amoebocyte Lysate) are very relevant in this context. They are derived from the blue blood (amoebocyts) of horseshoe crabs (*Carcinoscorpious rotundicauda* and *Tachypleus gigas*) found abundantly in the estuaries of West Bengal and Odisha, which are the two maritime states in the North-East coast of the Indian sub-continent. CAL and TAL are used for bacterial endotoxin test.
13. Estuarine fishes are the sources of omega-3 Fatty acids, which is good for heart. Coronary heart disease occurs due to atherosclerosis and thrombosis. Atherosclerosis is initiated by endothelial damage that may result from stress related events and may finally lead to thrombosis and arrhythmia, the main cause of sudden death. The omega-3 fatty acid synthesizes with the help of enzymes, eicosanoids. The antithrombotic effect of omega-3 fatty acid is due to these eicosanoids, which inhibit platelet aggregation, vasoconstriction and adhesion and induces vasodialation. Omega-3 fatty acid is also good for diabetic patients as it improves insulin resistance syndrome that leads to heart disease. It is also reported to be good for arthritis.

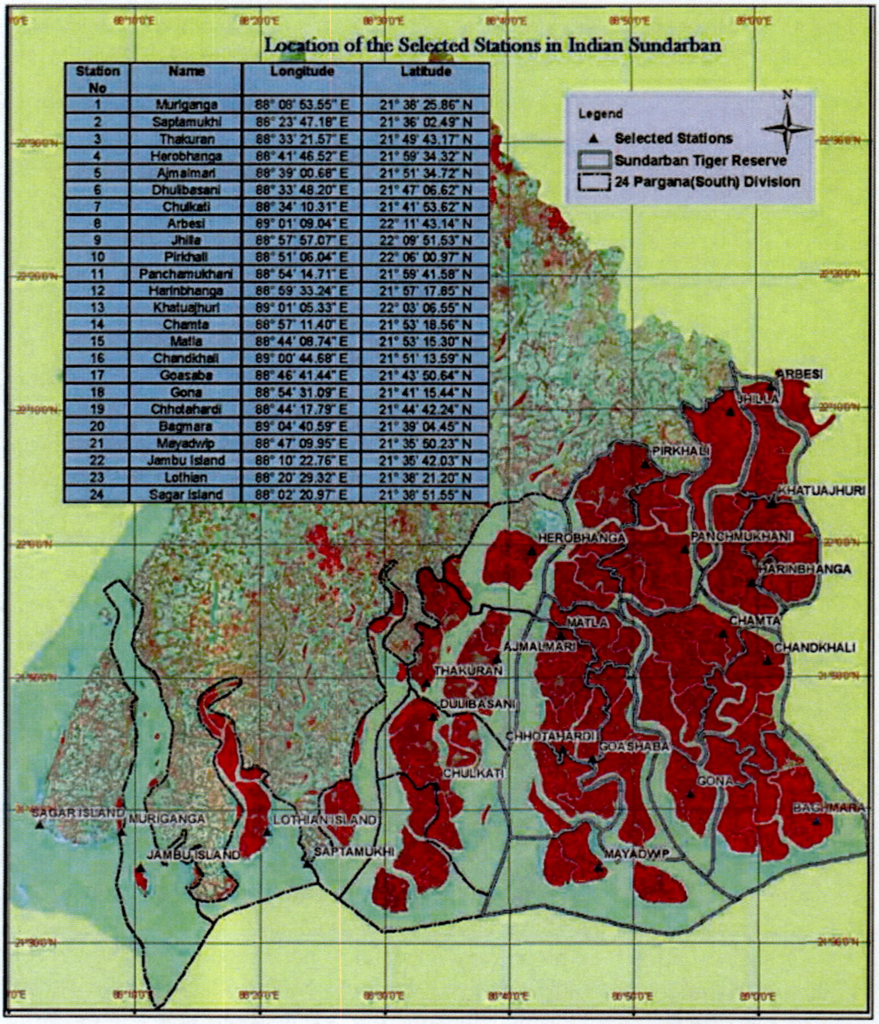


**Fig. 1.6** Stations in Indian Sundarbans selected for organic carbon study (*Source* Mitra and Zaman 2016)

14. Several medicinal plants widely distributed in the coastal and estuarine regions were tested for antiviral, antifungal and mosquito larvicidal activities. In the Parangipettai coast of India (South east coast), 113 medicinal plants have been recorded. Out of 36 plants, 3 plants showed high anti-HIV activity. The anti-HIV activity of coastal population was also tested in comparison with their terrestrial counterparts. In general, coastal populations exhibited better anti-HIV activity than terrestrial ones. 23 plants were screened for antibacterial activity against 13 human pathogenic bacterial strains, of which 23 extracts showed anti-bacterial activity against one or more bacteria. 25 plants

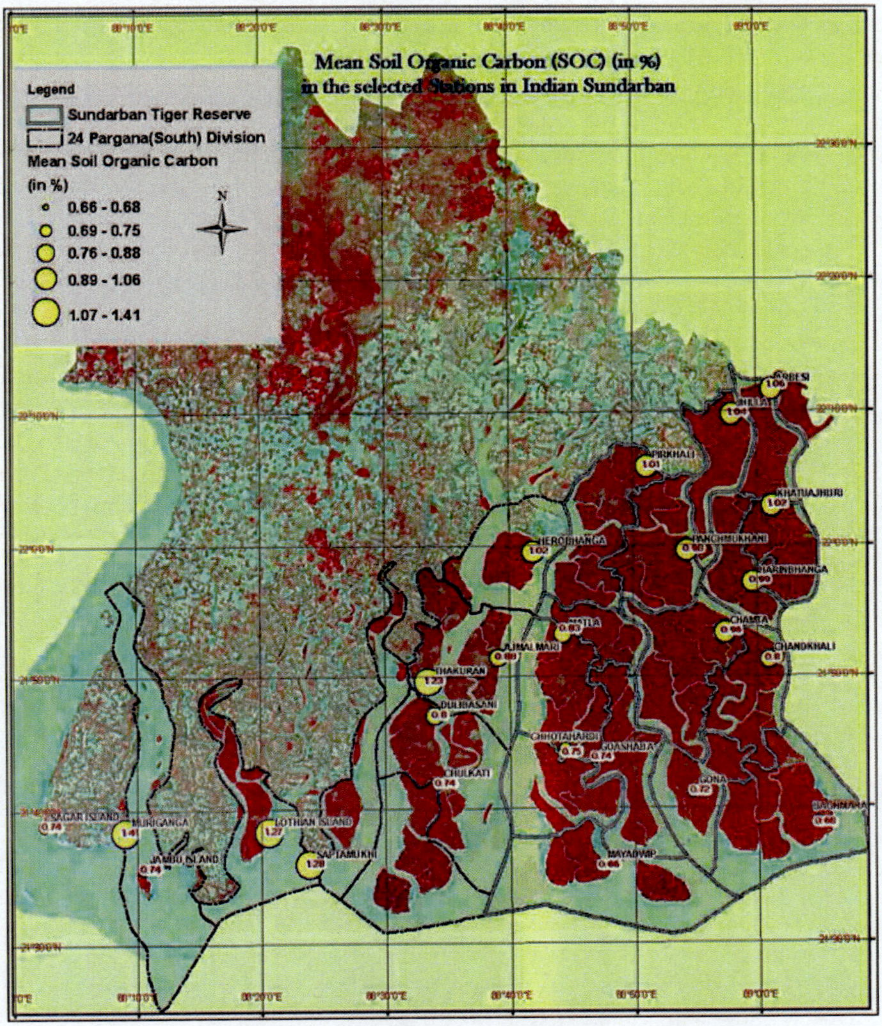


Fig. 1.7 Percentage of organic carbon in the intertidal mudflats of Indian Sundarbans (*Source* Mitra and Zaman 2016)

were screened for antifungal activity against 6 fungi. All the plant extracts examined showed activity against one or more fungi. 2 plant species exhibited high antifungal activity. In general, the coastal samples showed higher activity than terrestrial ones. 11 plants were tested for mosquito larvicidal activity against two mosquito species namely, *Culex quinquefasciatus* and *Aedes aegypti*, of which, 5 samples were effective. In addition to these services, the mangrove flora thriving in the coastal and estuarine region has unique medicinal properties (readers are advised to consult Annexure 1 for detailed information).

15. The estuarine ecosystem and surrounding island villages are unique sites of cultural and religious convergence unlike the urban areas. This may be attributed to extreme need for livelihood and adverse environmental conditions prevailing in the estuarine ecosystem that often break the barrier of caste and religion amongst the people. A relevant example in this context is the religious and cultural canvas of Indian Sundarbans. In this mangrove dominated deltaic complex, people of both religions—Islam and Hindu worship the same deities before entering into the forest for collecting their daily needs. The principal local folk deities are Dakshin Rai, Kalu Rai, Badar Sahib, Bakra Gazi, Sona Pir, Sawal Pir, Gazi Saheb and the first Goddess, Bonbibi (Fig. 1.8). Wood cutters, honey collectors and fishermen worship Dakshin Rai, Narayani Ma, Bonbibi, Gazi Saheb, Kalu Rai, Barkhan Gazi and Sa Janguli. In addition, the Goddess Manasa is worshipped with the belief to protect the people from venomous snakes, and Jagatguru from cobras. Makal Thakur and Biswalakshmi are treated as the God and Goddess of fish respectively. Manik Pir is worshipped for the welfare of cows and Olabibi for protection against cholera.

**Fig. 1.8** Bonbibi, 'the Goddess of forest' worshipped by the local people of Indian Sundarbans irrespective of religion and caste (*Source* Mitra and Zaman 2016)

For protection of the embankments, people refer to Ganga as the supreme Goddess.

Historically, people have depended on bays, estuaries and coastal regions for their living requirements; this has resulted in the establishment of major population centers along the shores of estuaries and coastal seas. In the United States today, over one-half of the population lives within 80 km (50 mi) of the coasts (including the Great Lakes). This population, with its necessary industries, energy-generating facilities, recreational activities, and waste-treatment plants has created a significant negative impact on the coastal and estuarine zones.

The ecosystem services of the estuaries are presently under threat due to rapid industrialization, urbanization and unplanned tourism. Added to the uncontrolled fishing and extraction of natural resources, estuaries and brackish waters are subjected to constant changes in their physico-chemical condition due to various anthropogenic activities. The wide spread reclamation of water areas, construction of barrages and other salt water extraction structure, dredging and aquatic pollution etc. cause substantial negative impact on the positive health of the estuaries. Days are not too far when the ecosystem services of the estuaries will be greatly squeezed.

In general, the total economic valuation of an estuarine system can be calculated as the sum of four components. These are (direct) use value, indirect use value, option value and non-use value (Table 1.12).

For monetary valuation many of these components are easier to measure, as for example, use values are easy to measure because they are observable quantities of products consumed as well as their standard market prices can be utilized to determine the economic value. Recreational use can also be measured from the log book entries maintained in the ecotourism camps, hotels, resorts, sanctuaries, eco parks, or Reserve Forest. The charges for visiting and staying in the heart of Indian Sundarbans are highlighted in Fig. 1.9 and Table 1.13.

**Table 1.12** Components of economic valuation of an estuarine ecosystem

| Type | Description | Example |
|---|---|---|
| Direct use value | Resources that provide direct benefits to the stakeholders | Fishery resources, irrigation facilities, aquaculture activities, existing tourism units etc. |
| Indirect use value | Users get indirect benefits and often some distance away from where they originate | Pollution filtering capacity of estuarine system, flood control, etc. |
| Option value | Users may be willing to pay for acquiring benefits from estuarine resources in future | May be a future source of drinking water or introduction of mangrove tourism |
| Non-use value | Users may be willing to pay for conservation of resources, which will never be used directly | Preservation or protection of endangered species arising out of a sense of environmental stewardship that is unrelated to the indirect and direct uses in the current or in the future |

**Fig. 1.9** Price for visiting and staying in a forest camp in the heart of Indian Sundarbans (*Source* Mitra and Zaman 2016)

**Table 1.13** Approximate price (in $) for traveling and staying in the heart of Indian Sundarbans

| Items | Price ($) |
|---|---|
| Staying charge per head per day | 1 |
| Air condition launch (per day) | 33.34 |
| General launch (per day) | 10 |
| Boat | 5.83 |
| Still camera | Free |
| Video camera | 3.33 |
| Cinema, documentary (per day) | 250 |
| Ethnic cottage | 20 |
| Round cottage (per day) | 16.67 |
| Govt. employee on duty (excluding forest department staffs) (per day) | 6.67 |
| Dormitory (of 10 beds) | 2.5 |
| Educational trip per student (per day) | 0.17 |
| Children under 5 years | Free |

*Note* Conversion to $ has been done on the rate of $1 = INR.60.00

**Fig. 1.10** Mangroves are potential controlling agents of erosion (*Source* Mitra and Zaman 2016)

Certain ecosystem services are extremely difficult to measure as for example the pollution control, erosion control and flood control potential of an estuarine system or the oxygen generation and carbon dioxide sequestration by the estuarine microphytes and macrophytes etc. The mangroves of estuarine system, for example, are noted for erosion control and combat sea level rise (Fig. 1.10). In such cases 'shadow value' are estimated in order to 'price' the produced services.

It is also difficult to measure the option values and non-use values of the estuary as they are not reflected as physical commodities or in any form of observable/measurable items. These values are estimated by using surveys that ask people a series of questions about their willingness to pay for ecosystem services they do not use (Table 1.14).

## 1.3 Acidification of Indian Estuaries

Indian estuaries are mostly subject to high degree of pollution that arises from industrial activities, sewage discharge, aquaculture farms, brick kilns, tourism units etc., and depending on the riverine discharge, the salinity and pollution load exhibit dynamic scenario.

We have made extensive studies on ten major estuaries of Indian sub-continent namely (i) Vellar estuary in southern Tamil Nadu, (ii) Ennore estuary in Tamil Nadu,

**Table 1.14** Economic valuation methods for water

| Method | Approach | Water service appropriate for method | Data needs | Limitations |
|---|---|---|---|---|
| Contingent valuation method | Ask people directly their willingness to pay (WTP) | All use values and non-use value (e.g., drinking water, fishing, protecting species) | Survey with scenario description and questions about WTP for specific services | Potential biases due to hypothetical nature of scenarios |
| Travel cost method | Estimate demand curve from data on travel expenditures | Recreation: boating, fishing, swimming | Survey on expenditures of time and money to travel to specific sites | Only captures recreational benefits; difficult to apply for multiple destination trips |
| Hedonic property value method | Identify contribution of environmental quality to land values | Water quality, wetland services | Property values and characteristics including environmental quality | Requires extensive information about ecosystem services of hundreds of specific sites |
| Change in productivity method | Assess impact of change in water service on produced goods | Commercial fisheries, agricultural uses | Impact of change in water service on production; net values of produced goods | Information on biological impacts of changes in ecosystem services often unavailable |

*Source* Pagiola et al. (2004)

(iii) Adyar estuary in Tamil Nadu, (iv) Rushikulya estuary in Odisha, (v) Kali estuary in Karnataka, (vi) Mandovi estuary in Goa, (vii) Kalapet estuary in Puducherry, (viii) Mahi estuary in Gujarat, (ix) Mindhola estuary in Gujarat, (x) Tapi estuary in Gujarat. The aquatic status of each of these estuaries in terms of pH is discussed here with data.

### *1.3.1 Vellar Estuary*

The Vellar estuary, in the Tamil Nadu State of India meets the Bay of Bengal and is approximately 200 km to the south from the city of Chennai (11° 29′ N, 79° 46′ E). The estuary is made by the Vellar River (which is more than 300 km long) and represents a narrow (200–500 m wide) and rather shallow water (the characteristic depths are 2–4 m) bay having no sharp constrictions or expansions, which is approximately 20 km long. A pronounced constriction is observed only at the entrance of the estuary due to a sandy bar deposited by the surf. The high tide in the mouth of the estuary is about 90 cm above the zero level; the littoral zone is approximately 30–100 m wide. The shores in the seaward part are low; the littoral zone is composed of sand or silted sand with sometimes occurring spots of mangrove bushes. It is pertinent to note that the restoration of mangroves in the estuary was initiated about 20 years ago; previously, the shores were open and the littoral zone was mostly sandy. In the lower part of the estuary, the coasts are populated. In the estuary and neighboring marine area, there is an intensive fishery using nets casted from fishing vessels and trawlers. We made an extensive study at the mouth of the Vellar estuary and observed a decrease in pH value in all the three seasons (Table 1.15).

**Table 1.15** Mean seasonal surface water pH values at the mouth of the Vellar estuary

| Year | Premonsoon high tide | Monsoon high tide | Postmonsoon high tide |
|---|---|---|---|
| 1984 | 8.32 | 8.31 | 8.32 |
| 1985 | 8.32 | 8.30 | 8.31 |
| 1986 | 8.33 | 8.31 | 8.32 |
| 1987 | 8.32 | 8.30 | 8.32 |
| 1988 | 8.33 | 8.31 | 8.33 |
| 1989 | 8.32 | 8.32 | 8.32 |
| 1990 | 8.32 | 8.29 | 8.32 |
| 1991 | 8.32 | 8.31 | 8.32 |
| 1992 | 8.31 | 8.30 | 8.31 |
| 1993 | 8.33 | 8.31 | 8.33 |
| 1994 | 8.32 | 8.31 | 8.32 |
| 1995 | 8.32 | 8.29 | 8.32 |
| 1996 | 8.33 | 8.32 | 8.31 |
| 1997 | 8.31 | 8.31 | 8.32 |
| 1998 | 8.32 | 8.30 | 8.32 |
| 1999 | 8.31 | 8.30 | 8.28 |
| 2000 | 8.31 | 8.29 | 8.31 |
| 2001 | 8.31 | 8.30 | 8.31 |

(continued)

**Table 1.15** (continued)

| Year | Premonsoon high tide | Monsoon high tide | Postmonsoon high tide |
|---|---|---|---|
| 2002 | 8.32 | 8.30 | 8.32 |
| 2003 | 8.31 | 8.30 | 8.30 |
| 2004 | 8.32 | 8.30 | 8.30 |
| 2005 | 8.30 | 8.30 | 8.32 |
| 2006 | 8.32 | 8.29 | 8.31 |
| 2007 | 8.30 | 8.29 | 8.30 |
| 2008 | 8.30 | 8.28 | 8.30 |
| 2009 | 8.34 | 8.29 | 8.28 |
| 2010 | 8.29 | 8.28 | 8.30 |
| 2011 | 8.30 | 8.28 | 8.30 |
| 2012 | 8.28 | 8.27 | 8.28 |
| 2013 | 8.27 | 8.26 | 8.27 |
| 2014 | 8.27 | 8.25 | 8.27 |
| 2015 | 8.26 | 8.23 | 8.25 |
| 2016 | 8.24 | 8.22 | 8.23 |
| 2017 | 8.24 | 8.21 | 8.23 |
| 2018 | 8.23 | 8.20 | 8.22 |
| 2019 | 8.24 | 8.17 | 8.21 |
| Mean | 8.3022 | 8.2833 | 8.2967 |
| Decrease | $-0.0023\ yr^{-1}$ | $-0.0040\ yr^{-1}$ | $-0.0031\ yr^{-1}$ |

### 1.3.2 Ennore Estuary

Ennore is located on the northeast coast of Chennai and Ennore coast consists of alluvial tracts, beach dunes, tidal flats and creek in the eastern part. Ennore comprises of lagoons, with salt marshes and backwaters, which are submerged under water during high tide and form an arm of the sea opening in to the Bay of Bengal. Ennore coast receives major amount of untreated domestic sewage from Royapuram area, untreated or treated industrial effluents from Manali Industrial Belt, which houses many refineries and chemical industries. The dredging activities in Ennore area result in changes in the landscape, sediment transport and dust pollution to the coast by quarrying process taking place. Southern arm of the creek is well developed with industries, utilities, suburban residential areas and fishing hamlets. Northern section of the creek is connected to the Pulicat lagoon and has two major developments North Chennai Thermal power Core (NCTPS) and Ennore satellite port has chocked the mouth of the Ennore creek. Raw municipal sewage, industrial trade effluents industrial cooling waters all of them make it through Buckingham canal, enters into Ennore estuary and eventually drains into the Bay of Bengal of Chennai coast. The surface water pH values at the mouth of Ennore estuary is presented in Table 1.16.

**Table 1.16** Mean seasonal surface water pH values at the mouth of the Ennore estuary

| Year | Premonsoon Hhigh tide | Monsoon high tide | Postmonsoon high tide |
|---|---|---|---|
| 1984 | 8.34 | 8.32 | 8.33 |
| 1985 | 8.33 | 8.33 | 8.34 |
| 1986 | 8.34 | 8.32 | 8.33 |
| 1987 | 8.33 | 8.31 | 8.33 |
| 1988 | 8.34 | 8.32 | 8.34 |
| 1989 | 8.33 | 8.33 | 8.33 |
| 1990 | 8.33 | 8.30 | 8.33 |
| 1991 | 8.33 | 8.32 | 8.33 |
| 1992 | 8.32 | 8.31 | 8.32 |
| 1993 | 8.34 | 8.32 | 8.34 |
| 1994 | 8.33 | 8.32 | 8.33 |
| 1995 | 8.33 | 8.30 | 8.33 |
| 1996 | 8.34 | 8.33 | 8.32 |
| 1997 | 8.32 | 8.32 | 8.33 |
| 1998 | 8.33 | 8.32 | 8.33 |
| 1999 | 8.32 | 8.31 | 8.29 |
| 2000 | 8.32 | 8.30 | 8.32 |
| 2001 | 8.32 | 8.31 | 8.32 |
| 2002 | 8.33 | 8.31 | 8.33 |
| 2003 | 8.32 | 8.31 | 8.31 |
| 2004 | 8.33 | 8.31 | 8.31 |
| 2005 | 8.31 | 8.31 | 8.33 |
| 2006 | 8.33 | 8.30 | 8.32 |
| 2007 | 8.31 | 8.30 | 8.31 |
| 2008 | 8.31 | 8.29 | 8.31 |
| 2009 | 8.35 | 8.30 | 8.29 |
| 2010 | 8.30 | 8.29 | 8.31 |
| 2011 | 8.31 | 8.29 | 8.31 |
| 2012 | 8.29 | 8.28 | 8.29 |
| 2013 | 8.28 | 8.27 | 8.28 |
| 2014 | 8.28 | 8.26 | 8.28 |
| 2015 | 8.27 | 8.24 | 8.26 |
| 2016 | 8.25 | 8.23 | 8.24 |

(continued)

Table 1.16 (continued)

| Year | Premonsoon Hhigh tide | Monsoon high tide | Postmonsoon high tide |
|---|---|---|---|
| 2017 | 8.25 | 8.22 | 8.24 |
| 2018 | 8.26 | 8.21 | 8.23 |
| 2019 | 8.27 | 8.21 | 8.23 |
| **Mean** | 8.3136 | 8.2950 | 8.3075 |
| **Decrease** | −0.0020 $yr^{-1}$ | −0.0031 $yr^{-1}$ | −0.0029 $yr^{-1}$ |

## *1.3.3 Adyar Estuary*

The present study was carried out for Adyar estuary in the coastal water of East coast of Tamil Nadu. The Adyar estuary is situated in the southern part of Chennai city at Lat. 13° 06′ N, Long 80° 18′ E of south east coast of India. The mean seasonal data of surface water pH at the mouth of the Adayar estuary is presented in Table 1.17.

**Table 1.17** Mean seasonal surface water pH values at the mouth of the Adyar estuary

| Year | Premonsoon high tide | Monsoon high tide | Postmonsoon high tide |
|---|---|---|---|
| 1984 | 8.34 | 8.31 | 8.33 |
| 1985 | 8.34 | 8.34 | 8.35 |
| 1986 | 8.33 | 8.33 | 8.34 |
| 1987 | 8.34 | 8.32 | 8.34 |
| 1988 | 8.33 | 8.33 | 8.35 |
| 1989 | 8.34 | 8.34 | 8.34 |
| 1990 | 8.34 | 8.31 | 8.34 |
| 1991 | 8.34 | 8.33 | 8.34 |
| 1992 | 8.33 | 8.32 | 8.33 |
| 1993 | 8.35 | 8.33 | 8.35 |
| 1994 | 8.34 | 8.33 | 8.34 |
| 1995 | 8.34 | 8.31 | 8.34 |
| 1996 | 8.35 | 8.34 | 8.33 |
| 1997 | 8.33 | 8.33 | 8.34 |
| 1998 | 8.34 | 8.33 | 8.34 |
| 1999 | 8.33 | 8.32 | 8.30 |
| 2000 | 8.33 | 8.31 | 8.33 |
| 2001 | 8.33 | 8.32 | 8.33 |
| 2002 | 8.34 | 8.32 | 8.34 |
| 2003 | 8.33 | 8.32 | 8.32 |
| 2004 | 8.34 | 8.32 | 8.32 |

(continued)

**Table 1.17** (continued)

| Year | Premonsoon high tide | Monsoon high tide | Postmonsoon high tide |
|---|---|---|---|
| 2005 | 8.32 | 8.32 | 8.34 |
| 2006 | 8.34 | 8.31 | 8.33 |
| 2007 | 8.32 | 8.31 | 8.32 |
| 2008 | 8.32 | 8.30 | 8.32 |
| 2009 | 8.36 | 8.31 | 8.30 |
| 2010 | 8.31 | 8.30 | 8.32 |
| 2011 | 8.32 | 8.30 | 8.32 |
| 2012 | 8.30 | 8.29 | 8.30 |
| 2013 | 8.29 | 8.28 | 8.29 |
| 2014 | 8.29 | 8.27 | 8.29 |
| 2015 | 8.28 | 8.25 | 8.27 |
| 2016 | 8.26 | 8.24 | 8.25 |
| 2017 | 8.26 | 8.23 | 8.25 |
| 2018 | 8.25 | 8.22 | 8.24 |
| 2019 | 8.26 | 8.19 | 8.23 |
| **Mean** | 8.3211 | 8.3036 | 8.3169 |
| **Decrease** | $-0.0023\ yr^{-1}$ | $-0.0034\ yr^{-1}$ | $-0.0029\ yr^{-1}$ |

### *1.3.4 Rushikulya Estuary*

River Rushikulya originates from the Rushimala mountains of Eastern Ghat. Geographically it located in the state of Odisha between 19° 22′–19° 24′ N and 85° 02′–85° 05′ E and discharges into Bay of Bengal near Ganjam. The estuary has a depth of 3.2 m and it is influenced by the semidiurnal tide. The nutrient supply in the estuary is mainly from freshwater flow. The estuarine ecosystem is facing a serious warning in recent time due to natural hazards and mean sea level rise phenomenon. The frequent storms at Odisha Coast have driven wave further inland into the deeper estuarine zone. At Rushikulya estuary, a long sand spit which was running parallel to the coast, separating the estuary from the sea, was eroded significantly due to the strong surge exerted by cyclone Phailin, which affected many species of fishes, shrimps and birds. Hence the cyclonic surge-induced coastal morphological changes could be one of the major reasons for decline in the number (failure of mass nesting) of these migratory endangered species. The surface water pH values at the confluence of the estuary and Bay of Bengal are presented in Table 1.18.

**Table 1.18** Mean seasonal surface water pH values at the mouth of the Rushikulya estuary

| Year | Premonsoon High tide | Monsoon High tide | Postmonsoon High tide |
|---|---|---|---|
| 1984 | 8.29 | 8.26 | 8.29 |
| 1985 | 8.29 | 8.27 | 8.28 |
| 1986 | 8.30 | 8.26 | 8.29 |
| 1987 | 8.29 | 8.27 | 8.29 |
| 1988 | 8.30 | 8.28 | 8.30 |
| 1989 | 8.29 | 8.29 | 8.29 |
| 1990 | 8.29 | 8.26 | 8.29 |
| 1991 | 8.29 | 8.28 | 8.29 |
| 1992 | 8.28 | 8.27 | 8.28 |
| 1993 | 8.30 | 8.28 | 8.30 |
| 1994 | 8.29 | 8.28 | 8.29 |
| 1995 | 8.29 | 8.26 | 8.29 |
| 1996 | 8.30 | 8.29 | 8.28 |
| 1997 | 8.28 | 8.28 | 8.29 |
| 1998 | 8.29 | 8.28 | 8.29 |
| 1999 | 8.28 | 8.27 | 8.25 |
| 2000 | 8.28 | 8.26 | 8.28 |
| 2001 | 8.28 | 8.27 | 8.28 |
| 2002 | 8.29 | 8.27 | 8.29 |
| 2003 | 8.28 | 8.27 | 8.27 |
| 2004 | 8.29 | 8.27 | 8.27 |
| 2005 | 8.27 | 8.27 | 8.29 |
| 2006 | 8.29 | 8.26 | 8.28 |
| 2007 | 8.27 | 8.26 | 8.27 |
| 2008 | 8.27 | 8.25 | 8.27 |
| 2009 | 8.31 | 8.26 | 8.25 |
| 2010 | 8.26 | 8.25 | 8.27 |
| 2011 | 8.27 | 8.25 | 8.27 |
| 2012 | 8.25 | 8.24 | 8.25 |
| 2013 | 8.24 | 8.23 | 8.24 |
| 2014 | 8.24 | 8.22 | 8.24 |
| 2015 | 8.23 | 8.20 | 8.22 |
| 2016 | 8.21 | 8.19 | 8.20 |
| 2017 | 8.21 | 8.18 | 8.21 |
| 2018 | 8.22 | 8.17 | 8.22 |

(continued)

**Table 1.18** (continued)

| Year | Premonsoon High tide | Monsoon High tide | Postmonsoon High tide |
|---|---|---|---|
| 2019 | 8.23 | 8.14 | 8.21 |
| **Mean** | 8.2733 | 8.2525 | 8.2686 |
| **Decrease** | $-0.0017\ yr^{-1}$ | $-0.0034\ yr^{-1}$ | $-0.0023\ yr^{-1}$ |

### *1.3.5 Kali Estuary*

The River Kali is northern most important estuarine system in Uttara Kannada district of Karnataka state. The estuary has considerable impact on adjoining coastal waters and also influences the fishery, fish and shellfish seeds and molluscan resources. The estuary spreads approximately about 23 km in axis until the edge of the elevated plateau of Sahyadri (14° 54′ 25" N and 74° 19′ 30" E) and drains into the Karwar Bay (14° 50′ 21" N and 74° 10′ 05" E) on the central west coast of India. An extensive survey made on the aquatic pH at estuarine mouth shows a decreasing trend (Table 1.19).

**Table 1.19** Mean seasonal surface water pH values at the mouth of the Kali estuary

| Year | Premonsoon high tide | Monsoon high tide | Postmonsoon high tide |
|---|---|---|---|
| 1984 | 8.35 | 8.33 | 8.34 |
| 1985 | 8.34 | 8.34 | 8.35 |
| 1986 | 8.35 | 8.33 | 8.34 |
| 1987 | 8.34 | 8.32 | 8.34 |
| 1988 | 8.35 | 8.33 | 8.35 |
| 1989 | 8.34 | 8.34 | 8.34 |
| 1990 | 8.34 | 8.31 | 8.34 |
| 1991 | 8.34 | 8.33 | 8.34 |
| 1992 | 8.33 | 8.32 | 8.33 |
| 1993 | 8.35 | 8.33 | 8.35 |
| 1994 | 8.34 | 8.33 | 8.34 |
| 1995 | 8.34 | 8.31 | 8.34 |
| 1996 | 8.35 | 8.34 | 8.33 |
| 1997 | 8.33 | 8.33 | 8.34 |
| 1998 | 8.34 | 8.33 | 8.34 |
| 1999 | 8.33 | 8.32 | 8.30 |

(continued)

**Table 1.19** (continued)

| Year | Premonsoon high tide | Monsoon high tide | Postmonsoon high tide |
|---|---|---|---|
| 2000 | 8.33 | 8.31 | 8.33 |
| 2001 | 8.33 | 8.32 | 8.33 |
| 2002 | 8.34 | 8.32 | 8.34 |
| 2003 | 8.33 | 8.32 | 8.32 |
| 2004 | 8.34 | 8.32 | 8.32 |
| 2005 | 8.32 | 8.32 | 8.34 |
| 2006 | 8.34 | 8.31 | 8.33 |
| 2007 | 8.32 | 8.31 | 8.32 |
| 2008 | 8.32 | 8.30 | 8.32 |
| 2009 | 8.36 | 8.31 | 8.30 |
| 2010 | 8.31 | 8.30 | 8.32 |
| 2011 | 8.32 | 8.30 | 8.32 |
| 2012 | 8.30 | 8.29 | 8.30 |
| 2013 | 8.29 | 8.28 | 8.29 |
| 2014 | 8.29 | 8.27 | 8.29 |
| 2015 | 8.28 | 8.25 | 8.27 |
| 2016 | 8.26 | 8.24 | 8.25 |
| 2017 | 8.26 | 8.23 | 8.25 |
| 2018 | 8.25 | 8.22 | 8.24 |
| 2019 | 8.27 | 8.20 | 8.22 |
| **Mean** | 8.3228 | 8.3044 | 8.3169 |
| **Decrease** | $-0.0023\ yr^{-1}$ | $-0.0037\ yr^{-1}$ | $-0.0034\ yr^{-1}$ |

## 1.3.6 *Mandovi Estuary*

The Mandovi River/Mahadayi River is described as the lifeline of the Indian state of Goa. The river has a length of ~80 km, 29 km in Karnataka and 52 km in Goa. It originates from a cluster of 30 springs at Bhimgad in the Western Ghats in the Belgaum district of Karnataka. The river has a 2032 $km^2$ catchment area in Karnataka and a 1580 $km^2$ catchment area in Goa. With its cerulean waters, Dudhsagar Falls and Varapoha Falls, it is also known as the Gomati in a few places. The Mandovi and the Zuari are the two primary rivers in the state of Goa.

The Mandovi enters Goa from the north via the Sattari Taluka in Goa and from Uttara Kannada District of Karnataka near the Castle Rock Railway station. The Mandovi flows through Belgaum, Uttara Kannada in Karnataka and Cumbarjua, Divadi and Chodné in Goa, eventually pouring into the Arabian Sea. Mandovi joins with the Zuari at a common point at Cabo Aguada, forming the Mormugao harbour. The river Mapusa is a tributary of the Mandovi. We made several visits at Mandovi

**Table 1.20** Mean seasonal surface water pH values at the mouth of the Mandovi estuary

| Year | Premonsoon high tide | Monsoon high tide | Postmonsoon high tide |
|---|---|---|---|
| 1984 | 8.33 | 8.31 | 8.32 |
| 1985 | 8.32 | 8.32 | 8.33 |
| 1986 | 8.33 | 8.31 | 8.32 |
| 1987 | 8.32 | 8.30 | 8.32 |
| 1988 | 8.33 | 8.31 | 8.33 |
| 1989 | 8.32 | 8.32 | 8.32 |
| 1990 | 8.32 | 8.29 | 8.32 |
| 1991 | 8.32 | 8.31 | 8.32 |
| 1992 | 8.31 | 8.30 | 8.31 |
| 1993 | 8.33 | 8.31 | 8.33 |
| 1994 | 8.32 | 8.31 | 8.32 |
| 1995 | 8.32 | 8.29 | 8.32 |
| 1996 | 8.33 | 8.32 | 8.31 |
| 1997 | 8.31 | 8.31 | 8.32 |
| 1998 | 8.32 | 8.31 | 8.32 |
| 1999 | 8.31 | 8.30 | 8.28 |
| 2000 | 8.31 | 8.29 | 8.31 |
| 2001 | 8.31 | 8.30 | 8.31 |
| 2002 | 8.32 | 8.30 | 8.32 |
| 2003 | 8.31 | 8.30 | 8.30 |
| 2004 | 8.32 | 8.30 | 8.30 |
| 2005 | 8.30 | 8.30 | 8.32 |
| 2006 | 8.32 | 8.29 | 8.31 |
| 2007 | 8.30 | 8.29 | 8.30 |
| 2008 | 8.30 | 8.28 | 8.30 |
| 2009 | 8.34 | 8.29 | 8.28 |
| 2010 | 8.29 | 8.28 | 8.30 |
| 2011 | 8.30 | 8.28 | 8.30 |
| 2012 | 8.28 | 8.27 | 8.28 |
| 2013 | 8.27 | 8.26 | 8.27 |
| 2014 | 8.27 | 8.25 | 8.27 |
| 2015 | 8.26 | 8.23 | 8.25 |
| 2016 | 8.24 | 8.22 | 8.23 |

(continued)

**Table 1.20** (continued)

| Year | Premonsoon high tide | Monsoon high tide | Postmonsoon high tide |
|---|---|---|---|
| 2017 | 8.24 | 8.21 | 8.23 |
| 2018 | 8.23 | 8.20 | 8.22 |
| 2019 | 8.24 | 8.20 | 8.22 |
| **Mean** | 8.3025 | 8.2850 | 8.2975 |
| **Decrease** | $-0.0026\ yr^{-1}$ | $-0.0031\ yr^{-1}$ | $-0.0029\ yr^{-1}$ |

estuary and made an extensive survey on aquatic pH at the mouth of estuary. The seasonal data are highlighted in Table 1.20.

### *1.3.7 Kalapet Estuary*

The study area is the region along Puducherry situated on the east coast of India, between 79.87° E and 79.79° E longitudes and 12.05° N and 11.75° N latitudes. Nearly 40% of the land area is used for industrial, residential and commercial purposes. Fishing villages are scattered along the estuary.

The three major physiographic units generally observed are coastal plain (younger and older), alluvial plain and uplands. The entire area, except the northeastern corner, is mostly covered by sedimentary formations ranging in age from cretaceous to recent. The Kalapet coast of Puducherry has been selected in the present study and a declining trend of surface water pH is observed in the present study (Table 1.21).

**Table 1.21** Mean seasonal surface water pH values at the mouth of the Kalapet estuary/Kalapet coast

| Year | Premonsoon high tide | Monsoon high tide | Postmonsoon high tide |
|---|---|---|---|
| 1984 | 8.34 | 8.32 | 8.33 |
| 1985 | 8.33 | 8.33 | 8.34 |
| 1986 | 8.34 | 8.32 | 8.33 |
| 1987 | 8.33 | 8.31 | 8.33 |
| 1988 | 8.34 | 8.32 | 8.34 |
| 1989 | 8.33 | 8.33 | 8.33 |
| 1990 | 8.33 | 8.30 | 8.33 |
| 1991 | 8.33 | 8.32 | 8.33 |
| 1992 | 8.32 | 8.31 | 8.32 |
| 1993 | 8.34 | 8.32 | 8.34 |
| 1994 | 8.33 | 8.32 | 8.33 |
| 1995 | 8.33 | 8.30 | 8.33 |
| 1996 | 8.34 | 8.33 | 8.32 |

(continued)

**Table 1.21** (continued)

| Year | Premonsoon high tide | Monsoon high tide | Postmonsoon high tide |
|---|---|---|---|
| 1997 | 8.32 | 8.32 | 8.33 |
| 1998 | 8.33 | 8.32 | 8.33 |
| 1999 | 8.32 | 8.31 | 8.29 |
| 2000 | 8.32 | 8.30 | 8.32 |
| 2001 | 8.32 | 8.31 | 8.32 |
| 2002 | 8.33 | 8.31 | 8.33 |
| 2003 | 8.32 | 8.31 | 8.31 |
| 2004 | 8.33 | 8.31 | 8.31 |
| 2005 | 8.31 | 8.31 | 8.33 |
| 2006 | 8.33 | 8.30 | 8.32 |
| 2007 | 8.31 | 8.30 | 8.31 |
| 2008 | 8.31 | 8.29 | 8.31 |
| 2009 | 8.35 | 8.30 | 8.29 |
| 2010 | 8.30 | 8.29 | 8.31 |
| 2011 | 8.31 | 8.29 | 8.31 |
| 2012 | 8.29 | 8.28 | 8.29 |
| 2013 | 8.28 | 8.27 | 8.28 |
| 2014 | 8.28 | 8.26 | 8.28 |
| 2015 | 8.27 | 8.24 | 8.26 |
| 2016 | 8.25 | 8.23 | 8.24 |
| 2017 | 8.25 | 8.22 | 8.24 |
| 2018 | 8.24 | 8.21 | 8.23 |
| 2019 | 8.25 | 8.19 | 8.22 |
| **Mean** | 8.3125 | 8.2944 | 8.3072 |
| **Decrease** | $-0.0026\ yr^{-1}$ | $-0.0037\ yr^{-1}$ | $-0.0031\ yr^{-1}$ |

### *1.3.8 Mahi Estuary*

Mahi River is one of the major rivers of Gujarat. The estuarine stretch extends up to 50 km upstream (Fajalpur latitude 22° 26′ N and longitude 73° 04′ E) and opens into Gulf of Khambhat at Kamboi (latitude 22° 12′ N and longitude 72° 36′ E). Being funnel shaped (width 200 km at mouth of Gulf terminating to 6 km at the extreme end of Gulf, the Mahi estuary); the Gulf of Khambhat shows high tidal amplitude (up to 12 m), extreme water current and churning of bed material. Moreover, sediment input from major rivers like Narmada, Tapi and Mahi contribute to high turbidity. Owning to these peculiarities, the geomorphology and hydrodynamics of the Gulf of Khambhat makes the estuary very specific in terms of sedimentology as well as water quality. The downstream area can be recognized as estuarine mudflats with silty-clayey composition while, the upstream area has coarse sand and gravel bed. The seasonal variation of surface water pH showed a declining trend (Table 1.22).

**Table 1.22** Mean seasonal surface water pH values at the mouth of the Mahi estuary

| Year | Premonsoon high tide | Monsoon High tide | Postmonsoon High tide |
|---|---|---|---|
| 1984 | 8.35 | 8.33 | 8.34 |
| 1985 | 8.34 | 8.34 | 8.35 |
| 1986 | 8.35 | 8.33 | 8.34 |
| 1987 | 8.34 | 8.32 | 8.34 |
| 1988 | 8.35 | 8.33 | 8.35 |
| 1989 | 8.34 | 8.34 | 8.34 |
| 1990 | 8.34 | 8.31 | 8.34 |
| 1991 | 8.34 | 8.33 | 8.34 |
| 1992 | 8.33 | 8.32 | 8.33 |
| 1993 | 8.35 | 8.33 | 8.35 |
| 1994 | 8.34 | 8.33 | 8.34 |
| 1995 | 8.34 | 8.31 | 8.34 |
| 1996 | 8.35 | 8.34 | 8.33 |
| 1997 | 8.33 | 8.33 | 8.34 |
| 1998 | 8.34 | 8.33 | 8.34 |
| 1999 | 8.33 | 8.32 | 8.30 |
| 2000 | 8.33 | 8.31 | 8.33 |
| 2001 | 8.33 | 8.32 | 8.33 |
| 2002 | 8.34 | 8.32 | 8.34 |
| 2003 | 8.33 | 8.32 | 8.32 |
| 2004 | 8.34 | 8.32 | 8.32 |
| 2005 | 8.32 | 8.32 | 8.34 |
| 2006 | 8.34 | 8.31 | 8.33 |
| 2007 | 8.32 | 8.31 | 8.32 |
| 2008 | 8.32 | 8.30 | 8.32 |
| 2009 | 8.36 | 8.31 | 8.30 |
| 2010 | 8.31 | 8.30 | 8.32 |
| 2011 | 8.32 | 8.30 | 8.32 |
| 2012 | 8.30 | 8.29 | 8.30 |
| 2013 | 8.29 | 8.28 | 8.29 |
| 2014 | 8.29 | 8.27 | 8.29 |

(continued)

Table 1.22 (continued)

| Year | Premonsoon high tide | Monsoon High tide | Postmonsoon High tide |
|---|---|---|---|
| 2015 | 8.28 | 8.25 | 8.27 |
| 2016 | 8.27 | 8.24 | 8.25 |
| 2017 | 8.28 | 8.23 | 8.25 |
| 2018 | 8.28 | 8.22 | 8.24 |
| 2019 | 8.28 | 8.20 | 8.22 |
| **Mean** | 8.3247 | 8.3044 | 8.3169 |
| **Decrease** | $-0.0020\ yr^{-1}$ | $-0.0037\ yr^{-1}$ | $-0.0034\ yr^{-1}$ |

### *1.3.9 Mindhola Estuary*

Mindhola Estuary which originates in the Western Ghats follows a westward and maundering course joining the Arabian Sea at Danti. It broadens into a wide estuary below Nanod, 25 km inland. The estuarine region has extensive sand banks and mudflats. The estuary receives large quantities of mostly untreated sewage and industrial effluents from urban areas and industrial units respectively. The estuary is under considerable tidal influence (tide 0.5–5.5 m) with extended ebb periods. Currents are 0.8–1.9 m/s with the excursion of 3–17 km inside the estuary. There is a general tendency of increase in BOD associated with low DO content which reverses during high tide. Nitrite and ammonia are often higher than expected for the estuaries along the coast. Chlorophyll *a* and zooplankton vary widely. Macrobenthos is low with few species. The aquatic pH also shows a decreasing trend through seasons (Table 1.23).

**Table 1.23** Mean seasonal surface water pH values at the mouth of the Mindhola estuary

| Year | Premonsoon high tide | Monsoon high tide | Postmonsoon high tide |
|---|---|---|---|
| 1984 | 8.33 | 8.32 | 8.33 |
| 1985 | 8.33 | 8.33 | 8.33 |
| 1986 | 8.34 | 8.32 | 8.33 |
| 1987 | 8.33 | 8.31 | 8.33 |
| 1988 | 8.34 | 8.32 | 8.32 |
| 1989 | 8.33 | 8.33 | 8.33 |
| 1990 | 8.33 | 8.30 | 8.33 |
| 1991 | 8.33 | 8.32 | 8.33 |
| 1992 | 8.32 | 8.31 | 8.32 |
| 1993 | 8.34 | 8.32 | 8.34 |
| 1994 | 8.33 | 8.32 | 8.33 |
| 1995 | 8.33 | 8.30 | 8.33 |

(continued)

Table 1.23 (continued)

| Year | Premonsoon high tide | Monsoon high tide | Postmonsoon high tide |
|---|---|---|---|
| 1996 | 8.34 | 8.33 | 8.32 |
| 1997 | 8.32 | 8.32 | 8.33 |
| 1998 | 8.33 | 8.32 | 8.33 |
| 1999 | 8.32 | 8.31 | 8.29 |
| 2000 | 8.32 | 8.30 | 8.32 |
| 2001 | 8.32 | 8.31 | 8.32 |
| 2002 | 8.33 | 8.31 | 8.33 |
| 2003 | 8.32 | 8.31 | 8.31 |
| 2004 | 8.33 | 8.31 | 8.31 |
| 2005 | 8.31 | 8.31 | 8.33 |
| 2006 | 8.33 | 8.30 | 8.32 |
| 2007 | 8.31 | 8.30 | 8.31 |
| 2008 | 8.31 | 8.29 | 8.31 |
| 2009 | 8.35 | 8.30 | 8.29 |
| 2010 | 8.30 | 8.29 | 8.31 |
| 2011 | 8.31 | 8.29 | 8.31 |
| 2012 | 8.31 | 8.28 | 8.29 |
| 2013 | 8.31 | 8.27 | 8.28 |
| 2014 | 8.31 | 8.26 | 8.28 |
| 2015 | 8.31 | 8.24 | 8.28 |
| 2016 | 8.30 | 8.23 | 8.27 |
| 2017 | 8.31 | 8.22 | 8.27 |
| 2018 | 8.31 | 8.21 | 8.28 |
| 2019 | 8.31 | 8.15 | 8.27 |
| **Mean** | 8.3222 | 8.2933 | 8.3114 |
| **Decrease** | $-0.0006\ yr^{-1}$ | $-0.0049\ yr^{-1}$ | $-0.0017\ yr^{-1}$ |

### *1.3.10 Tapi Estuary*

Tapi, a major perennial river of the west coast of India, is an important source of freshwater to the region. The 720 km long river originates near Multai in the Betoul district of Madhya Pradesh with a catchment area of 65,145 $km^2$ of which around 4000 $km^2$ is in Gujarat. During seaward course, the river meanders through the hilly terrain of the Western Ghats before entering the coastal alluvial plains of Gujarat to meet the Arabian Sea near Hazira. The shallow and wide lower segment of the river exhibits characteristics of a typical estuary with strong currents associated with significantly high tidal influence upto 25 km upstream. Further inland however, the seawater excursion is restricted due to creation of a causeway at Rander which also hinders the freshwater outflow. The estuary and nearshore areas of Hazira exhibit a typical character of South Gujarat coast with (a) vast intertidal regions composed of poorly sorted sediment made up sand silt and clay with isolated rocky outcrops, (b) supra-littoral region either barren or dominated by salt tolerant plant species of P*rosopis, Acacia and Zyphus*, (c) gently sloping continental shelf with uneven seafloor often strewn with sand bars, and (d) high tidal influence due to the proximity to the Gulf of Khambhat.

The riverine discharge to the sea is controlled by the Ukai and Kakrapar dams constructed on the river at 141 and 115 km upstream respectively. Mean runoff of Tapi was $1.7982 \times 10^{10}$ $m^3$/y in 1975. After the construction of dams, it has reduced to an average of $7.301 \times 10^9$ $m^3$/y during 1982–91. In 1995 a weir-cum-causeway was constructed across the river at Rander that prevented seawater incursion further inland. In recent years the river discharge of 8000 $m^3$ /s during monsoon decreases to 10 to 45 $m^3$/s during November–May leading to stagnation in the riverine and inner estuarine segments, during the dry season. Due to proximity of the Gulf, the region experiences significantly high tidal influence with mean spring and neap tidal ranges of 5.7 and 4.3 m respectively at Hazira. The tidal influence however decreases with distance into the estuary with spring and neap tidal ranges of 2.3 and 0.4 m respectively at Surat. During the period of freshwater dominance the flood duration of 6 h in the open shore reduces to 4 to 5 h in the mouth segment of the estuary and decreases to barely 2 h at Surat with corresponding increase in the ebb period. Decrease in flood period though to a lesser extent occurs in the landward direction during the dry season also. The Tapi River is subjected to sporadic floods associated with heavy rainfall in the catchment area during July–September. When the flood coincides with spring tide, the water level increases substantially inundating vast areas. A level of 10 m with respect to CD has been recorded at Magdalla during one such flood in the past. The construction of two dams on the river has however reduced the flood fury to a great extent. The mean seasonal surface water pH at the estuarine mouth shows a declining trend with time (Table 1.24).

The decrease of pH follows a unique trend with highest value in monsoon followed by postmonsoon and premonsoon (Figs. 1.11, 1.12, and 1.13). We presume that run-off during monsoon brings lots of land based wastes in the estuary, which dips the pH value to a minimum. The process of estuarine acidification is primarily linked to land based activities like emission from industries, wastes from urban settlements, aqua-cultural farms etc. The spatial variation as observed in Tables 1.15, 1.16, 1.17, 1.18, 1.19, 1.20, 1.21, 1.22, 1.23, and 1.24 is because of the variation in the magnitude of anthropogenic activities in the coastal, upstream and downstream zones surrounding the estuaries.

Coastal upwelling can also lead to formation of hotspots of coastal pH change because of naturally high levels of $CO_2$ coupled with the increased anthropogenic $CO_2$ content of these waters. In some regions, an increase in the intensity of upwelling, which amplifies the effects on pH, is associated with anthropogenic climate change. However, the upwelling phenomenon is not detailed as the book primarily addresses estuarine acidification.

**Table 1.24** Mean seasonal surface water pH values at the mouth of the Tapi estuary

| Year | Premonsoon high tide | Monsoon high tide | Postmonsoon high tide |
|---|---|---|---|
| 1984 | 8.35 | 8.34 | 8.35 |
| 1985 | 8.35 | 8.35 | 8.36 |
| 1986 | 8.34 | 8.34 | 8.35 |
| 1987 | 8.35 | 8.33 | 8.35 |
| 1988 | 8.33 | 8.34 | 8.36 |
| 1989 | 8.35 | 8.35 | 8.35 |
| 1990 | 8.35 | 8.32 | 8.35 |
| 1991 | 8.35 | 8.34 | 8.35 |
| 1992 | 8.34 | 8.33 | 8.34 |
| 1993 | 8.34 | 8.34 | 8.36 |
| 1994 | 8.35 | 8.34 | 8.35 |
| 1995 | 8.35 | 8.32 | 8.35 |
| 1996 | 8.36 | 8.35 | 8.34 |
| 1997 | 8.34 | 8.34 | 8.35 |
| 1998 | 8.35 | 8.34 | 8.35 |
| 1999 | 8.34 | 8.33 | 8.31 |
| 2000 | 8.34 | 8.32 | 8.34 |
| 2001 | 8.34 | 8.33 | 8.34 |
| 2002 | 8.35 | 8.33 | 8.35 |
| 2003 | 8.34 | 8.33 | 8.33 |
| 2004 | 8.35 | 8.33 | 8.33 |
| 2005 | 8.33 | 8.33 | 8.35 |
| 2006 | 8.35 | 8.32 | 8.34 |
| 2007 | 8.33 | 8.32 | 8.33 |
| 2008 | 8.33 | 8.31 | 8.33 |
| 2009 | 8.37 | 8.32 | 8.31 |
| 2010 | 8.32 | 8.31 | 8.33 |
| 2011 | 8.33 | 8.31 | 8.33 |
| 2012 | 8.31 | 8.30 | 8.31 |
| 2013 | 8.30 | 8.29 | 8.30 |
| 2014 | 8.30 | 8.28 | 8.30 |
| 2015 | 8.29 | 8.26 | 8.28 |
| 2016 | 8.27 | 8.25 | 8.26 |
| 2017 | 8.27 | 8.24 | 8.26 |

(continued)

**Table 1.24** (continued)

| Year | Premonsoon high tide | Monsoon high tide | Postmonsoon high tide |
|---|---|---|---|
| 2018 | 8.26 | 8.23 | 8.25 |
| 2019 | 8.25 | 8.20 | 8.24 |
| **Mean** | 8.3297 | 8.3142 | 8.3272 |
| **Decrease** | $-0.0029\ yr^{-1}$ | $-0.0040\ yr^{-1}$ | $-0.0031\ yr^{-1}$ |

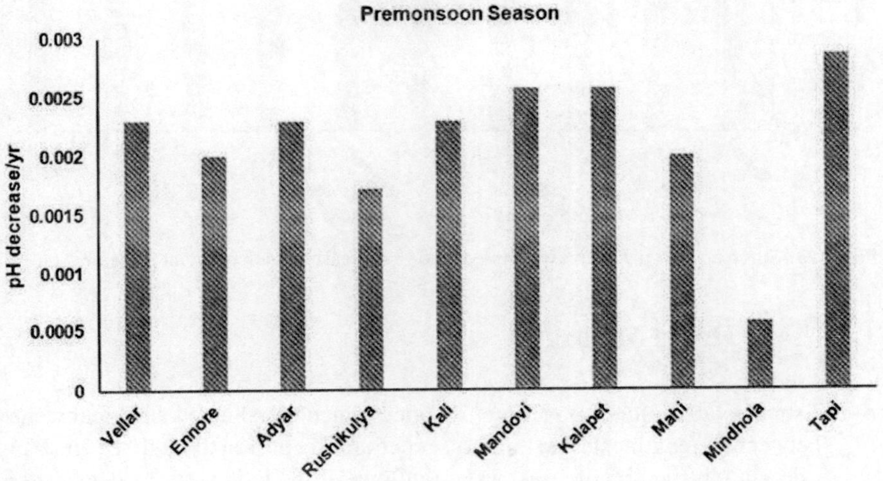


**Fig. 1.11** Surface water pH decrease/yr in premonsoon season in ten imortant Indian Estuaries

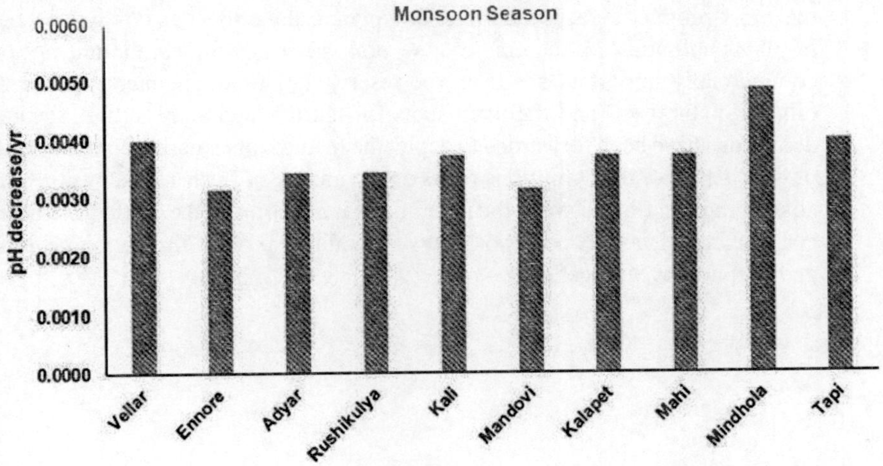


**Fig. 1.12** Surface water pH decrease/yr in monsoon season in ten imortant Indian Estuaries

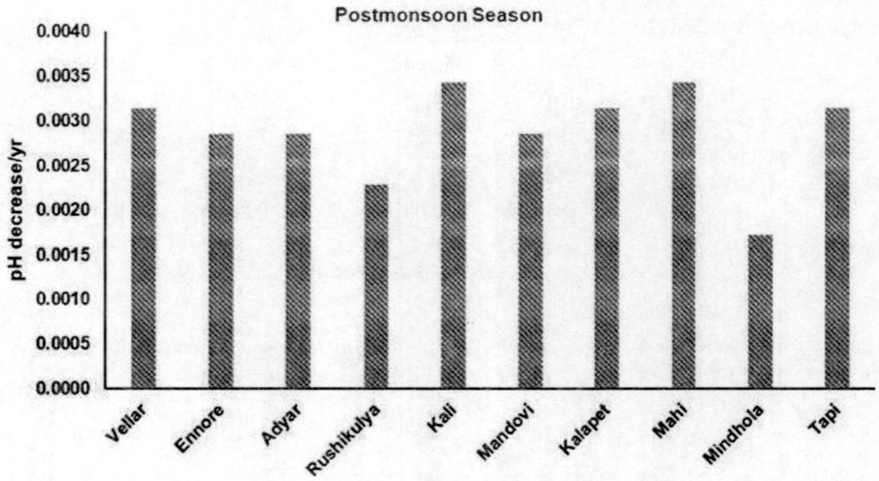


**Fig. 1.13** Surface water pH decrease/yr in postmonsoon season in ten imortant Indian Estuaries

## 1.4 Take Home Messages

A. Estuaries are unique spots on our planet, which are situated and sandwiched between the continents and the seas. Penetrated by the sea through the recurring tides and flushed by the freshwater outflows of the lotic system, an estuary is a dynamic system where the freshwater meets seawater. Blessed by the fertile flows of both the seas and the rivers, these fascinating biotopes are by far the most productive ecosystem on our planet, the abode of unique species of plants and animals, the cradle of several species of finfish, the nursery of commercially important shellfish, the reservoir of food, chemicals, mineral, oils and natural gas and the ideal spots for tourism and aquaculture. Several definitions have been forwarded to depict the features of estuarine compartment of which most of the definitions focus on the mixing of fresh and salt water as a zone of intersection between the river and sea. Estuaries of the world have been classified from various points of views like salinity profile, circulation pattern, geomorphology, energetic etc.

B. Estuaries are noted for unique taxonomic diversity of species, biological productivity and aesthetic value. The water, sediment and biotic communities of the estuaries offer several ecosystem services that provide both direct and indirect benefits to run the wheel of human civilization. They provide ecosystem services in the domains of provisioning services (fishes, minerals, mangrove based plants etc.), regulating services (water quality maintenance, biogeochemical cycles, carbon sink, flood control etc.), supporting services (biodiversity, life cycle stages of several species of shell fish and finfish, route for various rare fauna like turtles, dolphins etc.) and cultural services (religious convergence and rituals).

C. Indian estuaries are mostly subject to high degree of pollution that arises from industrial activities, sewage discharge, aquaculture farms, brick kilns, tourism units etc., and depending on the riverine discharge, the salinity and pollution load exhibit dynamic scenario. Extensive studies have been conducted on ten major estuaries of Indian sub-continent namely (i) Vellar estuary in southern Tamil Nadu, (ii) Ennore estuary in Tamil Nadu, (iii) Adyar estuary in Tamil Nadu, (iv) Rushikulya estuary in Odisha, (v) Kali estuary in Karnataka, (vi) Mandovi estuary in Goa, (vii) Kalapet estuary in Puducherry, (viii) Mahi estuary in Gujarat, (ix) Mindhola estuary in Gujarat, (x) Tapi estuary in Gujarat. In all the estuaries a decreasing trend of pH has been observed, which can be considered as foot print of acidification.

## Annexure 1: List of True Mangrove Flora with Medicinal Properties

The mangrove dominated Indian Sundarbans is the survival ground of a variety of endemic plants, which have great medicinal potential (Table 1.25).

The present annexure is an attempt to add another colored feather on the provisioning service cap of Sundarban estuarine mangrove floral species in the field of pharmacology, especially relevant in the COVID phase. Knowledge on the medicinal property of mangroves can help to cure diseases of the island dwellers locally, whose access to the city hospitals and clinics have totally been cut-off in the COVID period due to irregular transport system. Even the passenger vessel/boat, which is the only mode of transport for the Sundarban people has posed restrictions on the number of passengers to avoid the probability of contamination.

Table 1.25 Common true mangrove flora of Indian Sundarbans with medicinal values

| Mangrove species | Identifying features | Medicinal Values |
|---|---|---|
| <br>*Acanthus ilicifolius*<br>Common name: Hargoja | 1. Shrub like morphology with prominent stilt roots, usually found in sheltered mangrove areas; thrive on the supralittoral zone<br>2. Leaves lanceolate with serrated margins armed with spines<br>3. Flowers with long spike inflorescence, light blue or violet in colour | 1. The root is expectorant, and is used in coughs and asthma<br>2. The root, boiled in milk, is largely used in leucorrhoea and general debility<br>3. The plant extract has anti- inflammatory property |
| <br>*Aegiceros corniculatum*<br>Common name: Khalsi | 1. Shrub tolerant to high saline areas<br>2. Bark reddish brown with leaves elliptical, leaf-tip notched, cuneate at base<br>3. Fruit green to reddish in maturation, sharply curved<br>4. Fragrant white flowers, curved yellow | 1. It is used for curing asthma<br>2. It has anti diabetic property<br>3. It is used to cure rheumatism |

(continued)

**Table 1.25** (continued)

| Mangrove species | Identifying features | Medicinal Values |
|---|---|---|
| <br>*Avicennia alba*<br>Common name: Kalo baen | 1. Trees are tolerant to high salinity, pneumatophores spongy, narrowly pointed with slender stilt roots<br>2. Bark dark brown or even black<br>3. Leaves lanceolate to elliptical, leaf-tip acute, lower surface silver grey to white<br>4. Curved fruit with relatively long beak | 1. Paste prepared from seeds helps to relieve small pox ulceration<br>2. Wood ash after burning wood is given in dysentery<br>3. The plant extract is used to cure snake bites |
| <br>*Avicennia marina*<br>Common name: Piara baen | 1. Trees are tolerant to high salinity<br>2. Pneumatophores pencil like<br>3. Bark yellowish brown<br>4. Leaves elliptical, leaf-tip rolling, lower surface white to light grey<br>5. Inflorescence terminal or axillary, orange yellow in colour | 1. Bark decoction applied on "Haza" (a kind of sore due to wet sand and saline water) for curing purpose<br>2. Paste prepared from seeds helps to relieve small pox ulceration<br>3. Resinous exude is used for birth control purposes |

(continued)

**Table 1.25** (continued)

| Mangrove species | Identifying features | Medicinal Values |
|---|---|---|
|  *Avicennia officinalis* Common name: Sada baen | 1. Trees are tolerant to high salinity, pneumatophores pencil like<br>2. Bark yellowish brown. Leaves elliptical, leaf-tip rolling, lower surface white to light grey<br>3. Inflorescence terminal or axillary, orange yellow | 1. Ash obtained after burning the dried bark is used as antacid<br>2. Paste of green (raw) fruits is applied to cure boils and skin diseases<br>3. It is used to cure hepatitis |
|  *Sonneratia apetala* Common name: Keora | 1. Trees are with long, corky, forked pneumatophores<br>2. Stem light brown in colour<br>3. Leaves thick, coriaceous, narrowly elliptic oblong tapering towards apex<br>4. Flowers are cream coloured in axilliary cymes with globose berry seated in flattened calyx tube | 1. Fruit contains 28 g/100gm of Vitamin C and hence helps to fight cough and cold<br>2. Consumption of *S. apetala* fruit reduces the risk of cardiovascular disease<br>3. It is used to cure neurodegenerative diseases, such as Parkinson's and Alzheimers's |

(continued)

**Table 1.25** (continued)

| Mangrove species | Identifying features | Medicinal Values |
|---|---|---|
|  *Bruguiera gymnorrhiza* Common name: Kankra | 1. Trees generally found on elevated interior parts of mangrove forest with prominent buttress roots<br>2. Bark dark grey. Leaves simple, elliptical-oblong, leathery and leaf-tip acuminate<br>3. Flowers axillary, single with red calyx, red in colour and almost 16 lobed; fruits are cigar shaped, stout and dark green | 1. Bark decoction with coconut oil is applied to cure dermatological problems<br>2. The extract is used to cure eye diseases<br>3. Leaves and stems are used to treat diarrhea, fever, diabetes and pain |
|  *Ceriops decandra* Common name: Goran | 1. Stilt roots arising from pyramidal stem base<br>2. Light grey barked stem<br>3. Leaves elliptic-oblong, emarginate at apex, cuneate at base<br>4. Flowers axillary in condensed cymes<br>5. Fruit is berry, dark red when mature, warty towards tip, ridged, not hanging down | 1. Obstetric and haemorrhage cases are treated with an infusion of *Ceriops* bark<br>2. It is used to cure hepatitis<br>3. It is used to cure ulcers |

(continued)

**Table 1.25** (continued)

| Mangrove species | Identifying features | Medicinal Values |
|---|---|---|
|  *Excoecaria agallocha*<br>Common name: Genwa | 1. Prominent main root absent, many laterally spreading snake like roots producing elbo-shaped pegs<br>2. Bark greyish Poisonous milky latex highly irritating to eyes<br>3. Leaves light green with wavy margin<br>4. Catkin inflorescence terminal or axillary, orange yellow | 1. Latex applied on leprotic wounds<br>2. Leaf decoction is claimed as medicine in epilepsy<br>3. It is used to treat conjunctivitis |
|  *Heritiera fomes* (Sterculiaceae)<br>Common name: Sundari | 1. Trees with numerous peg-like pneumatophores and blind root suckers<br>2. Young branches covered with shining golden-brown scales<br>3. Leaves elliptic with lower surface shining with silvery scales<br>4. Flowers golden yellow with reddish tinge inside and fruits sub-globose, woody, indehiscent with longitudinal and transverse ridges | 1. It has antiviral property<br>2. It has antibacterial property<br>3. It has anticarcinogen property |

## References


Banerjee K, Mitra A, Bhattacharyya DP (2003) Phytopigment level of the aquatic subsystem of Indian Sundarbans at the apex of Bay of Bengal. Sea Explorers 6:39–46

Banerjee K, Sengupta K, Raha A, Mitra A (2013) Salinity based allometric equations for biomass estimation of Sundarban mangroves. Biomass Bioenerg 56:382–391

Banerjee K, Mitra A, Bhattacharyya DP, Choudhury A (2002) Role of nutrients on phytoplankton diversity in the North–East coast of the Bay of Bengal. In: Kumar A (ed) Ecology and ethology of aquatic biota. Daya Publishing House, pp 102–109

Biswas AN (1985) Geohydro-morphometry of Hooghly estuary. J Inst Eng (India) 66:61–73

Cameron WM, Pritchard DW (1963) Estuaries pp 306–324 In: Hill MN (ed) The Sea, 2. John Wiley and Sons, pp 306–324

Chakraborty SK, Choudhury A (1985). Distribution of fiddler crabs in Sundarbans mangrove estuarine complex. In: India proceedings of national symposium on biology, utilization and conservation of mangroves, pp 467–472

Mitra A (1998) Status of coastal pollution in West Bengal with special reference to heavy metals. J Ind Ocean Stud 5(2):135–138

Mitra A, Banerjee K, Sengupta K, Gangopadhyay A (2009) Pulse of climate change in Indian Sundarbans: a myth or reality. Nat Acad Sci Lett 32:1–7

Mitra A, Choudhury A (1994) Heavy metal concentrations in oyster *Crassostrea cucullata* of Henry's Island. India J Eco Biol 6(2):157–159

Mitra A, Choudhury A, Zamaddar YA (1992) Effects of heavy metals on benthic molluscan communities in Hooghly estuary. Proc. Zool. Soc. 45:481–496

Mitra A, Ghosh PB, Choudhury AA (1987) Marine bivalve *Crassostrea cucullata* can be used as an indicator species of marine pollution. In: Proceedings of national seminar on estuarine management, pp 177–180

Mitra A, Trivedi S, Choudhury A (1994). Inter-relationship between trace metal pollution and physico-chemical variables in the frame work of Hooghly estuary. Indian Ports, pp 27–35

Mitra, A, Zaman, S (2015) Blue carbon reservoir of the blue planet. Springer, ISBN 978-81-322-2106-7

Mitra A, Zaman S (2016). Basics of marine and estuarine ecology. Springer, ISBN 978-81-322-2705-2

Mondal K, Mukhopadhyay SK, Biswas H, De TK, Jana TK (2006) Fluxes of nutrients from the tropical river Hooghly at the land–ocean boundary of Sundarbans, NE Coast of Bay of Bengal, India. J of Mar Syst 62:9–21

Pagiola S, Konrad VR, Joshua B (2004) Assessing the value of ecosystem conservation (World Bank environmental department Paper No. 101), The World Bank Environmental Department, in collaboration with the nature conservancy and IUCN—The World Conservation Union

Perillo GME (1995). Definitions and geomorphological classifications of estuaries. In: Perillo GME (ed) Geomorphology and sedimentology of estuaries. Developments in Sedimentology 53, Elsevier, Amsterdam, pp 17–47

Pritchard (1967). What is an estuary: physical stand point? In: GH Lauff (ed) Estuaries. American Association for the Advancement of Science. Publication 83 Washington, DC, pp 3–5

Saclauso CA (1989) Brackish water aquaculture: threat to the environment? Naga 12:6–8

Sadhuram Y, Sarma VV, Murthy RTV, Rao PB (2005) Seasonal variability of physico–chemical characteristics of the Haldia channel of Hooghly estuary. J Earth Syst Sci 114(1):37–49

Saha SB, Mitra A, Bhattacharyya SB, Choudhury A (1999) Heavy metal pollution in Jagannath canal, an important tidal water body of the North Sundarbans aquatic ecosystem of West Bengal. Ind J Environ Prot 801–804

Sengupta K, Roy Chowdhury M, Bhattacharya SB, Raha A, Zaman S, Mitra A (2013) Spatial variation of stored carbon in *Avicennia alba* of Indian Sundarbans discovery. Nature 23(8):19–24

Tabuchi R (2003) The rehabilitation of mangroves in Southeast Asia. Retrieved 1 June 2007 from:http://www.unu.edu/env/plec/marginal/proceedings/TabuchiCH12.pdf

Trivedi S, Mitra A, Gupta A, Chaudhuri A, Neogi S, Ghosh I, Choudhury A (1994) Inter-relationship between physico-chemical parameters and uptake of pollutants by estuarine plants *Ipomea pescarpes* . In: Proceedings of the seminar: on our environment: its challenges to development projects. American Society of Civil Engineers—India Section, pp SC-1–SC-6

# Chapter 2
# Estuaries of Indian Sundarbans: A Test Bed for Acidification Study

**Abstract** Indian Sundarban delta, at the apex of Bay of Bengal is noted for the presence of seven dynamic estuaries. The estuaries in the western sector of the delta complex are hyposaline in nature with relatively low surface water pH (mean pH value = ~7.90) as they receive the industrial effluents from multifarious industrial units. On contrary, the estuaries of central Indian Sundarbans are hypersaline with relatively high pH as the freshwater discharge from the mighty River Ganga is blocked due to massive siltation. These hypersaline estuaries receive saline water from Bay of Bengal and exhibit a pH value around 8.28 throughout the year. Considering the contrasting nature of the estuaries of western and central Indian Sundarbans, the impacts of aquatic pH on several physico-chemical and biological components have been highlighted in this chapter on a comparative scale. The significant spatio-temporal variation of aquatic pH can be attributed to factors like seawater intrusion into the estuary from Bay of Bengal, sewage discharge (from point and nonpoint sources) and photosynthetic activity by the mangrove vegetation that exhibit variable biomass and distribution in and around the experimental sites.

## 2.1 Estuaries of Indian Sundarbans

An estuary may be considered as the zone of intersection of fresh and marine waters with unique physico-chemical and biological characteristics. Estuaries generally occupy those areas of the coasts, which are least subject to marine features/activities and wave actions and thus are major sites for development of harbours, recreational activities and aquacultural farms. In estuaries, fresh water collected over vast regions of the land pours into the ocean, which sends salt water upstream far beyond the river mouth. According to Odum (1971), estuaries belong to different class of "fluctuating water level ecosystems".

Each estuary has its respective identity in terms of the amount of river discharge, depth and general topography, specific circulation pattern, climatic regime and vertical tide range that influence its ecology and biodiversity, e.g. the estuaries of Indian Sundarbans has a salinity range from 4.0 to 30.0 psu depending on season



A. Mitra and S. Zaman, *Estuarine Acidification*,
https://doi.org/10.1007/978-3-030-84792-0_2

and locations unlike the Mandovi and Zuari estuaries of Goa, where the salinity oscillation is not so significant.

The estuaries are the live matrix of deltaic complex, on which the unique spectrum of biological diversity is embedded. In Indian Sundarbans, approximately 2069 $km^2$ area is occupied by tidal river system or estuaries, which finally end up in the Bay of Bengal. The seven main estuaries, from west to east are highlighted in Table 2.1 along with their important characteristics.

The major estuaries of lower Gangetic delta are depicted in Fig. 2.1.

The estuaries of Indian Sundarbans exhibit distinct variation in salinity in the same time period. We observed that salinity in the Matla estuary flowing through central Indian Sundarbans exhibit higher salinity than Hooghly estuary in the western Indian Sundarbans (*Vide* Tables 2.2 and 2.3 and Figs. 2.2 and 2.3).

It is clearly visualized from the data sets that the central sector experiences a high saline aquatic sub-system, which is the result of complete obstruction of the fresh water flow from the upstream region owing to Bidyadhari siltation since the late fifteenth century (Chaudhuri and Choudhury 1994; Banerjee et al. 2013; Mitra 2013; Sengupta et al. 2013). The Matla estuary in the central Indian Sundarban cannot be referred to as an ideal estuary as there is no head on discharge or dilution of the system with fresh water. Thus Matla can be designated as a tidal channel, whose survival depends on the tidal flow from Bay of Bengal.

The hyposaline environment of western Indian Sundarbans may be attributed to Farakka barrage discharge situated in the upstream region of Ganga-Bhagirathi-Hooghly river system. 10-year surveys (1999–2008) on water discharge from Farakka dam revealed an average discharge of $(3.7 \pm 1.15) \times 10^3\ m^3s^{-1}$. Higher discharge values were observed during the monsoon with an average of $(3.81 \pm 1.23) \times 10^3\ m^3s^{-1}$, and the maximum of the order 4524 $m^3s^{-1}$ during freshet (September). Considerably lower discharge values were recorded during premonsoon with an average of $(1.18 \pm 0.08) \times 10^3\ m^3s^{-1}$, and the minimum of the order 846 $m^3s^{-1}$ during May. During postmonsoon discharge, values were moderate with an average of $(1.98 \pm 0.97) \times 10^3\ m^3s^{-1}$ (Mitra 2013).

An attempt has also been taken to visualize the salinity profile of the region after a span of 30 years (forecasting) on the basis of present trend. The exponential smoothing method that produces maximum-likelihood estimate of the variable predicts a salinity value of 13.05 psu in 2043, which is a decrease of 38.4% since 1984 in western Indian Sundarban estuary (over a span of 60 years).

Considering the observed data set of 30 years (1984–2013), it can be predicted that salinity will be around 36 psu after a period of 30 years in the central sector of Indian Sundarbans, which is an indication of alarming hypersaline condition (a rise by 67.1%) in 2043 in this sector.

**Table 2.1** Important tidal rivers of Indian Sundarbans

| Estuary | Description |
|---|---|
| Hooghly | • It forms the western border of Indian Sundarbans and is exposed to considerable anthropogenic activities<br>• It is the main river of West Bengal and is a direct continuation of the River Ganges<br>• Most of the coastal industries of West Bengal are concentrated along the western bank of this river |
| Muriganga | • It is a branch of the Hooghly River<br>• It flows along the east of Sagar Island, the largest island in the deltaic complex<br>• Unique mangrove vegetation with dominancy of *Sonneratia apetala*, *Avicennia* spp. and *Excoecaria agallocha* is found along the bank of this river |
| Saptamukhi | • It has relatively high salinity compared to Hooghly estuary<br>• It is connected with the Muriganga (Bartala) branch of the Hooghly River through Hatania-Duania canal |
| Thakuran | • It begins near Jayanagar in South 24 Parganas and has a number of connections with the Saptamukhi<br>• It was connected in the earlier times with the Kolkata canal through the Kultali and the Piyali rives, which exist today in a dying state |
| Matla | • This river originates at the confluence of Bidyadhari, Khuratya and the Rampur Khal close to the town of Canning in 24 Parganas (South)<br>• Matla is connected to Bidya and ultimately flows to the Bay of Bengal. The fresh water connection and discharge to this river has been lost in the recent times<br>• Salinity of the river water is relatively high (in comparison to Hooghly or Muriganga) owing to fresh water cut-off from the upstream region |
| Bidyadhari* | • This was flourishing branch of the Bhagirathi during the fifteenth and sixteenth century, but now serves only as a sewage and excess rainwater outlet from the city of Kolkata<br>• The river bed is completely silted and presently it is almost in dying condition |

(continued)

**Table 2.1** (continued)

| Estuary | Description |
|---|---|
| Goashaba | • The waters of Matla and Harinbhanga (Raimangal) through a large number of canals form the estuary<br>• The estuary and its numerous creeks flow through the reserve forests |
| Harinbhanga | • The salinity of Harinbhanga estuary is relatively low compared to Matla estuary in the central Indian Sundarbans<br>• The Harinbhanga (also known as Ichamati and Raimangal) forms a natural demarcation between India and Bangladesh |

* Presently a dying estuary and not considered within the seven major types

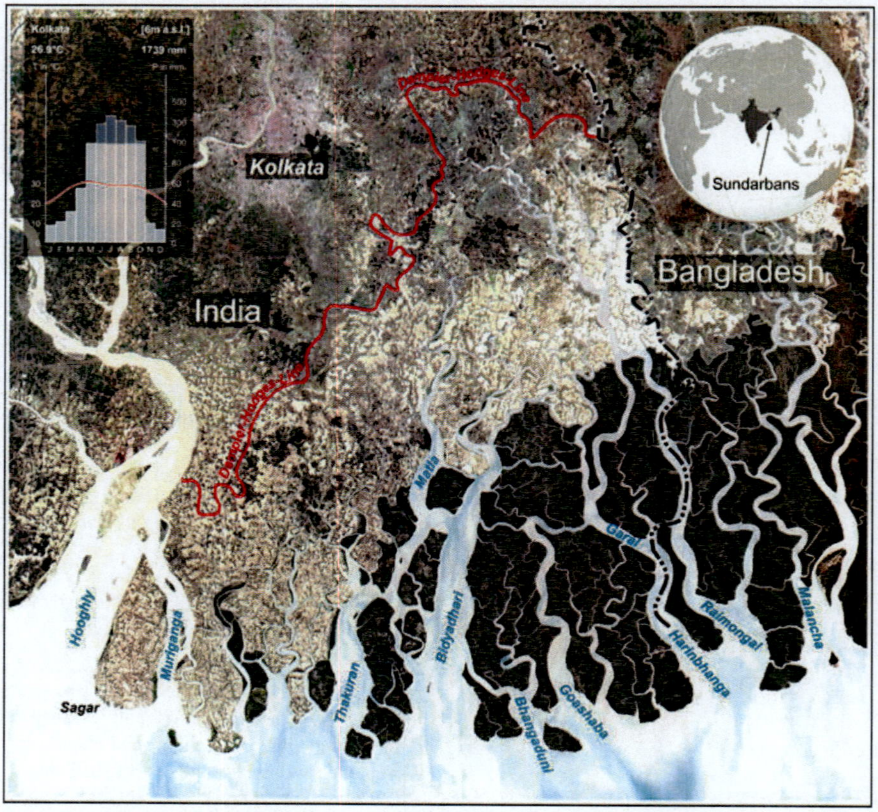


**Fig. 2.1** Major estuaries in the lower Gangetic delta complex

**Table 2.2** Surface water salinity (psu) in the Jharkhali station along the Matla Estuary of central Indian Sundarbans

| Year | Premonsoon | Monsoon | Postmonsoon |
|---|---|---|---|
| 1984 | 22.07 | 20.67 | 21.81 |
| 1985 | 23.59 | 20.74 | 22.08 |
| 1986 | 24.34 | 20.80 | 22.02 |
| 1987 | 24.22 | 20.14 | 22.67 |
| 1988 | 25.04 | 20.03 | 22.40 |
| 1989 | 25.69 | 20.35 | 23.08 |
| 1990 | 25.84 | 21.02 | 23.54 |
| 1991 | 25.01 | 21.67 | 23.57 |
| 1992 | 24.72 | 21.73 | 23.96 |
| 1993 | 25.60 | 21.86 | 23.77 |
| 1994 | 26.00 | 21.74 | 24.21 |
| 1995 | 26.04 | 21.99 | 24.99 |
| 1996 | 26.08 | 21.61 | 25.05 |
| 1997 | 26.83 | 22.01 | 24.92 |
| 1998 | 26.81 | 22.06 | 25.43 |
| 1999 | 26.75 | 20.64 | 25.41 |
| 2000 | 26.98 | 21.37 | 25.51 |
| 2001 | 27.04 | 22.60 | 25.96 |
| 2002 | 27.13 | 22.92 | 26.42 |
| 2003 | 27.45 | 22.98 | 26.48 |
| 2004 | 27.20 | 23.04 | 26.45 |
| 2005 | 27.18 | 23.16 | 27.56 |
| 2006 | 27.86 | 23.78 | 27.38 |
| 2007 | 27.69 | 23.83 | 27.69 |
| 2008 | 28.34 | 23.92 | 27.72 |
| 2009 | 31.68 | 22.87 | 28.30 |
| 2010 | 29.87 | 22.96 | 28.54 |
| 2011 | 30.13 | 23.37 | 28.89 |
| 2012 | 30.08 | 23.52 | 28.72 |
| 2013 | 30.46 | 23.69 | 28.92 |
| 2014 | 30.73 | 24.04 | 29.28 |
| 2015 | 30.81 | 23.89 | 29.80 |
| 2016 | 31.09 | 24.03 | 29.88 |
| 2017 | 30.80 | 25.03 | 29.30 |
| 2018 | 30.96 | 25.41 | 30.68 |

**Table 2.3** Surface water salinity (psu) in the Diamond Harbour station along the Hooghly Estuary adjacent to theWestern Indian Sundarbans

| Year | Premonsoon | Monsoon | Postmonsoon |
|---|---|---|---|
| 1984 | 11.68 | 7.14 | 9.17 |
| 1985 | 11.04 | 7.10 | 9.11 |
| 1986 | 10.87 | 6.97 | 9.13 |
| 1987 | 10.99 | 6.00 | 8.17 |
| 1988 | 11.55 | 5.84 | 8.81 |
| 1989 | 11.02 | 5.56 | 8.33 |
| 1990 | 10.85 | 6.02 | 7.67 |
| 1991 | 10.32 | 5.13 | 7.14 |
| 1992 | 10.07 | 5.07 | 7.37 |
| 1993 | 9.97 | 4.99 | 7.96 |
| 1994 | 9.45 | 4.84 | 6.83 |
| 1995 | 9.63 | 5.04 | 6.40 |
| 1996 | 8.86 | 4.97 | 6.44 |
| 1997 | 8.31 | 4.66 | 6.67 |
| 1998 | 7.70 | 4.44 | 6.12 |
| 1999 | 7.95 | 3.96 | 6.14 |
| 2000 | 7.04 | 3.78 | 6.19 |
| 2001 | 7.43 | 3.42 | 5.43 |
| 2002 | 6.51 | 3.29 | 5.41 |
| 2003 | 6.25 | 3.55 | 5.77 |
| 2004 | 6.20 | 2.99 | 5.84 |
| 2005 | 6.39 | 2.47 | 5.23 |
| 2006 | 5.77 | 2.76 | 5.35 |
| 2007 | 5.17 | 2.34 | 5.43 |
| 2008 | 5.07 | 2.99 | 4.40 |
| 2009 | 11.04 | 3.35 | 4.20 |
| 2010 | 5.00 | 2.23 | 4.17 |
| 2011 | 4.66 | 2.06 | 4.00 |
| 2012 | 5.01 | 1.99 | 4.09 |
| 2013 | 4.93 | 1.95 | 3.93 |
| 2014 | 4.26 | 1.92 | 3.86 |
| 2015 | 5.07 | 1.87 | 3.71 |
| 2016 | 4.09 | 1.08 | 3.87 |
| 2017 | 4.12 | 1.01 | 3.55 |
| 2018 | 3.06 | 0.85 | 2.17 |

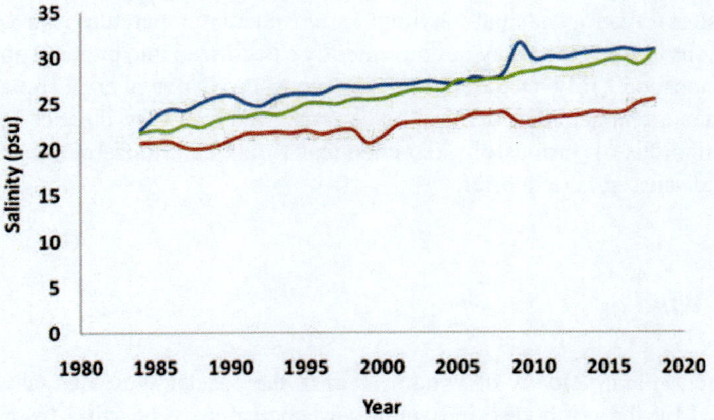


**Fig. 2.2** Surface water salinity (psu) in the Jharkhali station along the Matla Estuary of central Indian Sundarbans; the blue, green and red lines represent the salinity values during premonsoon, postmonsoon and monsoon seasons respectively and the peak in blue line is the extreme salinity witnessed during the supercyclone Aila that hit Indian Sundarbans on 25th May, 2009

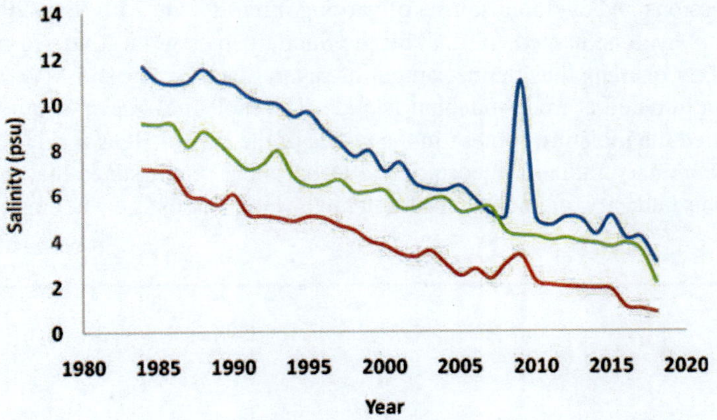


**Fig. 2.3** Surface water salinity (psu) in the Diamond Harbour station along the Hooghly Estuary of western Indian Sundarbans; the blue, green and red lines represent the salinity values during premonsoon, postmonsoon and monsoon seasons respectively and the peak in blue line is the extreme salinity witnessed during the supercyclone Aila that hit Indian Sundarbans on 25th May, 2009

## 2.2 Climate of Indian Sundarbans

The deltaic lobe of Indian Sundarbans experiences a moderate type of climate because of its location adjacent to the Bay of Bengal as well as due to regular tidal flushing in the estuaries. Wave actions, micro and macro tidal cycles, long shore currents are recorded in most of the islands of the ecosystem. Coastal processes are very dynamic and are accelerated by tropical cyclones, which is locally called "Kal Baisakhi"

(Norwesters). Average annual maximum atmospheric temperature is around 35 °C. The seasons in study area may be conveniently categorized into premonsoon (March-June), monsoon (July–October) and postmonsoon (November- February). Each season has a characteristic feature of its own, which is very distinct and unique. The oscillations of various physico-chemical variables in different seasons of the year are discussed here in brief.

### 2.2.1 Wind

The direction and velocity of wind system in the coastal West Bengal are mainly controlled by the north-east and south-west monsoons. The wind from the north and north-east commences at the beginning of October and continues till the end of March. The month of January and February are relatively calm with an average wind speed around 3.5 km/h. Violent wind speed recommences from the south-west around the middle of March and continues till September. During this period, several low pressure systems occur in this region, a number of which takes the form of depressions and cyclonic storms of varying intensity. On 25th May, 2009, a type of tropical cyclone named "AILA" hit the Sundarbans region. It was formed in the central Bay of Bengal as the net output of several factors. Around May 20th, 2009 monsoon initiated at mid Andaman Islands. The moisture—laden south westerlies accelerated the moisture content in the winds of the Bay of Bengal (Fig. 2.4).

A preliminary Indian Meteorological Department report stated that the cyclone retained its intensity for about 15 h after it hit the landmasses as it was close to the Bay

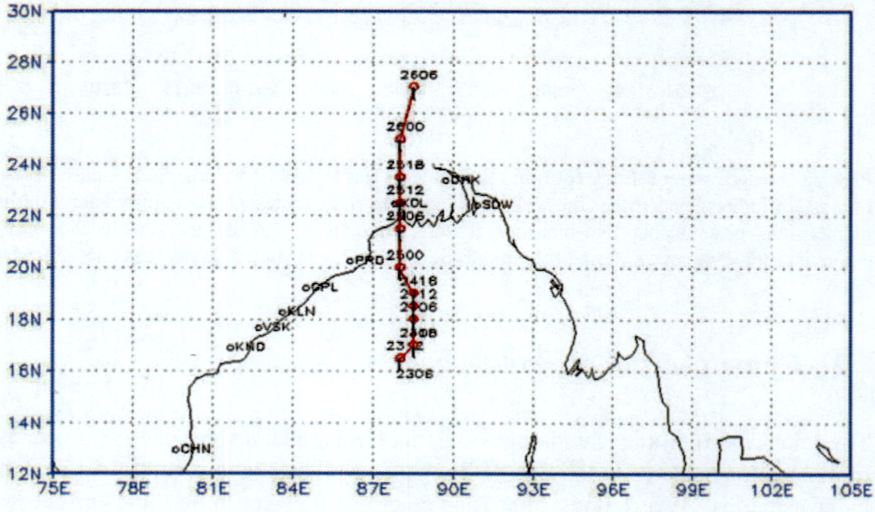


**Fig. 2.4** Track of AILA during 23rd to 26th May, 2009. *Source* Regional Specialized Meteorological Center—Tropical Cyclone, New Delhi

of Bengal. It lay centered over the Gangetic delta for quite some time, ascertaining the availability of moisture. This is peculiar nevertheless because premonsoon storms rarely hit the maritime state of West Bengal with such severity. A large number of mangroves were uprooted during the AILA period, but this vegetation acted as bioshield and reduced the vulnerability of island villages.

The variations in the wind speed in the northern and southern sector of Bay of Bengal had led to the curling of winds, which is known as positive relative vortices. Basically an area of depression developed in the Bay of Bengal on May 20, 2009 which transformed into a cyclone on May 23, 2009 and hit the deltaic complex of Indian Sundarbans on May 25, 2009 destroying the lives and properties of island dwellers. It lay centered over the Gangetic delta for quite some time, ascertaining the availability of moisture. This is peculiar nevertheless because premonsoon storms rarely hit the maritime state of West Bengal with such severity.

Natural hazards are keenly related to the lives and livelihoods of Sundarban people. Bulbul, one of the most recent tropical cyclonic storms, wreaked havoc in southern West Bengal along with parts of Odisha on 8th and 9th November 2019. The storm is said to have triggered losses between INR 15,000 and 19,000 crores, as per official projections.

The storm surges primarily affected the districts of North and South Parganas and East Midnapore in West Bengal, claiming several lives, destroying houses, physical assets and sweeping away ripe agricultural harvests. Despite the adequacy of pre-arranged assistance, reliefs and contingencies provisioned by the state, Bulbul led to significant devastations for marginal rural communities in the districts—primarily subsistence agriculturalists.

Extreme climate-change events and frequent cyclones like Bulbul and Fani calls for a conscious recognition of the need of productive ecosystems like mangroves and an effective governance mechanism working towards long term environmental sustainability.

In recent time, another natural disaster in the form of cyclone Amphan, hit West Bengal on May 20, 2020 claiming 86 lives and affecting over 10 million people in the eastern Indian state. The powerful storm ripped through eastern India and neighbouring Bangladesh. The storm in West Bengal caused massive damage to standing crops, thousands of trees were uprooted, and power and water supply was interrupted in the state capital Kolkata. Many in the state have lost their entire homes as well.

Cyclone Amphan made its landfall in West Bengal between Digha and Hatiya on the afternoon of May 20, as a very severe cyclonic storm with sustained wind speeds of 155–165 kms per hour spiraling up to 185 kms per hour. The coastal state of Odisha and the neighbouring country of Bangladesh were also impacted by the cyclone.

It is estimated that about 3900 $km^2$ in the Sundarbans suffered damage, with over 10 lakh homes and shops being devastated, trees being uprooted, and roads being cracked open. The storm also damaged creeks and embankments, bringing saline water into villages and destroying 109,577 hectares of cultivable land and impacting 443,144 farmers.

It has been observed that saline inundation in agricultural fields has destroyed standing crops and set back any prospects of cultivation to at least two years. The incoming monsoon also brings with it new challenges for rehabilitation. Freshwater and potable water availability has been affected and it is now up to the residents to begin picking up from what's left due to the damage.

Basically the inshore region of Bay of Bengal is prone to several categories of storms in frequent intervals of time. The table here reflects the frequency of occurrence of cyclonic storm in the Indian Sundarbans during 1990–2020 (Table 2.4).

**Table 2.4** A brief summary of extremely severe cyclonic storms (ESCSs) over the Bay of Bengal region during 1990–2020

| S. no. | Year | Name year-wise | Lifetime | MSW (knots) | Duration during ESCS (h) |
|---|---|---|---|---|---|
| | 1990[a] | 1990 | 6 days 6 h | 127 | 60 |
| | 1991[a] | 1991 | 5 days 12 h | 127 | 36 |
| | 1992[b] | 1992 | 5 days 12 h | 102 | 48 |
| | 1993[b] | 1993 | 3 days 6 h | 90 | 3 |
| | 1994[a] | 1994 | 3 days 3 h | 102 | 18 |
| | 1995[b] | 1995 | 3 days 21 h | 102 | 15 |
| | 1997[a] | 1997 | 4 days 6 h | 90 | 12 |
| | 1999[b] | 1999 | 4 days | 90 | 3 |
| | 1999[b] | Orissa SuCS | 5 days 21 h | 140 | 24 |
| | 2000[b] | 2000 | 4 days 3 h | 102 | 12 |
| | 2000[b] | 2000 | 5 days 9 h | 90 | 3 |
| | 2004[a] | 2004 | 3 days | 90 | 6 |
| | 2006[a] | Mala | 4 days 15 h | 100 | 24 |
| | 2007[b] | Sidr | 4 days 18 h | 115 | 63 |
| | 2008[a] | Nargis | 6 days 6 h | 90 | 9 |
| | 2009[a] | Aila | 2 days | 67 | 24 |
| | 2010[b] | Giri | 2 days 18 h | 105 | 6 |
| | 2013[b] | Phailin | 6 days 3 h | 115 | 54 |
| | 2014[b] | Hudhud | 7 days 6 h | 100 | 27 |
| | 2019 | Bulbul | 3 days | 72 | 2 |
| | 2019[a] | Fani | 8 days 9 h | 116 | 27 |
| | 2020[a] | Amphan SuCS | 5 days 12 h | 102 | 60 |

VSCS very severe cyclonic storm (wind speed between 64 and 89 knots); ESCS extremely severe cyclonic storm (wind speed between 90 and 119 knots); SuCS super cyclonic storm (wind speed more than 120 knots); [a]During premonsoon period (March–May); [b]During post-monsoon period (October–December); One knot is equal to 1.8 kmph

### 2.2.2 Waves and Tides

The wind is the basic driving force for generating surface waves in the coastal zone of West Bengal. Sea waves in this region rarely become destructive except during cyclonic storms. During Norwesters', the wind speed rises above 100 km/h and is usually accompanied by huge tidal waves. When the cyclonic incidences coincide with the spring tides, wave height can rise over 5 m above the mean sea level. Ripple waves appear in the month of October, November and December when wind generated wave height varies approximately between 0.20 and 0.35 m. In the month of April to August, large wavelets are formed in the shelf region and they start breaking when they approach towards the coastal margin.

Wave height rises up to 2 m during this period, which causes maximum scouring of land masses. Wave actions, micro and macro-tidal cycles and long shore currents are recorded in most of the islands in this ecosystem. With the change in seasons, tidal pattern in the estuarine systems of coastal West Bengal also changes. During the monsoon month, the effect of flood tide is more or less countered and nullified by freshets and there is a strong predominance of ebb tide. The strength of flood tide over ebb tide is at a minimum during the postmonsoon season. Conversely, during the premonsoon season, the effect of flood tide is considerably stronger than that of the ebb tide.

### 2.2.3 Water Temperature

In coastal West Bengal, the seasonal variation of surface water temperature is not so drastic between premonsoon and monsoon seasons. The premonsoon period (March to June) is characterized by a mean surface water temperature around 34 °C. The monsoon period (July to October) shows a mean surface water temperature around 32 °C and the postmonsoon period (November to February) is characterized by cold weather with a mean surface water temperature around 23 °C (Mitra 2000).

### 2.2.4 Rainfall

Rainfall is usually maximum during the month of August/September and then monsoon period lasts from July to October. The south-west wind triggers precipitation in the monsoon period with an average rainfall of about 165 mm. The postmonsoon period (November to February) is characterized by negligible rainfall and the premonsoon period (March to June) is basically dry, but occasionally accompanied by rains and thunderstorms. Rough weather with frequent cyclonic depressions lasts from mid March to mid September. Average annual rainfall is 1920 mm. Average humidity is about 82% and is more or less uniform throughout the year.

Table 2.5 highlights the average rainfall data for three decades (1981–2018). The average monthly data of rainfall has been computed on the basis of secondary data available from Canning station (Source: IMD—Indian Meteorological Department) and it is observed that monsoon in the lower Gangetic delta region starts from June and continues till October.

## 2.3 Pulse of Acidification in Hooghly and Matla Estuaries of Indian Sundarbans

Over the last century, the atmospheric concentration of carbon dioxide has risen at a rate 100 times faster than any change observed during the past 650,000 years (Siegenthaler et al. 2005). There is broad consensus that this ongoing change is a direct result of human activity, principally by fossil fuel burning, cement production and changing land use (Hansen et al. 2007). Atmospheric levels of carbon dioxide have consequently increased from pre-industrial levels of 280 ppm to a concentration of approximately 380 ppm (Feely et al. 2004). Almost 50% of all anthropogenic carbon dioxide emitted to the atmosphere has diffused passively into the ocean, thereby significantly decreasing the rate of global warming (Sabine et al. 2004). Concentrations of atmospheric carbon dioxide are rising at a rate of 3.3% per year and will continue this rising trend (Canadell et al. 2007). Hydrological models predict that, based on proposed future emissions of carbon dioxide, the average oceanic pH will decline by 0.3–0.5 by the year 2100 and by 0.7 within the next 300 years (Caldeira and Wickett 2003). Leakage from carbon dioxide seabed storage would create locally faster and stronger acidification than that induced by atmospheric carbon dioxide (Hawkins 2004). Estuaries, being the arms of oceans and seas are also exposed to the process of acidification.

On the basis of the global trend of acidification, we have tried to focus on the issue at the local level in the Hooghly and Matla estuaries situated in the western and central Indian Sundarbans respectively in the lower Gangetic delta region.

We conducted a study in these two estuaries of Indian Sundarbans during 1984 to 2014 using aquatic pH as proxy in three stations of Indian Sundarbans namely Diamond Harbour and Namkhana in western Indian Sundarbans along the Hooghly estuary and Ajmalmari in the central Indian Sundarbans along the Matla estuary.

The pH ranged from 8.10 (during 2014 low tidal phase) to 8.33 (during 1984 high tide condition) at Diamond Harbour (Fig. 2.5a, b).

The pH ranged from 8.18 (during 2014 low tide phase) to 8.34 (during 1984 high tide condition) (Fig. 2.6a, b).

The pH ranged from 8.20 (during 2014 low tide phase) to 8.33 (during 1984 high tide condition (Fig. 2.7a, b).

The surface water pH is an important environmental parameter on which the life of any aquatic organism depends. It shows more or less a uniform value throughout the seasons with minor fluctuations mainly in the monsoon months, when, dilution

**Table 2.5** Average monthly rainfall data (in mm) collected for Canning region, 24 Parganas, South (West Bengal)

| Year | Jan | Feb | Mar | Apr | May | Jun | Jul | Aug | Sep | Oct | Nov | Dec |
|---|---|---|---|---|---|---|---|---|---|---|---|---|
| 1981 | – | – | – | – | – | – | – | – | – | – | 0 | 0.38 |
| 1982 | 0 | 1.69 | 1.65 | 2.77 | 0.85 | 5.69 | 8.49 | 10.4 | 4.23 | 0.36 | 0.87 | 0 |
| 1983 | 0.14 | 1.67 | 1.2 | 1.08 | 2.51 | 11.22 | 6.42 | 17.79 | 9.96 | 3.61 | 0.25 | 0.18 |
| 1984 | 0.57 | 0.03 | 0 | 2.29 | 3.19 | 23.81 | 7.26 | 15.55 | 6.69 | 3.28 | 0 | 0 |
| 1985 | 0.88 | 0.21 | 0.2 | 1.12 | 5.11 | 9.25 | 9.25 | 12.2 | 7.69 | 3.86 | 0 | 0 |
| 1986 | 0.33 | 0 | 0.05 | 1.04 | 6.2 | 6.34 | 10.66 | 3.35 | 27.91 | 8.55 | 7.54 | 0.05 |
| 1987 | 0 | 0.7 | 0.6 | 5.8 | 4.21 | 4.73 | 11.55 | 13.41 | 9.58 | 0.81 | 2 | 0.32 |
| 1988 | 0 | 0.92 | 0.12 | 0.71 | 5.49 | 20.29 | 17.4 | 9.28 | 6.38 | 4.89 | 4.66 | 0 |
| 1989 | 0.07 | 0.25 | 0.32 | 0 | 5.15 | 2.83 | 9.1 | 10.35 | 11.34 | 6.63 | 0.39 | 0.08 |
| 1990 | 0 | 2.48 | 6.35 | 3.27 | 4.79 | 10.18 | 16.6 | 10.77 | 10.4 | 5.72 | 2.41 | 0.06 |
| 1991 | 2 | 0.64 | 0.55 | 1.13 | 0.49 | 16.21 | 8.64 | 8.07 | 8.02 | 4.56 | 0.02 | 0.57 |
| 1992 | 0.92 | 1.73 | 0 | 0.59 | 6.1 | 13.31 | 12.13 | 13.51 | 16.1 | 3.8 | 0 | 0 |
| 1993 | 0.03 | 0 | 2.75 | 0.77 | 4.61 | 10.34 | 11.34 | 11.55 | 17.73 | 3.97 | 0.43 | 0 |
| 1994 | 0.11 | 0.84 | 0.25 | 3.82 | 2.35 | 8.21 | 14.93 | 9.12 | 6.46 | 2.52 | 1.92 | 0 |
| 1995 | 0.23 | 1.36 | 0.08 | 0.61 | 7.39 | 11.37 | 17.6 | 10 | 17.84 | 3.08 | 6.68 | 0 |
| 1996 | 0.29 | 0.3 | 0.07 | 0.89 | 2.56 | 18.97 | 11.88 | 16.72 | 5.09 | 9.94 | 0 | 0 |
| 1997 | 0.57 | 0.49 | 1.36 | 6.64 | 3.41 | 7.12 | 12.98 | 15.78 | 9.36 | 0.16 | 0.49 | 0.48 |
| 1998 | 0.98 | 0.14 | 5.97 | 2.13 | 5.86 | 4.94 | 12.07 | 12.09 | 10.41 | 5.31 | 5.84 | 0 |
| 1999 | 0.08 | 0 | 0.01 | 0.05 | 6.72 | 8.94 | 16.47 | 13.38 | 14.91 | 5.96 | 0.07 | 0 |
| 2000 | 0.05 | 1.96 | 0 | 2.29 | 10.79 | 7.79 | 15.21 | 6.34 | 11.69 | 4.22 | 0.06 | 0.06 |
| 2001 | 0.09 | 0.01 | 0.8 | 3.43 | 6.8 | 14.54 | 10.63 | 6.87 | 5.84 | 5.14 | 0.62 | 0 |
| 2002 | 0.85 | 0 | 0.53 | 2.66 | 4.01 | 19.81 | 8.78 | 10.05 | 8.31 | 3.97 | 4.62 | 0 |
| 2003 | 0 | 0.51 | 0.9 | 2.26 | 4.36 | 10.65 | 11.9 | 11.49 | 9.37 | 17.09 | 0.67 | 1.29 |
| 2004 | 0.15 | 0.25 | 0.13 | 2.26 | 2.42 | 8.96 | 13.35 | 14.93 | 7.75 | 8.77 | 0 | 0 |
| 2005 | 1.47 | 0 | 4.25 | 1.24 | 2.27 | 8.3 | 17.89 | 9.17 | 11.73 | 16.74 | 0 | 0.03 |
| 2006 | 0 | 0 | 0.11 | 2.06 | 3.9 | 4.08 | 21.16 | 14.55 | 17.35 | 1.045 | 0.03 | 0 |
| 2007 | – | 2.8 | 0.07 | 1.74 | 5.63 | 5.32 | 20.25 | 8.97 | 21.39 | 3.9 | 2.13 | 0 |
| 2008 | 3.46 | 0.34 | 0.15 | 1.4 | 2.26 | 12.22 | 7.16 | 8.96 | 9.62 | 6 | – | – |
| 2009 | 0 | 0 | 0.53 | 0.03 | 6.52 | 1.79 | 13.19 | 11.47 | 10.05 | 4.93 | 0.34 | 0 |
| 2010 | 0 | 0.13 | 0 | 0.72 | 4.25 | 8.44 | 8.53 | 7.56 | 7.24 | 4.51 | 0.12 | 0.36 |
| 2011 | 0 | 0.24 | 1.2 | 7.31 | 3.34 | 14.09 | 8.98 | 22.56 | 7.86 | 3.1 | – | 0 |
| 2012 | 1.57 | 1.01 | 0.17 | 3.15 | 1.52 | 6.08 | 10.2 | 14.78 | 8.35 | 4.28 | 1.86 | 1.15 |
| 2013 | 0.12 | 0.18 | 0.15 | 0.63 | 6.36 | 8.25 | 8.65 | 22.08 | 14.84 | 9.3 | 0 | 0 |
| 2014 | 0 | 2.33 | 1.77 | 0 | – | 5.78 | 7.35 | – | 11.37 | 1.7 | 0 | 0 |
| 2015 | 0.14 | 0.19 | 1.83 | 0.74 | 4.24 | 8.21 | 6.80 | 13.14 | 20.14 | 2.85 | 0 | 0 |

(continued)

**Table 2.5** (continued)

| Year | Jan | Feb | Mar | Apr | May | Jun | Jul | Aug | Sep | Oct | Nov | Dec |
|---|---|---|---|---|---|---|---|---|---|---|---|---|
| 2016 | 0 | 0.07 | 1.09 | 0 | 5.10 | 4.98 | 5.93 | 9.68 | 12.65 | 3.10 | 0.12 | 0 |
| 2017 | 0 | 0.14 | 0 | 0 | 1.26 | 11.02 | 7.80 | 12.65 | 9.89 | 4.02 | 0 | 0.14 |
| 2018 | 0.10 | 1.02 | 1.21 | 0.35 | 3.58 | 13.65 | 11.66 | 14.20 | 8.42 | 5.17 | 0 | 1.02 |

*Source* IMD (Indian Meteorological Department; '–' means data not available; Jan—January; Feb—February; Mar—March; Apr—April; May—May; Jun—June; Jul—July; Aug—August; Sep—September; Oct—October; Nov—November; Dec—December

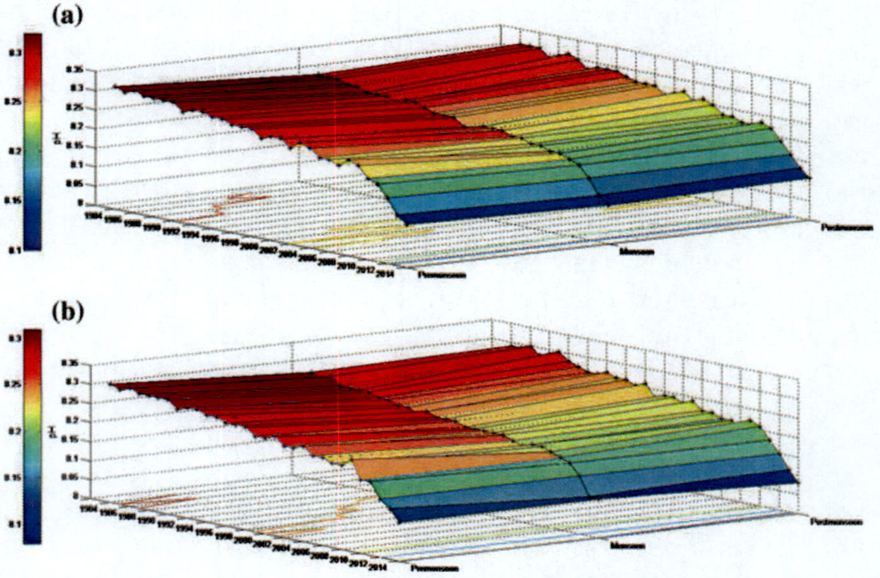


**Fig. 2.5** **a** Surface water pH at Diamond Harbour during high tide condition for three decades (1984–2014). **b** Surface water pH at Diamond Harbour during low tide condition for three decades (1984–2014)

by run-off, precipitation and water discharge suppress the surface water pH. Looking on the time series data of 31 years (1984–2014), the trend of acidification becomes quiet pronounced. In all the selected stations, an almost uniform decrease in surface water pH was observed. At Diamond Harbour, the pH value has decreased at the rate of 0.081 unit/decade in premonsoon, 0.074 unit/decade in monsoon and 0.077 unit/decade in postmonsoon (Fig. 2.5a, b). At Namkhana, the rate of decrease is 0.024 unit/decade in premonsoon and monsoon and 0.019 unit/decade in postmonsoon (Fig. 2.6a, b). The decrease of surface water pH at Ajmalmari is 0.012 unit/decade in premonsoon, 0.023 unit/decade in monsoon and 0.019 unit/decade in postmonsoon (Fig. 2.5a, b). This decrease is quite close to the observations recorded in several oceans and estuaries. For further research work in this vertical, we present here the backup data of Figs. 2.5, 2.6 and 2.7 as Table 2.6.

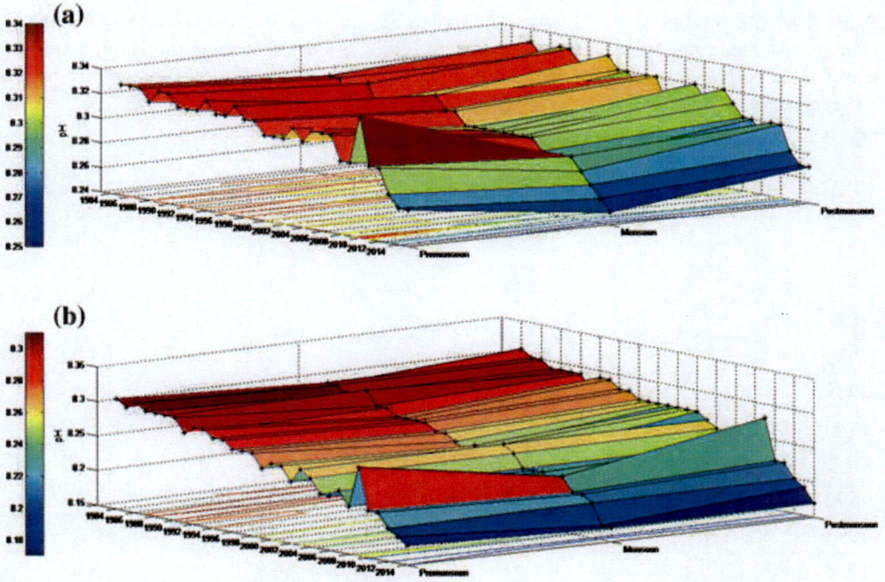

**Fig. 2.6** **a** Surface water pH at Namkhana during high tide condition for three decades (1984–2014). **b** Surface water pH at Namkhana during low tide condition for three decades (1984–2014)

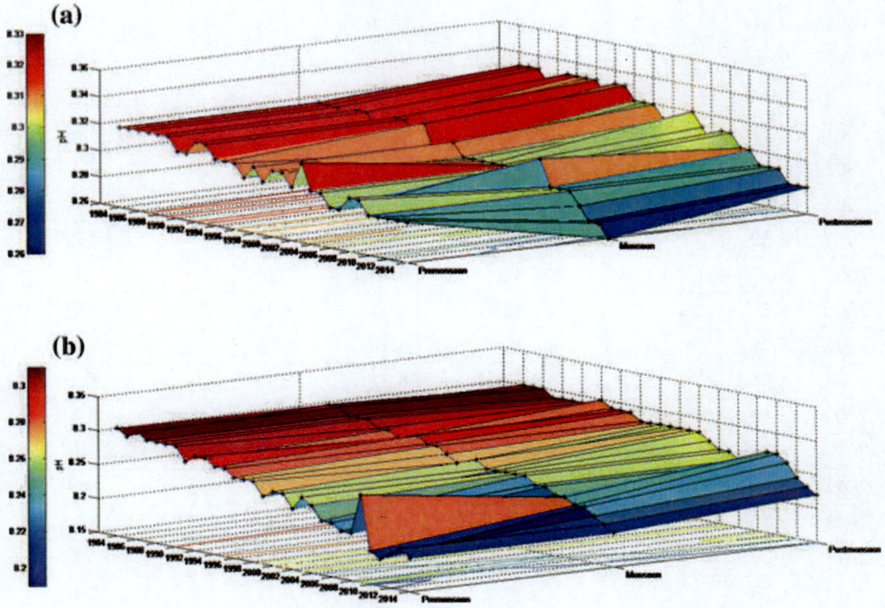

**Fig. 2.7** **a** Surface water pH at Ajmalmari during high tide condition for three decades (1984–2014). **b** Surface water pH at Ajmalmari during low tide condition for three decades (1984–2014)

**Table 2.6** Mean value of surface water pH at three sampling stations in Indian Sundarbans. Diamond Harbour and Namkhana are in western Indian Sundarbans along the Hooghly estuary and Ajmalmari is in the central Indian Sundarbans along the Matla estuary; Diamond Harbour is adjacent to western Indian Sundarbans

| Year | Ajmalmari | | Namkhana | | Diamond harbour | |
|---|---|---|---|---|---|---|
| | High tide | Low tide | High tide | Low tide | High tide | Low tide |
| 1984 | 8.33 | 8.31 | 8.34 | 8.31 | 8.33 | 8.31 |
| 1985 | 8.32 | 8.30 | 8.33 | 8.30 | 8.31 | 8.30 |
| 1986 | 8.32 | 8.31 | 8.33 | 8.31 | 8.32 | 8.31 |
| 1987 | 8.32 | 8.30 | 8.32 | 8.30 | 8.31 | 8.30 |
| 1988 | 8.32 | 8.30 | 8.33 | 8.30 | 8.31 | 8.30 |
| 1989 | 8.32 | 8.30 | 8.33 | 8.30 | 8.31 | 8.30 |
| 1990 | 8.31 | 8.30 | 8.32 | 8.30 | 8.31 | 8.30 |
| 1991 | 8.31 | 8.28 | 8.32 | 8.28 | 8.29 | 8.28 |
| 1992 | 8.32 | 8.29 | 8.32 | 8.29 | 8.30 | 8.29 |
| 1993 | 8.32 | 8.29 | 8.33 | 8.29 | 8.30 | 8.29 |
| 1994 | 8.31 | 8.28 | 8.32 | 8.28 | 8.29 | 8.28 |
| 1995 | 8.31 | 8.28 | 8.32 | 8.28 | 8.29 | 8.28 |
| 1996 | 8.31 | 8.27 | 8.33 | 8.27 | 8.28 | 8.27 |
| 1997 | 8.30 | 8.27 | 8.32 | 8.27 | 8.28 | 8.27 |
| 1998 | 8.31 | 8.27 | 8.32 | 8.27 | 8.28 | 8.27 |
| 1999 | 8.30 | 8.25 | 8.31 | 8.25 | 8.26 | 8.25 |
| 2000 | 8.31 | 8.26 | 8.31 | 8.26 | 8.27 | 8.25 |
| 2001 | 8.31 | 8.26 | 8.31 | 8.26 | 8.27 | 8.26 |
| 2002 | 8.30 | 8.24 | 8.32 | 8.24 | 8.25 | 8.24 |
| 2003 | 8.32 | 8.26 | 8.31 | 8.26 | 8.27 | 8.24 |
| 2004 | 8.30 | 8.25 | 8.32 | 8.25 | 8.26 | 8.25 |
| 2005 | 8.30 | 8.23 | 8.32 | 8.23 | 8.24 | 8.23 |
| 2006 | 8.29 | 8.23 | 8.32 | 8.23 | 8.24 | 8.23 |
| 2007 | 8.29 | 8.22 | 8.30 | 8.22 | 8.23 | 8.22 |
| 2008 | 8.30 | 8.22 | 8.30 | 8.22 | 8.25 | 8.22 |
| 2009 | 8.30 | 8.28 | 8.34 | 8.28 | 8.24 | 8.24 |
| 2010 | 8.29 | 8.20 | 8.30 | 8.20 | 8.21 | 8.19 |
| 2011 | 8.29 | 8.19 | 8.30 | 8.19 | 8.20 | 8.19 |
| 2012 | 8.29 | 8.20 | 8.28 | 8.20 | 8.17 | 8.15 |
| 2013 | 8.29 | 8.21 | 8.27 | 8.19 | 8.13 | 8.13 |
| 2014 | 8.29 | 8.20 | 8.27 | 8.18 | 8.11 | 8.10 |

We also made a seasonal survey for more than three decades at Sagar Island and Bali Island to confirm our previous data in spatio-temporal scale in a different anthropological scenario. (Tables 2.7 and 2.8). Sagar Island, in western Indian Sundarbans is noted for pilgrims (Figs. 2.8 and 2.9), tourism and fish landing activities, where as Bali Island in central Indian Sundarbans is adjacent to Reserve forest with minimum anthropogenic activities, except occasional tourism.

The Kapil Muni temple in Sagar South of western Indian Sundarban attracts about 10 lakhs devotees every year during Makar Sankranti (in mid-January) for taking holy bath in the estuarine water (Mitra and Zaman 2020), which also plays a major role in alteration of aquatic characteristics and can be considered as 'noise' in the domain of acidification of Sundarban estuaries. Apart from such sporadic anthropogenic impact, the western sector is also exposed to continuous discharge of wastes from industries, fish landing stations and busy markets that mix wastes of complex characters and alter the pH of the aquatic phase.

The seasonal behavior and decreasing trend of surface water pH are depicted in Figs. 2.10 and 2.11.

This decreasing trend is in alignment with the global trend. It has been documented that surface ocean pH has fallen by about 0.11 pH units from pre-industrial times till date. As pH is a measure of $H^+$ concentration and the pH scale is logarithmic, therefore for a decrease of 1 pH unit, the $H^+$ concentration increases by a factor of 10. Thus the decrease of 0.11 pH units is equivalent to about 29% increase in ocean $H^+$ concentration. It has been forecasted that if the burning of fossil fuels follows the present trend resulting in the rise of atmospheric $CO_2$, then the pH is likely to drop by 0.3–0.4 units by the end of the twenty-first century and increase ocean $H^+$ concentration by 100–150% above what it was in the pre-industrial times (http://www.us-ocb.org).

A small drop in pH value has great implication in the domain of biotic community. A very simple example may illustrate the fact. In human being, the arterial blood flow normally lies within the pH range of 7.35–7.45. A drop of 0.2–0.3 pH units can result in several health hazards like seizures, heart arrhythmia or coma. Similar modification has high probability to affect the marine organisms. Fundamental physiological processes of marine biota like photosynthesis, respiration, reproduction, calcification etc. are influenced by alternation of pH and carbonate ion concentration of the ambient aquatic phase.

The formation of calcareous skeletons or shells in the marine and estuarine molluscan species is an internal process where most organisms appear to convert $HCO_3^-$ to $CO_3^=$ to form calcium carbonate. However, this conversion creates protons ($H^+$) and hence the organisms must exert energy to transport the $H^+$ into the external environment (ambient water). One hypothesis is that as the pH of the ambient aquatic phase decreases, there is building of protons in the external environment due to which the organisms have to exert more energy to get rid of the protons resulting into lowering of calcification rate.

The process of ocean acidification happens through a stepwise manner as stated below:

**Table 2.7** Seasonal variation of surface water pH in Sagar Island (88° 04′ 0.51″E and 21° 37′ 49.90″N), a station in the western sector of Indian Sundarbans

| Year | Premonsoon | Monsoon | Postmonsoon |
|---|---|---|---|
| 1984 | 8.33 ± 0.03 | 8.31 ± 0.03 | 8.32 ± 0.03 |
| 1985 | 8.33 ± 0.03 | 8.32 ± 0.03 | 8.33 ± 0.03 |
| 1986 | 8.33 ± 0.03 | 8.32 ± 0.03 | 8.32 ± 0.03 |
| 1987 | 8.32 ± 0.03 | 8.3 ± 0.03 | 8.32 ± 0.03 |
| 1988 | 8.33 ± 0.03 | 8.31 ± 0.03 | 8.33 ± 0.03 |
| 1989 | 8.32 ± 0.03 | 8.32 ± 0.03 | 8.33 ± 0.03 |
| 1990 | 8.32 ± 0.03 | 8.3 ± 0.03 | 8.32 ± 0.03 |
| 1991 | 8.32 ± 0.03 | 8.31 ± 0.03 | 8.32 ± 0.03 |
| 1992 | 8.32 ± 0.03 | 8.3 ± 0.03 | 8.32 ± 0.03 |
| 1993 | 8.33 ± 0.03 | 8.31 ± 0.03 | 8.32 ± 0.03 |
| 1994 | 8.32 ± 0.03 | 8.31 ± 0.03 | 8.32 ± 0.03 |
| 1995 | 8.32 ± 0.03 | 8.3 ± 0.03 | 8.32 ± 0.03 |
| 1996 | 8.33 ± 0.03 | 8.32 ± 0.03 | 8.31 ± 0.03 |
| 1997 | 8.32 ± 0.03 | 8.31 ± 0.03 | 8.32 ± 0.03 |
| 1998 | 8.32 ± 0.03 | 8.31 ± 0.03 | 8.32 ± 0.03 |
| 1999 | 8.31 ± 0.03 | 8.3 ± 0.03 | 8.3 ± 0.03 |
| 2000 | 8.31 ± 0.03 | 8.3 ± 0.03 | 8.31 ± 0.03 |
| 2001 | 8.31 ± 0.03 | 8.3 ± 0.03 | 8.31 ± 0.03 |
| 2002 | 8.32 ± 0.03 | 8.3 ± 0.03 | 8.30 ± 0.03 |
| 2003 | 8.31 ± 0.03 | 8.3 ± 0.03 | 8.31 ± 0.03 |
| 2004 | 8.31 ± 0.03 | 8.29 ± 0.03 | 8.3 ± 0.03 |
| 2005 | 8.32 ± 0.03 | 8.3 ± 0.03 | 8.32 ± 0.03 |
| 2006 | 8.31 ± 0.03 | 8.3 ± 0.03 | 8.31 ± 0.03 |
| 2007 | 8.3 ± 0.03 | 8.29 ± 0.03 | 8.3 ± 0.03 |
| 2008 | 8.3 ± 0.03 | 8.28 ± 0.02 | 8.3 ± 0.03 |
| 2009 | 8.34 ± 0.04 | 8.29 ± 0.03 | 8.29 ± 0.03 |
| 2010 | 8.3 ± 0.03 | 8.29 ± 0.03 | 8.3 ± 0.03 |
| 2011 | 8.3 ± 0.03 | 8.28 ± 0.02 | 8.3 ± 0.03 |
| 2012 | 8.28 ± 0.02 | 8.27 ± 0.02 | 8.28 ± 0.02 |
| 2013 | 8.27 ± 0.02 | 8.26 ± 0.02 | 8.27 ± 0.02 |
| 2014 | 8.27 ± 0.02 | 8.25 ± 0.02 | 8.27 ± 0.02 |
| 2015 | 8.28 ± 0.02 | 8.25 ± 0.02 | 8.26 ± 0.02 |
| 2016 | 8.27 ± 0.02 | 8.23 ± 0.02 | 8.26 ± 0.02 |
| 2017 | 8.25 ± 0.03 | 8.21 ± 0.03 | 8.24 ± 0.01 |
| 2018 | 8.23 ± 0.02 | 8.17 ± 0.04 | 8.21 ± 0.02 |
| Mean | 8.30 ± 0.03 | 8.28 ± 0.02 | 8.30 ± 0.02 |
| Decrease | 0.003/yr | 0.004/yr | 0.003/yr |

**Table 2.8** Seasonal variation of surface water pH in Bali Island (88° 39′ 46″E and 22° 15′ 45.00″N), a station in the central sector of Indian Sundarbans

| Year | Premonsoon | Monsoon | Postmosoon |
|---|---|---|---|
| 1984 | 8.34 ± 0.04 | 8.32 ± 0.04 | 8.33 ± 0.04 |
| 1985 | 8.34 ± 0.04 | 8.32 ± 0.04 | 8.33 ± 0.04 |
| 1986 | 8.34 ± 0.04 | 8.32 ± 0.04 | 8.33 ± 0.04 |
| 1987 | 8.33 ± 0.04 | 8.31 ± 0.04 | 8.32 ± 0.04 |
| 1988 | 8.32 ± 0.04 | 8.32 ± 0.04 | 8.32 ± 0.04 |
| 1989 | 8.32 ± 0.04 | 8.32 ± 0.04 | 8.33 ± 0.04 |
| 1990 | 8.33 ± 0.04 | 8.31 ± 0.04 | 8.33 ± 0.04 |
| 1991 | 8.33 ± 0.04 | 8.31 ± 0.04 | 8.33 ± 0.04 |
| 1992 | 8.33 ± 0.04 | 8.32 ± 0.04 | 8.33 ± 0.04 |
| 1993 | 8.33 ± 0.04 | 8.32 ± 0.04 | 8.32 ± 0.04 |
| 1994 | 8.33 ± 0.04 | 8.31 ± 0.04 | 8.33 ± 0.04 |
| 1995 | 8.31 ± 0.04 | 8.32 ± 0.04 | 8.33 ± 0.04 |
| 1996 | 8.31 ± 0.04 | 8.3 ± 0.03 | 8.32 ± 0.04 |
| 1997 | 8.3 ± 0.03 | 8.3 ± 0.03 | 8.32 ± 0.04 |
| 1998 | 8.31 ± 0.04 | 8.31 ± 0.04 | 8.32 ± 0.04 |
| 1999 | 8.3 ± 0.03 | 8.3 ± 0.03 | 8.31 ± 0.04 |
| 2000 | 8.31 ± 0.04 | 8.3 ± 0.03 | 8.31 ± 0.04 |
| 2001 | 8.31 ± 0.04 | 8.3 ± 0.03 | 8.31 ± 0.04 |
| 2002 | 8.32 ± 0.04 | 8.3 ± 0.03 | 8.32 ± 0.04 |
| 2003 | 8.32 ± 0.04 | 8.3 ± 0.03 | 8.31 ± 0.04 |
| 2004 | 8.3 ± 0.03 | 8.3 ± 0.03 | 8.3 ± 0.03 |
| 2005 | 8.3 ± 0.03 | 8.3 ± 0.03 | 8.3 ± 0.03 |
| 2006 | 8.29 ± 0.03 | 8.3 ± 0.03 | 8.29 ± 0.03 |
| 2007 | 8.29 ± 0.03 | 8.31 ± 0.04 | 8.3 ± 0.03 |
| 2008 | 8.3 ± 0.03 | 8.29 ± 0.03 | 8.3 ± 0.03 |
| 2009 | 8.34 ± 0.04 | 8.3 ± 0.03 | 8.3 ± 0.03 |
| 2010 | 8.31 ± 0.04 | 8.29 ± 0.05 | 8.29 ± 0.03 |
| 2011 | 8.31 ± 0.04 | 8.28 ± 0.03 | 8.29 ± 0.03 |
| 2012 | 8.31 ± 0.04 | 8.27 ± 0.03 | 8.28 ± 0.03 |
| 2013 | 8.29 ± 0.04 | 8.27 ± 0.03 | 8.28 ± 0.03 |
| 2014 | 8.29 ± 0.04 | 8.26 ± 0.03 | 8.28 ± 0.03 |
| 2015 | 8.28 ± 0.03 | 8.25 ± 0.03 | 8.27 ± 0.03 |
| 2016 | 8.27 ± 0.03 | 8.24 ± 0.03 | 8.25 ± 0.02 |
| 2017 | 8.26 ± 0.04 | 8.21 ± 0.02 | 8.25 ± 0.02 |
| 2018 | 8.26 ± 0.03 | 8.19 ± 0.02 | 8.22 ± 0.01 |
| Mean | 8.31 ± 0.04 | 8.29 ± 0.03 | 8.30 ± 0.03 |
| Decrease | 0.002/yr | 0.004/yr | 0.003/yr |

**Fig. 2.8** Kapil Muni Ashram at Sagar Island

**Step 1**: In the first step, the atmospheric $CO_2$ dissolves in seawater to form aqueous $CO_2$ along with $H_2CO_3$

$$CO_2 + H_2O \leftrightarrow H_2CO_3$$

**Step 2**: In this step $H_2CO_3$ dissociates to produce $HCO_3^-$ and protons ($H^+$)

$$H_2CO_3 \leftrightarrow HCO_3^- + H^+$$

**Step 3**: In this step $HCO_3^-$ has the tendency to dissociate into $CO_3^=$ and protons ($H^+$), which results in the lowering of pH

$$HCO_3^- \leftrightarrow CO_3^= + H^+$$

It is to be noted in this context that the total dissolved inorganic carbon (DIC) in seawater is the sum total of aqueous $CO_2$, $HCO_3^-$ and $CO_3^=$. Thus,

$$DIC = CO_2(aq) + HCO_3^- + CO_3^=$$

The distribution of DIC between the species varies with seawater pH. Typically surface water of today's ocean have a pH around 8.1, which means that $HCO_3^-$ is the

**Fig. 2.9** Inside view of Kapil Muni Ashram at Sagar Island. About 10–15 lakhs people visit this place every year at Makar Sankranti (during mid January)

dominant species representing about 90% of DIC whereas, $CO_2$ (aq) represents about <1% of the DIC. DIC in the marine and estuarine system shifts with season. The three major seasons in the tropical estuaries are premonsoon (with lowest discharge), monsoon (with highest discharge) and postmonsoon (with moderate discharge). It has been documented that freshwater input in the coastal and estuarine ecosystem lowers the pH value. This hypothesis is confirmed after the observation of Trenton, Schuylkill and Christina rivers in the Delaware Bay (Table 2.9).

In the deltaic complex of Sundarbans similar results were seen when the data of Tables 2.7 and 2.8 were analyzed. In both Sagar Island (along the Hooghly estuary) and Bali Island (along the Matla estuary) the pH level decreased maximum in monsoon (0.004/yr), when the dilution factor is maximum in the Sundarban estuary. These two major estuaries have significantly different physico-chemical characteristics. The Hugli estuary in the western sector of Indian Sundarbans receives freshwater from the Farakka discharge. The Matla estuary in the central sector of the Indian Sundarbans practically receives no freshwater due to blockage of the Bidyadhari River by excessive siltation since the late fifteenth century (Chaudhuri and Chaudhury 1994; Mitra 2013).

The hyposaline environment of western Indian Sundarbans is primarily the result of Farakka barrage discharge situated in the upstream region of Ganga-Bhagirathi-Hooghly river system. 10-year surveys (1999–2008) on water discharge from Farakka

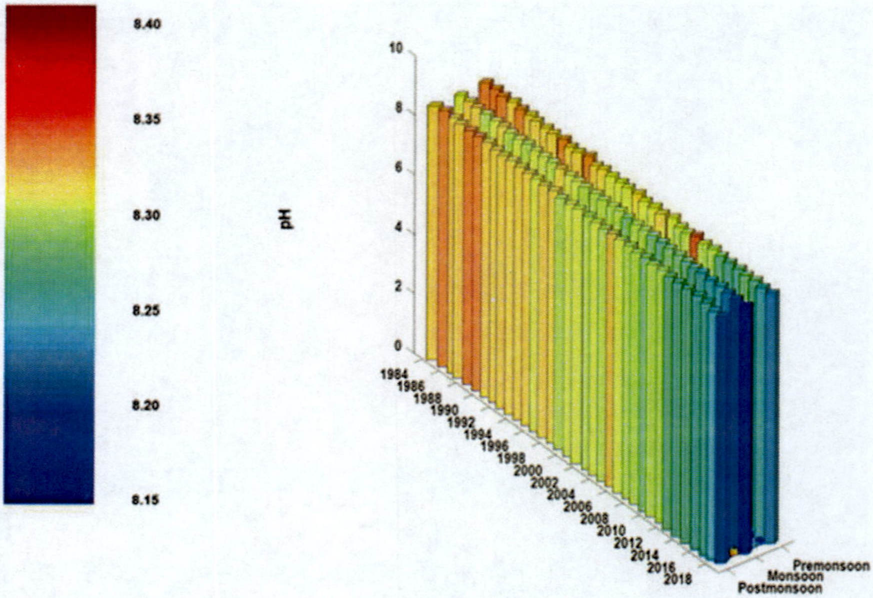


**Fig. 2.10** Seasonal variation of surface water pH in Sagar Island (88° 04′ 0.51″E and 21° 37′ 49.90″N), a station in the western sector of Indian Sundarbans

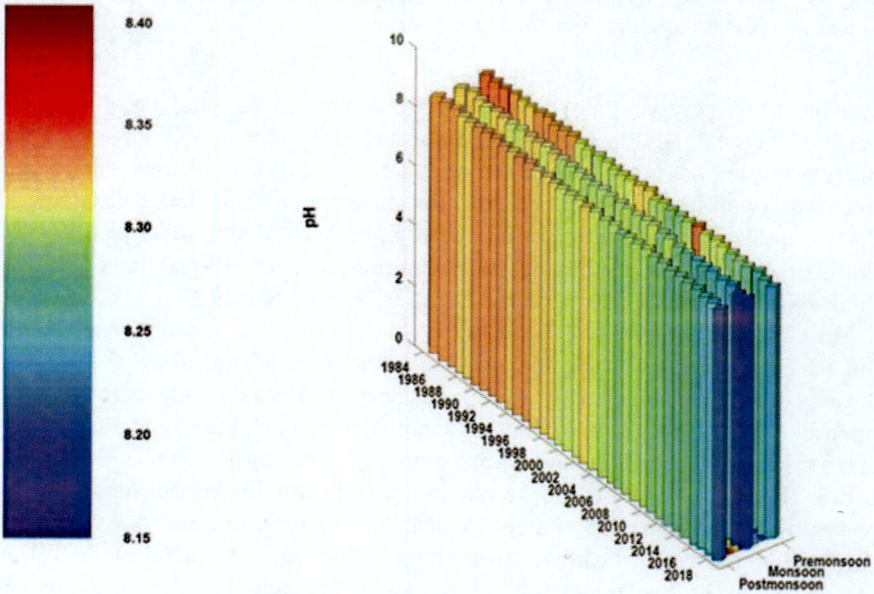


**Fig. 2.11** Seasonal variation of surface water pH in Bali Island (88° 39′ 46″E and 22° 15′ 45.00″N), a station in the central sector of Indian Sundarbans

**Table 2.9** Seasonal variability of physico-chemical parameters in three river estuaries in the Delaware Bay

| Location | Month | Discharge ($m^3s^{-1}$) | pH (at 25 °C) | DIC ($\mu$molkg$^{-1}$) |
|---|---|---|---|---|
| Trenton | March | 182 | 8.8 | 973.4 |
| | April | 442 | 7.8 | 745.2 |
| | May | 190 | 8.8 | 856.5 |
| | June | 199 | 8.2 | 857.8 |
| | July | 1127 | 7.2 | 783.5 |
| | September | 183 | 8.2 | 454.2 |
| | October | 170 | 8.1 | 1095.2 |
| Schuylkill | March | 52 | 8.9 | 1421.2 |
| | April | 32 | 8.1 | 1682.9 |
| | May | 105 | 7.9 | 1400.1 |
| | June | 271 | 7.7 | 1095.3 |
| | July | 60 | 7.8 | 1506.1 |
| | September | 19 | 8.1 | 2071.3 |
| | October | 34 | 8.3 | 1869.3 |
| Christina | March | 14 | 7.7 | 1056.5 |
| | April | 15 | 7.5 | 1076.4 |
| | May | 11 | 7.7 | 1134.1 |
| | June | 32 | 7.5 | 1089.4 |
| | July | 6 | 7.9 | 1326.9 |
| | September | 7 | 8.0 | 1188.6 |
| | October | 7 | 8.0 | 1199.6 |

dam revealed an average discharge of $(3.7 \pm 1.15) \times 10^3$ $m^3s^{-1}$. Higher discharge values were observed during the monsoon with an average of $(3.81 \pm 1.23) \times 10^3$ $m^3s^{-1}$, and the maximum of the order 4524 $m^3s^{-1}$ during freshet (September). Considerably lower discharge values were recorded during premonsoon with an average of $(1.18 \pm 0.08) \times 10^3$ $m^3s^{-1}$, and the minimum of the order 846 $m^3s^{-1}$ during May. During postmonsoon discharge, values were moderate with an average of $(1.98 \pm 0.97) \times 10^3$ $m^3s^{-1}$ (Mitra 2013).

The central sector represents a hypersaline environment due to complete obstruction of the fresh water flow from the upstream region owing to Bidyadhari siltation since the late fifteenth century (Chaudhuri and Choudhury 1994; Banerjee et al. 2013; Mitra 2013; Sengupta et al. 2013). The Matla estuary in the central Indian Sundarban cannot be referred to as an ideal estuary as there is no head on discharge or dilution of the system with fresh water. Thus Matla can be designated as a tidal channel, whose survival depends on the tidal flow from Bay of Bengal.

The salinity profiles of these two estuaries stand as proxy to freshwater discharge. The average salinity of the stations along the Hooghly estuarine stretch in western

Indian Sundarbans is less compared to the stations in central Indian Sundarbans along the Matla estuary. We have recorded the surface water salinity of ten stations, five along each of the two estuaries (Table 2.10 and Fig. 2.12).

It is observed from the data sets of aquatic salinity that during 2016, 2017 and 2018 the average salinity in western Indian Sundarbans are 20.67 psu, 18.6 psu and 17.18 psu respectively. In case of central Indian Sundarbans the average salinity during 2016, 2017 and 2018 are 21.66 psu, 22.98 psu and 24.37 psu respectively (Table 2.11). It is to be noted in this context that in western Indian Sundarbans the salinity has decreased with the passage of time, but the order is reverse in case of central Indian Sundarbans where the salinity showed an increase with time.

The comparatively higher inflow of freshwater in the western sector than central sector has decreased the pH also, which implies that apart from climate induced acidification freshwater also regulates the pH in an estuarine system, which is not the case in high seas and oceans. The regulatory role of freshwater in pH oscillation can be confirmed while critically analyzing the lowering of pH in eastern Indian Sundarbans, which is in the Reserve Forest zone with minimum anthropogenic influence. The region is hyposaline as it receives the fresh water due to inter-connection with several creeks and channels of Harinbhanga estuary (the aquatic border of India and Bangladesh Sundarbans) with the tributaries of Bangladesh Sundarbans that arise from Padma Meghna river system. Also the quantum of coastal and estuarine vegetation as mostly concentrated in the Reserve Forest area of eastern Indian Sundarbans do play a great role in altering the pH level of the aquatic phase.

The direct influence of anthropogenic activities on the pace of acidification is confirmed through an interesting study conducted by the present authors in three stations in the lower Gangetic delta region during the pre-COVID-19 lockdown (18th–22nd March, 2020) and COVID-19 lockdown phases (26th March–30th April, 2020) as discussed in details in Annexure 2. The results of this study are the outcomes of an acid test on the confirmatory role of anthropogenic activities in the process of acidification.

## 2.4 Take Home Messages

A. Estuaries are the live matrix of Indian Sundarban deltaic complex, on which the unique spectrum of biological diversity is embedded. In Indian Sundarbans, approximately 2069 sq. km area is occupied by tidal river system or estuaries, which finally end up in the Bay of Bengal. The seven main estuaries, from west to east are Hooghly, Muriganga, Saptamulki, Thakuran, Matla, Goashaba and Haribhanga. The estuaries of Indian Sundarbans exhibit distinct variation in salinity in the same time period.

B. The deltaic lobe of Indian Sundarbans experiences a moderate type of climate because of its location adjacent to the Bay of Bengal as well as due to regular tidal flushing in the estuaries. Wave actions, micro and macro tidal cycles, long shore currents are recorded in most of the islands of the ecosystem. Coastal

**Table 2.10** Sampling stations in western (stations 1–5) and central Indian (6–10) Sundarbans with salient features

| Study site | Longitude and latitude | Site description |
|---|---|---|
| Harinbari (Stn. 1) | 88° 04′ 22.88″E 21° 46′ 53.07″N | Situated in the western region of Indian Sundarbans almost in the middle of the Sagar Island; receives the water of the Hooghly River |
| Chemaguri (Stn. 2) | 88° 08′ 49.01″E 21° 39′ 42.88″N | Situated on the south-eastern side of Sagar Island and receives the water of the Mooriganga River |
| Sagar South (Stn. 3) | 88° 04′ 0.51″E 21° 37′ 49.90″N | Situated on the south-western part of the Sagar Island at the confluence of the River Hooghly and the Bay of Bengal. Anthropogenically stressed zone due to presence of passenger jetties, fishing activities and pilgrimage |
| Lothian Island (Stn. 4) | 88° 19′ 8.47″E 21° 39′ 08.04″N | Situated east of Bakkhali island; a Wildlife Sanctuary; faces the River Saptamukhi |
| Prentice Island (Stn. 5) | 88° 17′ 3.62″E 21° 42′ 43.31″N | Situated north of Lothian island; receives the water of the Saptamukhi River |
| Canning (Stn. 6) | 88° 41′ 04.43″E 22° 19′ 03.20″N | Situated in the central part of the Indian Sundarbans and faces the mighty River Matla, a tide-fed river. Due to presence of fish landing stations, passenger jetties and busy market, the area is anthropogenically stressed |
| Sajnekhali (Stn. 7) | 88° 48′ 15.78″E 22° 06′ 34.19″N | A Wildlife Sanctuary and a part of Sundarban Tiger Reserve; adjacent to River Bidhya and Gomor. Tourism pressure is extremely high in this station particularly during postmonsoon |
| Chotomollakhali (Stn. 8) | 88° 54′ 26.71″E 22° 10′ 40.00″N | Situated in the upper portion of Central Indian Sundarban adjacent to Jhila forest; receives the water of Rangabelia and Korankhali Rivers |
| Satjelia (Stn. 9) | 88° 52′ 49.51″E 22° 05′ 17.86″N | Situated adjacent to River Duttar in the upper region of Central Indian Sundarban facing western part of the Jhilla forest |
| Pakhiralaya (Stn. 10) | 88° 48′ 29.00″E 22° 07′ 07.23″N | Situated adjacent to River Gomor; opposite to Sajnekhali Wild Life Sanctuary |

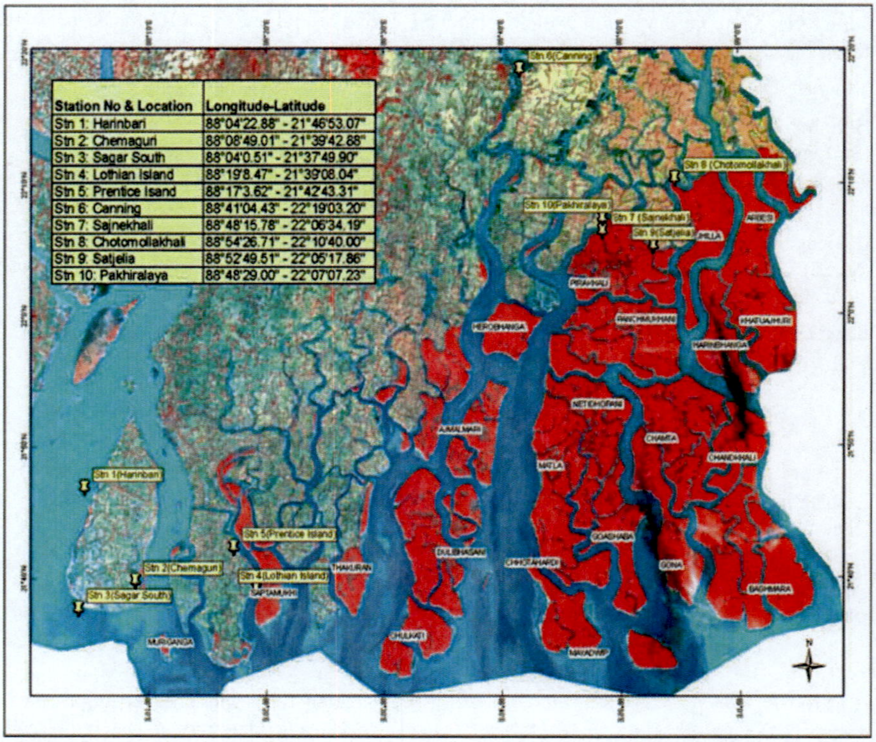


Fig. 2.12 Selected stations in Indian Sundarbans

processes are very dynamic and are accelerated by tropical cyclones, which is locally called "Kal Baisakhi" (Norwesters). Average annual maximum atmospheric temperature is around 35 °C. The seasons in study area may be conveniently categorized into premonsoon (March–June), monsoon (July–October) and postmonsoon (November–February). Each season has a characteristic feature of its own, which is very distinct and unique.

C. The surface water pH is an important environmental parameter on which the life of any aquatic organism depends. It shows more or less a uniform value throughout the seasons with minor fluctuations mainly in the monsoon months, when, dilution by run-off, precipitation and water discharge suppress the surface water pH. Looking on the time series data of 31 years (1984–2014), the trend of acidification becomes quiet pronounced in Indian Sundarban estuaries. In all the selected stations, an almost uniform decrease in surface water pH was observed.

D. The comparatively higher inflow of freshwater in the western sector than central sector of Indian Sundarbans has decreased the pH, which implies that apart from climate induced acidification freshwater also regulates the pH in an estuarine system, which is not the case in high seas and oceans. In the deltaic

**Table 2.11** Water salinity (in psu) in the selected stations of western and central Indian Sundarbans

| S. no. | Name | 2016 | 2017 | 2018 | Average | Sector |
|---|---|---|---|---|---|---|
| | Harinbari (Stn. 1) | 8.34 | 6.49 | 5.24 | 6.69 | Western sector average water salinity = 18.84 psu |
| | Chemaguri (Stn. 2) | 18.75 | 15.38 | 14.94 | 16.36 | |
| | Sagar South (Stn. 3) | 24.69 | 22.88 | 20.91 | 22.83 | |
| | Lothian island (Stn. 4) | 25.74 | 24.68 | 22.42 | 24.28 | |
| | Prentice island (Stn. 5) | 25.81 | 23.99 | 22.39 | 24.06 | |
| Average salinity | | 20.67 | 18.68 | 17.18 | | |
| | Canning (Stn. 6) | 9.54 | 11.63 | 12.51 | 11.23 | Central sector average water salinity = 23.05 psu |
| | Sajnekhali (Stn. 7) | 24.90 | 26.31 | 28.42 | 26.54 | |
| | Chotomollakhali (Stn. 8) | 23.11 | 25.79 | 26.03 | 24.98 | |
| | Satjelia (Stn. 9) | 25.95 | 26.10 | 28.75 | 26.93 | |
| | Pakhiralaya (Stn. 10) | 24.79 | 25.08 | 26.13 | 25.57 | |
| Average salinity | | 21.66 | 22.98 | 24.37 | | |

complex of Sundarbans similar results were seen when the data of stations along two major estuaries, Hooghly and Matla were analyzed. In both Sagar Island (along the Hooghly estuary) and Bali Island (along the Matla estuary) the pH level decreased maximum in monsoon (0.004/yr), when the dilution factor is maximum in the Sundarban estuary.

E. A small drop in pH value has great implication in the domain of biotic community. Fundamental physiological processes of marine biota like photosynthesis, respiration, reproduction, calcification etc. are influenced by alternation of pH and carbonate ion concentration of the ambient aquatic phase. The formation of calcareous skeletons or shells in the marine and estuarine molluscan species is an internal process where most organisms appear to convert $HCO_3^-$ to $CO_3^=$ to form calcium carbonate. However, this conversion creates protons ($H^+$) and hence the organisms must exert energy to transport the $H^+$ into the external environment (ambient water). One hypothesis is that as the pH of the ambient aquatic phase decreases, there is building of protons in the external environment due to which the organisms have to exert more energy to get rid of the protons resulting into lowering of calcification rate.

## Annexure 2: How the COVID Lockdown Phase Impacted the Pace of Acidification?

### Introduction

The mighty River Ganges flowing through the city of Kolkata serves as the life line of the people of the megacity. The River provides several ecosystem services like production of fishes, mode of transportation, recreational activities, sports, performance of religious rituals etc. (Mitra 2019). However, the river is still treated as the bin of all the wastes arising from domestic and industrial activities. In addition to this, wastes from automobile repairing units, emission of Green House Gases (GHGs) from industrial units and air-condition machines have made the city atmosphere highly polluted. Increased emission of $CO_2$ from anthropogenic sources (Mitra et al. 2020) has enhanced the entry of atmospheric $CO_2$ to the river water resulting in the formation of carbonic acid, thereby shifting the pH to a lower value. This is commonly referred as acidification and has been reported in several estuaries in the state of West Bengal (Mitra 2013; Mitra and Zaman 2014; Mitra and Zaman 2016; Mitra 2019). In this study, we have carried out a comparative analysis on surface water pH of the River Ganges between the pre-COVID-19 (18th March–22nd March, 2020) and COVID-19 lockdown phase (25th March–30th April, 2020). The COVID-19 pandemic provided an unique opportunity to venture in to such comparative picture as there is no movements of vessels, boats and trawlers in the river and also all the industrial operations have been ceased during the lockdown phase of the state—a condition of retrieving the natural parameters with the withdrawal of anthropogenic influences. This paper, therefore, can be treated as a comparative account of natural versus human induced factors using River Ganges as the test bed.

### Materials and Methods

### Study Site

Kolkata, Capital city of the maritime state of West Bengal, is the third largest city in India and situated in the east bank of the Hooghly River with an area of 187.33 $km^2$. Three sites along the bank of the River Ganges were selected for the present study namely Ramkrishna Ghat (22° 34′ 19.8″N; 88° 20′ 17.0″E), Botanical Garden (22° 33′ 06.4″N; 88° 18′ 06.6″E) and Babughat (22° 34′ 10.3″N; 88° 20′ 28.5″E).

### Measurement of Aquatic pH

pH of the surface water in the selected sampling station was measured during high tide condition with a portable pH meter (sensitivity = ± 0.02). The measurement was carried out during pre-COVID-19 lockdown (18th–22nd March, 2020) and COVID-19 lockdown phases (26th March–30th April, 2020).

### Statistical Analysis

ANOVA was carried out to know whether significant variation of aquatic pH exists between sites and time phase (pre-COVID-19 and COVID-19 lockdown phases).

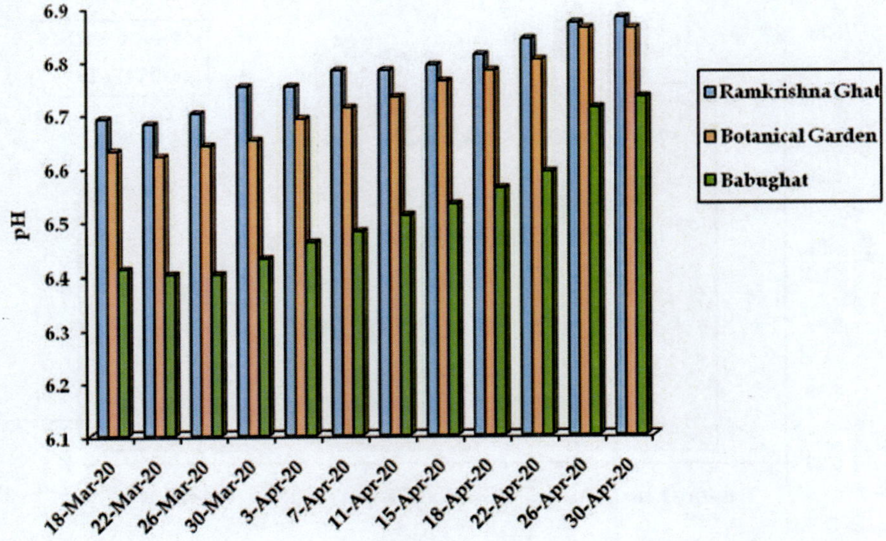


**Fig. 2.13** Spatio-temporal variation of pH duirng the study period

**Results**

The spatio-temporal variations of aquatic pH in the three study sites are highlighted in Fig. 2.13. In the lockdown phase due to COVID-19, the pH value increases in three selective sites (Fig. 2.14).

Results of ANOVA show significant variations between stations and between pre-COVID-19 and COVID-19 lockdown phases ($p < 0.01$) (Table 2.12).

**Discussion**

The increase of atmospheric carbon dioxide in West Bengal, a maritime state in northeast coast of India has touched almost 51% since 1980 (Mitra 2019). The gradual increase of carbon dioxide coupled with unplanned expansion of shrimp culture in places like Sundarbans (adjacent to the city), unplanned urban development and industrial activities has lowered the aquatic pH considerably (Mitra and Zaman 2016). It is interesting to note that in all the sampling stations selected in the present study, the pH has increased by 2.84%, 3.46% and 4.99% at Ramkrishna Ghat, Botanical garden and Babughat respectively. This may be due to complete closure of all industrial operations along the bank of the River Ganges. Also the movements of vessels, floatels (a boat or ship that serves as a hotel, sometimes permanently moored to a dock) and other recreational activities have also been ceased to abide by the rules of mass gathering and social distancing during the COVID—19 lock down phase. The $CO_2$ level of the atmosphere also showed a considerable dip in the value (Mitra et al. 2020). All these have posed a joint impact on pH level of the aquatic phase (synergistic effect) due to which a sudden turn in the trend of acidification has taken place.

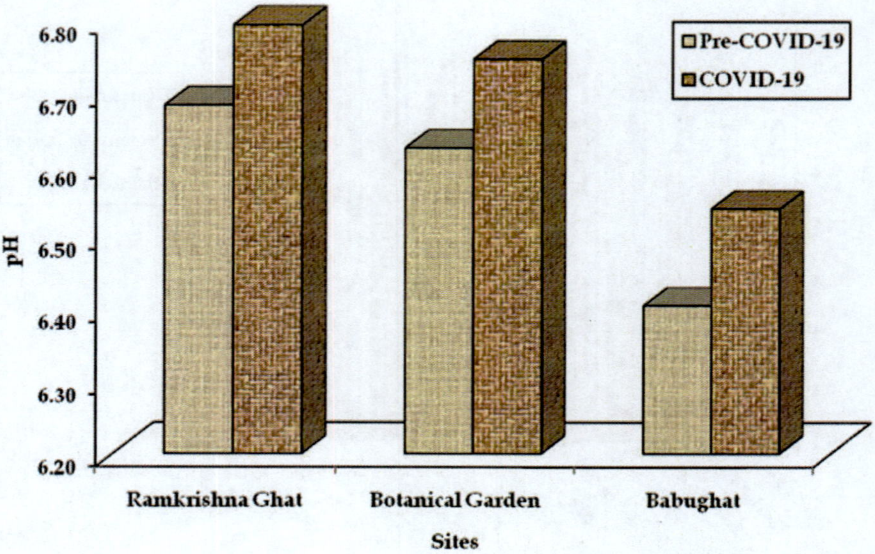


**Fig. 2.14** pH variation in the three selected sampling sites between pre-COVID-19 and COVID-19 lockdown period

**Table 2.12** ANOVA of the pH value between sites and between pre-COVID-19 and COVID-19 period

| Source of variation | SS | df | MS | F | P-value | F crit |
|---|---|---|---|---|---|---|
| Between Pre-COVID-19 and COVID-19 | 0.022571 | 1 | 0.022571 | 288.7505 | 0.003445 | 18.51282 |
| Between Sites | 0.080143 | 2 | 0.040071 | 512.6418 | 0.001947 | 19 |
| Error | 0.000156 | 2 | 7.82E-05 | | | |
| Total | 0.10287 | 5 | | | | |

The phenomenon of acidification due to climate change is never supportive and congenial for the aquatic organisms preferably for the survival of molluscs and other aquatic organisms with calcareous shell. Thus COVID-19 lockdown phase served as a boon to these faunal communities by providing a reverse swing in the process of acidification.

## References

Banerjee K, Sengupta K, Raha A, Mitra A (2013) Salinity based allometric equations for biomass estimation of Sundarban mangroves. Biomass Bioenerg 56:382–391

Caldeira K, Wickett ME (2003) Anthropogenic carbon and ocean pH. Nature 425:365

Canadell JG, Le Quéré C, Raupach M, Field CB, Buitenhuis ET, Ciais P, Conway TJ, Gillet NP, Hougton RA, Marland G (2007) Contributions to accelerating atmospheric $CO_2$ growth from economic activity, carbon intensity, and efficiency of natural sinks. Proc Natl Acad Sci USA 104:18866–18870

Chaudhuri AB, Choudhury A (1994) Mangroves of the Sundarbans India. Published by IUCN

Feely RA, Sabine CL, Lee K, Berelson W, Kleypas JA, Fabry VJ, Millero FJ (2004) Impact of anthropogenic $CO_2$ on the $CaCO_3$ system in the oceans. Science 281:237–240

Hansen J, Sato M, Ruedy R, Kharecha P, Lacis A, Miller R, Nazarenko L, Lo K, Schmidt GA, Russell G, Aleinov I, Bauer S, Baum E, Cairns B, Canuto V, Chandler M, Cheng Y, Cohen A, Del Genio A, Faluvegi G, Fleming E, Friend A, Hall T, Jackman C, Jonas J, Kelley M, Kiang NY, Koch D, Labow G, Lerner J, Menon S, Novakov T, Oinas V, Perlwitz J, Perlwitz J, Rind D, Romanou A, Schmunk R, Shindell D, Stone P, Sun S, Streets D, Tausnev N, Thresher D, Unger N, Yao M, Zhang S (2007) Dangerous human-made interference with climate: a GISS model E study. Atmos Chem Phys 7:2287–2312

Hawkins DG (2004) No exit: thinking about leakage from geologic carbon storages sites. Energy 29:1571–1578

Mitra A (2000) The Northeast coast of the Bay of Bengal and deltaic Sundarbans. In: Sheppard C (ed) Seas at the Millennium—an environmental evaluation. Elsevier Science, UK, pp 143–157

Mitra A (2013) Sensitivity of mangrove ecosystem to changing climate. Published by Springer, ISBN: 978-81-322-1508-0 (Print) 978-81-322-1509-7 (Online)

Mitra A (2019). Estuarine pollution in the Lower Gangetic Delta, published by Springer International Publishing, ISBN 978-3-319-93305-4, XVI, 371

Mitra A, Zaman S (2016) Basics of Marine and Estuarine Ecology. Springer ISBN 978-81-322-2705-2

Mitra A, Zaman S (2020) Environmental science—a ground zero observation on the Indian Subcontinent. Springer International Publishing, Switzerland

Mitra A, Ray Chadhuri T, Mitra A, Pramanick P, Zaman S (2020) Impact of COVID-19 related shutdown on atmospheric carbon dioxide level in the city of Kolkata. Parana J Sci Edu 6(3):84–92

Odum EP (1971) Fundamentals of ecology. Published by Philadelphia, Saunders, 3rd edn

Sabine CL, Feely RA, Gruber N, Key RM, Lee K, Bullister JL, Wanninkhof R, Wong CS, Wallace DWR, Tilbrook B, Millero FJ, Peng PH, Kozyr A, Ono T, Rios AF (2004) The oceanic sink for anthropogenic $CO_2$. Science 305:367–371

Sengupta K, Roy Chowdhury M, Bhattacharya SB, Raha A, Zaman S, Mitra A (2013) Spatial variation of stored carbon in *Avicennia alba* of Indian Sundarbans. Disc Nat 3(8):19–24

Siegenthaler U, Stocker TF, Monnin E, Luethi D, Schwander J, Stauffer B, Raynaud D, Barnola JM, Fisher H, Masson-Delmotte V, Jouzel J (2005) Stable carbon cycle–climate relationship during the last Pleistocene. Science 310:1313–1317

## Internet References

http://www.espl.noaa.gov/gmd/ccgg/trends
http://mcbi.marineconservation.org/publications/pub_pdfs/feely_etal_2008_pices.pdf.
http://www.us-ocb.org

# Chapter 3
# Impact of Acidification on Fishes

**Abstract** Fish protein is one of the primary sources that provide food security to the rapidly rising population of the planet Earth. It has been documented that several fish species exhibit a change in their sensory behavior during their stay in water bodies with low pH. The spawning pattern of fish is also affected if the pH level changes. A significant variation in finfish juvenile diversity and population size is observed in few experimental sites in the mangrove dominated Indian Sundarbans over a period of more than three decades (1984–2018), which is directly related to gradual lowering of pH as depicted through a time series analysis of aquatic pH in and around the deltaic complex. The chapter also presents an interesting change in the spawning zone of Indian shad (*Tenualosa ilisha*) which has shifted towards the downstream direction of the Hooghly estuary with the passage of time.

## 3.1 Taxonomic Diversity of Fish in Indian Sundarbans

Biodiversity refers to all life on the planet Earth and ranges from the microbes in the human gut to blue whale of the ocean. So far 1.7 millions species have been indentified and recorded on the Earth, but the total number is thought to be somewhat between 5 to 100 million. Biodiversity encompasses three important tiers or segments of biosphere, namely, habitat diversity, species diversity and genetic diversity. The major portion of the coastal zone of Bay of Bengal in the maritime state of West Bengal (India) is covered with luxuriant mangrove vegetation and the brackish-cum-saline aquatic phase of this environment nourishes the world's most famous mangrove chunk- the Sundarbans. In fact, mangroves forest, mangrove swamps and backwater of Sundarbans form a productive and protective margin of coastal West Bengal which is also the breeding ground and nursery of a wide spectrum of fish. We will be discussing in this section about the finfish variety of the estuarine waters of Indian Sundarbans.

Several researches have been carried out on the taxonomic diversity of fish species in the lower Gangetic delta. Pillay (1967) estimated the species number to be more than 120. Jhingran (1982) documented a total of 172 species and stated that the diversity is comparatively more in the high saline zone of the mangrove dominated



A. Mitra and S. Zaman, *Estuarine Acidification*,
https://doi.org/10.1007/978-3-030-84792-0_3

Indian Sundarbans. His estimate reveals 73 species of fresh water origin and 99 species of marine/higher salinity origin. Mandal and Nandi (1989) documented 141 species under 100 genera, while Chaudhuri and Choudhury (1994) recorded 250 species under 96 genera in the aquatic sub-system of Indian Sundarbans. Khan (2003) recorded 107 species from Sundarban Biosphere Reserve region, but this figure does not include the species restricted in the low saline upper zone of the Hooghly-Matla estuarine complex. The fish fauna of the estuarine waters around Sundarbans has been classified into residents and transients (migrants). The species which are present in all stages of their life cycle in all the seasons of the year in any zone of the estuary are referred to as resident species. Examples of resident species are *Mugil parsia, Mugil tade, Polynemus paradiseus, Polydactylus indicus, Otolithoides biauritus, Lates calcarifer, Hilsa toli, Arius jella, Harpodon nehereus, Setipinna taty, Ilisha elongata, Setipinna phasa, Coilia ramcarati, Pama pama* and *Sillaginopsis panijus.* The transient or migratory fishes enter and stay in the inshore region of Bay of Bengal and associated estuaries for a short period of time. Depending on their migratory pattern and direction, the migrants may be divided into three categories (Jhingran 1982).

(1) Marine forms that migrate downstream to upstream and spawn in fresh-water areas of the estuary (where there is about zero salinity) like *Tenualosa ilisha, Polynemus paradiseus, Sillaginopsis panijus and Pama pama* is one of the transient varieties (2) Freshwater species, which spawn in saline area of the estuary like *Pangasius pangasius* (3) Marine species, that spawns in less saline water of the estuary (salinity range 5.0–10.0 psu) like *Arius jella, Osteogeneious militaris* and *Polydactylus indicus.*

An updated checklist of fishes available in the mangrove creeks, estuaries and bays is highlighted here.

## A CHECKLIST OF SUNDARBAN FISH (BOTH CHONDRICHTHYES AND OSTEICHTHYES) IN BRACKISH WATER AND FRESH WATER PONDS.

### 1. ORDER: CARCHARHINIFORMES (GROUND SHARKS)

**Family: *Carcharhinidae* (Requiem sharks)**
White cheek shark *Carcharhinus dussumieri*
Bull shark *Carcharhinus leucas*
Black tip shark *Carcharhinus limbatus*
Black tip reef shark *Carcharhinus melanopterus*
Tiger shark *Galeocerdo cuvier*
Ganges shark *Glyphis gangeticus*
Milk shark *Rhizoprionodon acutus*
Indian dog shark or Spadenose shark *Scoliodon laticaudus* Whale shark *Rhincodon typus*
Grey bamboo shark *Chiloscyllium griseum*
Slender bambooshark *Chiloscyllium indicum*
**Family: *Sphyrnidae* (Hammerhead, bonnethead or scoophead sharks)**
Winghead shark, *Eusphyra blochii*

Scalloped hammerhead *Sphyrna lewini*
Great hammerhead *Sphyrna mokarran.*

2. **ORDER: PRISTIFORMES (SAWFISHES)**

**Family: *Pristidae***
Knifetooth sawfish *Anoxypristis cuspidata*
Largetooth sawfish *Pristis microdon*
Smalltooth sawfish *Pristis pectinata.*

3. **ORDER: RAJIFORMES (SKATES AND RAYS)**

**Family: *Dasyatidae* (Stingrays)**
Small-eye stingray *Dasyatis microps*
Pale-edged stingray *Dasyatis zugei*
Bleeker's whipray *Himantura bleekeri*
Ganges stingray *Himantura fluviatilis*
Scaly whipray *Himantura imbricata*
Pointed-nose stingray *Himantura jenkinsii*
Blackedge whipray *Himantura marginatus*
Honeycomb stingray *Himantura uarnak*
**Family: *Myliobatidae* (Eagle and manta rays)**
Spotted eagle ray *Aetobatus narinari*
Banded eagle ray *Aetomylaeus nichofii*
**Family: *Rhinobatidae* (Guitarfishes)**
Annandale's guitarfish *Rhinobatos annandalei*
Smoothback guitarfish *Rhinobatos lionotus*
Giant guitarfish *Rhynchobatus djiddensis.*

4. **ORDER: TORPEDINIFORMES (ELECTRIC RAYS)**

**Family: *Narcinidae* (Numbfishes)**
Brown numbfish *Narcine brunnea.*

5. **ORDER: OSTEOGLOSSIFORMES (BONY TONGUES)**

**Family: *Notopteridae* (Featherbacks or knifefishes)**
Clown knifefish *Chitala chitala*
Bronze featherback *Notopterus notopterus.*

6. **ORDER: ELOPIFORMES (TARPONS AND TENPOUNDERS)**

**Family: *Elopidae* (Tenpounders and ladyfishes)**
Tenpounder *Elops machnata*
**Family: *Megalopidae* (Tarpons)**
Indo-Pacific tarpon *Megalops cyprinoides.*

7. **ORDER: ALBULIFORMES (BONE FISHES) Family: *Albulidae* (Bone-fishes)** Roundjaw bonefish *Albula glossodonta*
8. **ORDER: ANGUILLIFORMES (EELS AND MORAYS)**

**Family: *Anguillidae* (Freshwater eels)**
Indian mottled eel *Anguilla bengalensis*
Indonesian shortfin eel *Anguilla bicolor*
Mottled eel *Anguilla nebulosa*
**Family: *Moringuidae* (Worm and spaghetti eels)**
*Moringua macrocephalus*
Purple spaghetti eel *Moringua raitaborua*
**Family: *Muraenidae* (Moray eels)**
Freshwater moray *Gymnothorax tile*
**Family: *Ophichthidae* (Snake eels and worm eels)**
Finny snake eel *Caecula pterygera*
Rice paddy eel *Pisodonophis boro*
Greenspot snake eel *Pisodonophis hijala*
Maimed snake eel *Muraenichthys schultzei*
**Family: *Muraenesocidae* (Pike conger eels)**
Yellow pike conger *Congresox talabon*
Indian pike conger *Congresox talabonoides*
Common pike conger *Muraenesox bagio*
Daggertooth pike conger *Muraenesox cinereus.*

9. **ORDER: CLUPEIFORMES (HERRINGS)**

**Family: *Clupeidae* (Herrings, shads, sardines and menhadens)**
Chacunda gizzard shad *Anodontostoma chacunda*
Thai gizzard shad *Anodontostoma thailandiae*
Ganges river sprat *Corica soborna*
White sardine *Escualosa thoracata*
Ganges river gizzard shad *Gonialosa manmina*
Indian river shad *Gudusia chapra*
Kelee shad *Hilsa kelee*
Bloch’s gizzard shad *Nematalosa nasus*
Tardoore *Opisthopterus tardoore*
Raconda *Raconda russeliana*
Indian oil sardine *Sardinella longiceps*
Hilsa shad, River shad *Tenualosa ilisha*
Toli shad *Tenualosa toli*
**Family: *Pristigasteridae***
Elongate ilisha *Ilisha elongata*
Coromandel ilisha *Ilisha filigera*
Kampen’s ilisha *Ilisha kampeni*
Bigeye ilisha *Ilisha megaloptera*
Indian ilisha *Ilisha melastoma*
Indian pellona *Pellona ditchela*
**Family: *Engraulidae* (Anchovies)**
Goldspotted grenadier anchovy *Coilia dussumieri*
Ramcarat grenadier anchovy *Coilia ramcarati*

Reynald's grenadier anchovy *Coilia reynaldi*
Shorthead hairfin anchovy *Setipinna breviceps*
Dusky hairfin anchovy *Setipinna melanochir*
Gangetic hairfin anchovy *Setipinna phasa*
Scaly hairfin anchovy *Setipinna taty*
Indian anchovy *Stolephorus indicus*
Dussumier's thryssa *Thryssa dussumieri*
Hamilton's thryssa *Thryssa hamiltonii*
Malabar thryssa *Thryssa malabarica*
Moustached thryssa *Thryssa mystax*
Oblique jaw thryssa *Thryssa purava*
Bengal thryssa *Thryssa spinidens*
Slender thryssa *Thryssa stenosoma*
Orangemouth anchovy *Thryssa vitrirostris*
**Family: *Chirocentridae* (Wolf herrings)**
Dorab wolf herring *Chirocentrus dorab*
Whitefin wolf herring *Chirocentrus nudus.*

10. **ORDER: GONORHYNCHIFORMES (MILK FISH)**

**Family: *Chanidae***
Milkfish White mullet *Chanos chanos.*

11. **ORDER: CYPRINIFORMES (CARPS)**

**Family: *Cyprinidae* (Minnows, carps, barbs)**
Mola carplet *Amblypharyngodon mola*
Mrigal *Cirrhinus cirrhosus*
Sind danio *Devario devario*
Silver razorbelly minnow *Salmostoma acinaces (Chela argentea)*
Large razorbelly minnow *Salmostoma bacaila*
Finescale razorbelly minnow *Salmostoma phulo*
Bengala barb *Bengala elanga*
Zebra danio *Danio rerio*
Flying rasbora (barb) *Esomus danricus*
Slender rasbora *Rasbora daniconius*
Silver hatchet danio *Chela cachius*
Indian glass barb *Chela laubuca*
Catla *Catla catla*
Reba *Labeo ariza*
Bata *Labeo bata*
Kalbosu (Orange-fin labeo) *Labeo calbasu*
Rohu *Labeo rohita*
Swamp barb *Puntius chola*
Rosy barb *Puntius conchonius*
Golden barb *Puntius gelius*
Olive barb *Puntius sarana*

Pool barb *Puntius sophore*
Onespot barb *Puntius terio*
Ticto barb *Puntius ticto.*

12. **ORDER: SILURIFORMES (CAT FISH)**

**Family: *Bagridae* (Bagrid cat fishes)**
Menoda cat fish *Hemibagrus menoda*
Day's mystus *Mystus bleekeri*
Gangetic mystus *Mystus cavasius*
Long whiskers cat fish *Mystus gulio*
Striped dwarf cat fish *Mystus vittatus*
Rita *Rita rita*
Long whiskered cat fish *Sperata aor*
Giant river cat fish *Sperata seenghala*
**Family: *Siluridae* (Sheat fishes)**
Butter cat fish *Ompok bimaculatus*
Pabdah cat fish *Ompok pabda*
Wallago *Wallago attu*
**Family: *Schilbeidae* (Schilbeid cat fishes)**
Gangetic ailia *Ailia coila*
Garua cat fish *Clupisoma garua*
Sharpnose cat fish *Eutropiichthys vacha*
Indian potasi *Pseudeutropius atherinoides*
Silond cat fish *Silonia silondia*
**Family: *Pangasiidae* (Shark cat fishes)**
Yellowtail cat fish, River pangus *Pangasius pangasius*
**Family: *Sisoridae* (Sisorid cat fishes)**
Dwarf goonch *Bagarius bagarius*
Goonch *Bagarius yarrelli*
Indian gagata *Gagata cenia*
*Gagata gagata*
*Glyptothorax botius*
*Glyptothorax telchitta*
*Gogangra viridescens*
Kosi nangra *Nangra nangra*
Sisor cat fish *Sisor rabdophorus*
**Family: *Clariidae* (Airbreathing cat fishes)**
Walking cat fish *Clarias batrachus*
**Family: *Chacidae* (Squarehead cat fishes)**
Squarehead cat fish *Chaca chaca*
**Family: *Ariidae* (Sea cat fishes)**
T hreadfin sea cat fish *Arius arius*
Engraved cat fish *Arius caelatus*
Blacktip sea cat fish *Arius dussumieri*
Gagora cat fish *Arius gagora*

Blackfin sea cat fish *Arius jella*
Spotted cat fish *Arius maculatus*
Smooth headed cat fish *Arius nenga*
Flat mouth cat fish *Arius platystomus*
Sagor cat fish *Arius sagor*
Sona sea cat fish *Arius sona*
Shovelnose sea cat fish *Arius subrostratus*
Beardless sea cat fish *Batrachocephalus mino*
River cat fish *Hemipimelodus jatius*
Soldier cat fish *Osteogeneiosus militaris*
**Family: *Heteropneustidae* (Airsac cat fishes)**
Stinging cat fish *Heteropneustes fossilis*
**Family: *Plotosidae* (Eeltail cat fishes)**
Gray eel cat fish *Plotosus canius*
Striped eel cat fish *Plotosus lineatus.*

13. **ORDER: AULOPIFORMES (GRINNERS)**

**Family: *Synodontidae* (Lizard fishes, Bombay duck)**
Greater lizard fish *Saurida tumbil*
Brushtooth lizard fish *Saurida undosquamis*
Bombay duck *Harpadon nehereus.*

14. **ORDER: BATRACHOIDIFORMES (TOAD FISHES)**

**Family: *Batrachoididae***
Grunting toad fish *Allenbatrachus grunniens.*

15. **ORDER: GADIFORMES (CODS)**

**Family: *Bregmacerotidae* (Codlets)**
Spotted codlet Bregmaceros mcclellandi.

16. **ORDER: CYPRINODONTIFORMES (RIVULINES, KILLI FISHES AND LIVE BEARERS)**

**Family: *Aplocheilidae* (Killi fishes)**
Blue panchax *Aplocheilus panchax.*

17. **ORDER: BELONIFORMES (NEEDLE FISHES)**

**Family: *Belonidae***
Banded needle fish *Strongylura leiura*
Spottail needle fish *Strongylura strongylura*
Freshwater gar fish *Xenentodon cancila*
**Family: *Hemiramphidae***
Gangetic half beak *Dermogenys brachynotopterus*
Wrestling half beak *Dermogenys pusilla*
Jumping half beak *Hemiramphus archipelagicus*
Congaturi half beak *Hyporhamphus limbatus*

Long billed half beak *Rhynchorhamphus georgii*
Ectuntio half beak *Zenarchopterus ectuntio*
**Family: *Adrianichthyidae* (Ricefishes)**
Rice fish *Oryzias carnaticus.*

18. **ORDER: SYNGNATHIFORMES (PIPEFISHES AND SEAHORSES)**

**Family: *Syngnathidae***
Freshwater pipefish *Ichthyocampus carce*
Crocodile tooth pipefish *Microphis cuncalus*
Deocata pipefish *Microphis deocata.*

19. **ORDER: SYNBRANCHIFORMES (SWAMP AND SPINY EELS)**

**Family: *Synbranchidae* (Swamp eels)**
Cuchia, Gangetic mud eel *Monopterus cuchia*
Bengal eel *Ophisternon bengalense*
Family: *Mastacembelidae* (Spiny eels)
Lesser spiny eel *Macrognathus aculeatus*
One-stripe spinyeel *Macrognathus aral*
Barred spiny eel *Macrognathus pancalus*
Zig-zag eel, Tire-track spiny eel *Mastacembelus armatus.*

20. **ORDER: SCORPAENIFORMES (SCORPION FISHES & FLAT HEADS)**

**Family: *Platycephalidae* (Flatheads)**
Bartail flathead *Platycephalus indicus.*

21. **ORDER: PERCIFORMES (PERCH-LIKE)**

**Family: *Latidae* (Perches)**
Barramundi, Giant seaperch *Lates calcarifer*
Waigieu seaperch *Psammoperca waigiensis*
**Family: *Ambassidae* (Glass fishes)**
Bald glassy Ambassis *gymnocephalus*
Elongate glass-perchlet *Chanda nama*
Highfin glassy perchlet *Parambassis lala*
Indian glassy fish *Parambassis ranga*
**Family: *Serranidae* (Sea basses: groupers and fairy basslets)**
Orange spotted grouper *Epinephelus coioides*
Giant grouper *Epinephelus lanceolatus*
**Family: *Terapontidae* (Grunters and tigerperches)**
Fourlined terapon *Pelates quadrilineatus*
Jarbua terapon *Terapon jarbua*
Small-scaled terapon *Terapon puta*
Largescaled therapon *Terapon theraps*
**Family**: ***Sillaginidae*** {**Sillagos (Smelt-whitings)**} Flathead sillago *Sillaginopsis panijus*

Clubfoot sillago *Sillago chondropus*
Silver sillago, *Sillago sihama*
Soringa sillago *Sillago soringa*
**Family: *Carangidae* (Jacks and pompanos)**
Indian threadfish *Alectis indicus*
Razorbelly scad *Alepes kleinii*
Longnose trevally *Carangoides chrysophrys*
Malabar trevally *Carangoides malabaricus*
Bigeye trevally *Caranx sexfasciatus*
Japanese scad *Decapterus maruadsi*
Golden trevally *Gnathanodon speciosus*
Torpedo scad *Megalaspis cordyla*
Black pomfret, Brown pomfret *Parastromateus niger*
Barred queen fish *Scomberoides tala*
Bigeye scad *Selar crumenophthalmus*
Yellowstripe scad *Selaroides leptolepis*
**Family: *Menidae* (Moonfishes, bat fishes)**
Moonfish *Mene maculata*
**Family: *Leiognathidae* (Pony fishes)**
Goldstripe pony fish *Leiognathus daura*
Common pony fish *Leiognathus equulus*
Striped pony fish *Leiognathus fasciatus*
Splendid pony fish *Leiognathus splendens*
Pugnose pony fish *Secutor insidiator*
Deep pugnose pony fish *Secutor ruconius*
**Family: *Lutjanidae* (Snappers)**
Mangrove red snapper *Lutjanus argentimaculatus* Humpback red snapper *Lutjanus gibbus*
John's snapper *Lutjanus johnii*
Malabar blood snapper *Lutjanus malabaricus*
Russell's snapper *Lutjanus russellii*
**Family: *Datnioididae***
Fourstripe perch *Datnioides polota*
**Family: *Lobotidae* (Tripletails)**
Atlantic tripletail *Lobotes surinamensis*
**Family: *Gerreidae* (Mojarras)**
Whipfin silver-biddy *Gerres filamentosus*
Saddleback silver-biddy *Gerres limbatus*
Slender silver-biddy *Gerres oblongus*
Common silver-biddy *Gerres oyena*
Strong spine silver-biddy *Gerres phaiya*
Small Bengal silver-biddy *Gerres setifer*
**Family: *Haemulidae* (Grunts)**
Bluecheek silver grunt *Pomadasys argyreus*
Silver bream *Pomadasys hasta*

**Family: *Sparidae* (Porgies and sea breams)**
Yellowfin seabream *Acanthopagrus latus*
King soldierbream *Argyrops spinifer*
Goldlined seabream *Rhabdosargus sarba*
**Family: *Nemipteridae* (Threadfin breams and spinycheeks)**
Japanese threadfin bream *Nemipterus japonicus*
**Family: *Sciaenidae* (Croakers and drums)**
Chaptis bahaba *Bahaba chaptis*
Reeve's croaker *Chrysochir aureus*
Bengal corvina *Daysciaena albida*
Goatee croaker *Dendrophysa russelii*
Belanger's croaker *Johnius belangerii*
Karut croaker *Johnius carutta*
Coitor croaker *Johnius coitor*
Cuja croaker *Macrospinosa cuja*
Soldier croaker *Nibea soldado*
Bronze croaker *Otolithoides biauritus*
Pama croaker *Otolithoides pama*
Hooghly croaker *Panna heterolepis*
*Pennahia ovata*
Blackspotted croaker *Protonibea diacanthus* Blotched tiger-tooth croaker *Pterotolithus maculatus*
**Family: *Polynemidae* (Threadfins)**
Fourfinger threadfin *Eleutheronema tetradactylum*
Indian threadfin *Leptomelanosoma indicum*
Striped threadfin *Polydactylus plebeius*
Sixfinger threadfin *Polydactylus sexfilis*
Paradise threadfin *Polynemus paradiseus*
**Family: *Mullidae* (Goat fishes)**
Sulphur goatfish *Upeneus sulphureus*
**Family: *Toxotidae* (Archerfishes)**
Largescale archerfish *Toxotes chatareus*
**Family: *Drepaneidae* (Sicklefishes)**
Concertina fish *Drepane longimana*
Spotted sicklefish *Drepane punctata*
**Family: *Monodactylidae* (Moonyfishes or fingerfishes)**
Silver moony *Monodactylus argenteus*
**Family: *Nandidae* (Asian leaf fishes)**
Gangetic leaf fish *Nandus nandus*
**Family: *Badidae***
Badis *Badis badis*
**Family: *Kurtidae* (Nurseryfishes)**
Indian hump head *Kurtus indicus*
**Family: *Mugilidae* (Mullets)**
Largescale mullet *Liza macrolepis*

Gold-spot mullet *Liza parsia*
Greenback mullet *Liza subviridis*
Tade mullet *Liza tade*
Flathead mullet *Mugil cephalus*
Corsula *Rhinomugil corsula*
Cascasia mullet, Yellowtail mullet *Sicamugil cascasia*
Bluetail mullet *Valamugil buchanani*
Bluespot mullet *Valamugil seheli*
Speigler's mullet *Valamugil speigleri*
**Family: *Cichlidae* (Cichlids)**
Green chromide *Etroplus suratensis*
**Family: *Uranoscopidae* (Stargazers)**
*Uranoscopus guttatus*
**Family: *Callionymidae* (Dragonets)**
River dragonet *Callionymus fluviatilis*
Arrow dragonet *Callionymus sagitta*
**Family: *Eleotridae* (Sleepers)**
Duckbill sleeper *Butis butis*
Gangetic sleeper *Odonteleotris macrodon*
Dusky sleeper *Eleotris fusca*
Lutea sleeper *Eleotris lutea*
**Family: *Gobiidae* (Gobies)**
Tropical sand goby *Acentrogobius caninus,*
*Acentrogobius cyanomos*
Spotted green goby *Acentrogobius viridipunctatus*
Dragon goby *Apocryptes bato*
Scribbled goby *Awaous grammepomus*
Largesnout goby *Awaous melanocephalus*
*Bathygobius ostreicola*
Bumblebee goby *Brachygobius nunus*
Boddart's goggle-eyed goby *Boleophthalmus boddarti* Mudskipper
*Boleophthalmus dussumieri*
Tank goby *Glossogobius giuris*
Glass goby *Gobiopterus chuno*
Rubicundus eelgoby *Odontamblyopus rubicundus*
Maned goby *Oxyurichthys microlepis*
Taileyed goby *Parachaeturichthys polynema*
Giant mudskipper *Periophthalmodon schlosseri*
*Periophthalmodon septemradiatus*
Atlantic mudskipper *Periophthalmus barbarus*
Pearse's mudskipper *Periophthalmus novemradiatus*
Elongate goby *Pseudapocryptes elongatus*
Many-finned eelgoby *Pseudotrypauchen multiradiatus*
Walking goby *Scartelaos histophorus*
Knight goby *Stigmatogobius sadanundio*

Eel worm goby *Taenioides anguillaris*
Burmese gobyeel *Taenioides buchanani*
Bearded worm goby *Taenioides cirratus*
Burrowing goby *Trypauchen vagina*
**Family: *Scatophagidae* (Scats)**
Spotted scat *Scatophagus argus*
**Family: *Siganidae* (Rabbit fishes)**
Streaked spinefoot *Siganus javus*
**Family: *Sphyraenidae* (Barracudas)**
Great barracuda *Sphyraena barracuda*
**Family: *Trichiuridae* (Cutlass fishes and scabbard fishes)**
Longtooth hairtail *Eupleurogrammus glossodon*
Smallhead hairtail *Eupleuroarammus muticus*
Coromandel hairtail *Lepturacanthus pantului*
Savalani hairtail *Lepturacanthus savala*
Gangetic hairtail *Trichiurus gangeticus*
Largehead hairtail *Trichiurus lepturus*
**Family: *Scombridae* (Mackerels, tunas and bonitos)**
Kawakawa *Euthynnus affinis*
Indian mackerel *Rastrelliger kanagurta*
Narrow-barred Spanish mackerel *Scomberomorus commerson* Indo-Pacific king mackerel *Scomberomorus guttatus*
**Family: *Stromateidae* (Butterfishes)**
Silver pomfret *Pampus argenteus*
Chinese silver pomfret *Pampus chinensis*
*Pampus cinereus*
**Family: *Anabantidae* (Climbing gouramies)**
Climbing perch *Anabas testudineus*
**Family: *Osphronemidae* (Gouramies)**
Banded gourami *Colisa fasciata*
Spiketail paradise fish *Pseudosphromenus cupanus* Dwarf gourami *Colisa lalia*
Frail gourami *Ctenops nobilis*
Honey gourami *Trichogaster chuna*
**Family: *Channidae* (Snakeheads)**
Barca snakehead
*Channa barca*
*Channa gachua*
Great snakehead *Channa marulius*
Walking snakehead *Channa orientalis*
Spotted snakehead *Channa punctata*
Snakehead murrel *Channa striata.*

22. **ORDER: PLEURONECTIFORMES (FLATFISHES)**

**Family: *Psettodidae* (Psettodids)**

Indian spiny turbot *Psettodes erumei*
**Family: *Paralichthyidae* (Largetooth flounders)**
Largetooth flounder *Pseudorhombus arsius*
Deep flounder *Pseudorhombus elevatus*
Malayan flounder *Pseudorhombus malayanus*
Three spotted flounder *Pseudorhombus triocellatus*
**Family: *Citharidae* (Chitarids)**
Yellow-dabbled flounder *Brachypleura novaezeelandiae*
**Family: *Soleidae* (Soles)**
Oriental sole *Brachirus orientalis*
Pan sole *Brachirus pan*
Eyed sole *Heteromycteris oculus*
Kaup's sole *Synaptura albomaculata*
Highfin sole *Zebrias altipinnis*
**Family: *Cynoglossidae* (Tongue fishes)**
Largescale tongue-sole *Cynoglossus arel*
Bengal tongue-sole *Cynoglossus cynoglossus*
Long tongue-sole *Cynoglossus lingua*
Malabar tongue-sole *Cynoglossus macrostomus*
Speckled tongue-sole *Cynoglossus puncticeps*
Bengal tongue-sole *Cynoglossus semifasciatus*
Doublelined tongue-sole *Paraplagusia bilineata.*

23. **ORDER: TETRADONTIFORMES (PUFFERS AND FILEFISHES)**

**Family: *Triacanthidae* (Triplespines)** Short-nosed tripodfish *Triacanthus biaculeatus*
**Family: *Tetraodontidae* (Puffers)**
Immaculate puffer *Arothron immaculatus*
Milkspotted puffer *Chelonodon patoca*
Green rough-backed puffer *Lagocephalus lunaris*
Lattice blaasop *Takifugu oblongus*
Ocellated pufferfish *Tetraodon cutcutia*
Green pufferfish *Tetraodon fluviatiles.*

## 3.2 Threats to Indian Sundarban Fishery in the Backdrop of Climate Change

Climate change is a bitter truth associated with the planet Earth in the present century and has both direct and indirect impacts on the fishery sector.

In this chapter we present the evidences of direct impact of climate change induced alteration of salinity and temperature on the fish community. Alteration of aquatic salinity and temperature have considerable impact on (i) spawning and larval recruitment, (ii) metabolic activity and growth and (ii) microbial load in fish tissues leading to several types of diseases.

### 3.2.1 Alteration of Spawning and Larval Recruitment

The metabolic activities, physiology, reproductive success, spawning and larval recruitment of fish are considerably governed by ambient environmental parameters like water temperature, salinity, pH etc. We have documented a study in the coastal zone of West Bengal at three different sites with contrasting salinity on the fish juvenile number.

West Bengal is a maritime state in the northeast part of the Indian sub-continent adjacent to Bangladesh. The coastal zone of this state spreads over an area of 10,158.22 sq. km. and is restricted within the latitude 21° 30′N to 22° 30′N and longitude 87° 25′E to 89° 10′E. The river Harinbhanga or Heronbhanga of the Indian Sundarbans (India- Bangladesh border) is the eastern border while the New Digha coast of Orissa-West Bengal border constitutes the western boundary of coastal West Bengal. With considerable degree of marine characteristics in the major portion of the ecosystem, the important morphotypes of coastal West Bengal are sand flats, coastal dunes, beaches, mudflats, estuaries, creeks, inlets and mangrove flats. There is a drastic variation of salinity and dilution factor between different horizontal transects of the ecosystem. The dilution factor, salinity and pH reveal significant variations at the same time in different locations of the coastal zone. The recent emergence of Haldia industrial complex and various fish landing stations in the coastal zone has also opened the gateway of input of various categories of wastes in this ecosystem.

Geographically, the study area encompasses three major districts of the state of West Bengal namely, 24 Parganas (North), 24 Parganas (South) and Medinipur (East). The Indian Sundarbans declared as the World Heritage Site by UNESCO in the year of 1989 fall within the North and South 24 Parganas districts. Three sampling stations were selected in and around this deltaic ecosystem namely, Diamond Harbour (station 1), Sagar light house (station 2) and Junput (station 3). Each of these sampling stations is markedly different from the other with respect to aquatic salinity due to their location and proximity to the Bay of Bengal.

Diamond Harbour (station 1) is situated in the low saline upper stretch of Hooghly estuary, just outside the northern boundary of Indian Sundarbans. The station is very near to the Haldia port-cum-industrial complex. Salinity of surface water is less around the station owing to its location in the extreme upstream region, far away from the Bay of Bengal and also due to huge fresh water discharge from the Hooghly River, which is perennial in nature.

Sagar light house (station 2) is situated in the south western tip of the Sagar Island and falls in the western sector of Indian Sundarbans. The station has rich mangrove vegetation and extensive mudflats. Although there are no industrial activities in this station, but the presence of sizeable number of shrimp farms (presently carrying

**Table 3.1** Mean seasonal variation in standing stock/fish juvenile count (N) of finfish species in three selected sampling stations during Jan' 2009 to Dec' 2018

| Year | Monthly average standing stock (N) of finfish species | | | | | | | | |
|---|---|---|---|---|---|---|---|---|---|
| | Stn 1 | | | Stn 2 | | | Stn 3 | | |
| | Prm | Mon | Pom | Prm | Mon | Pom | Prm | Mon | Pom |
| 2009 | 394 | 161 | 245 | 1598 | 134 | 522 | 1593 | 260 | 767 |
| 2010 | 322 | 133 | 291 | 1315 | 122 | 422 | 1303 | 213 | 625 |
| 2011 | 202 | 82 | 183 | 1273 | 99 | 416 | 1290 | 213 | 631 |
| 2012 | 197 | 79 | 175 | 1279 | 99 | 413 | 1229 | 201 | 596 |
| 2013 | 217 | 89 | 197 | 1090 | 84 | 361 | 1055 | 172 | 651 |
| 2014 | 219 | 89 | 198 | 716 | 56 | 233 | 679 | 123 | 333 |
| 2015 | 226 | 93 | 204 | 639 | 51 | 209 | 601 | 103 | 291 |
| 2016 | 223 | 92 | 201 | 545 | 42 | 180 | 521 | 89 | 255 |
| 2017 | 227 | 93 | 204 | 910 | 178 | 442 | 934 | 153 | 440 |
| 2018 | 392 | 38 | 67 | 936 | 73 | 304 | 1020 | 81 | 367 |

on shrimp culture by traditional method with a very low stocking density of prawn seeds) has enriched the surrounding water with nutrients and organic load.

Junput (Station 3) is situated in the Medinipur (East) district of coastal West Bengal, which is noted for its high aquatic salinity owing to its proximity to Bay of Bengal. The extremely high salinity has posed an inhibitory effect on the growth and survival of mangroves in this station. Although the station has no industry around its vicinity, but the presence of Digha tourist center and Shankarpur fishing harbour close to the station has multiplied the anthropogenic pressure around the zone.

Considering fish juvenile number as an indicator of reproductive success, egg development and spawning, we observed significant spatial difference of fish juvenile number in the study zone (Table 3.1).

Table 3.2 exhibits the abundance of finfish juvenile species in the three sampling stations. Members of genus *Coilia* sp., *Thryssa* sp., *Mugil* sp., *Tenualosa* sp., *Liza* sp., *Scatophagus* sp., *Stolephorus* sp., *Cynoglossus* sp., *Sillago* sp. are dominant (in terms of standing stock or biomass) in all the stations and in all the seasons of the study period, which suggests the wide range of tolerance of this genus in the present geographical locale. The study clearly depicts the regulation of finfish juveniles by aquatic salinity. The number of finfish juvenile is highest at station 3 (Junput) which is followed by station 2 (Sagar light house) and station 1 (Diamond Harbour). The highest juvenile number at Junput confirms the positive role of aquatic salinity in maintaining the fish germplasm in the study area. Diamond Harbour is an extremely low saline zone in the upstream area with an average aquatic salinity of 2.76 psu during premonsoon, 1.02 psu during monsoon and 2.32 during postmonsoon and hence chance of getting stenohaline fish juvenile species or saline water loving individuals is rare.

**Table 3.2** Distribution of finfish juvenile species in 10 gm composite wasted sample collected from three selected sampling stations during January, 2009 to December, 2018

| S. No | Species | Station 1 | Station 2 | Station 3 |
|---|---|---|---|---|
| 1 | *Coilia* sp. | ++ | ++ | +++ |
| 2 | *Thryssa hamiltonii (Gray)* | + | ++ | +++ |
| 3 | *Thryssa baelama* | + | ++ | +++ |
| 4 | *Torquigener oblongus* | + | + | ++ |
| 5 | *Rhinomugil corsula* | + | + | + |
| 6 | *Mugil cephalus* | + | + | + |
| 7 | *Sillaginopsis panijus* | + | + | + |
| 8 | *Sillago sihama* | + | + | + |
| 9 | *Sillago soringa* | + | + | + |
| 10 | *Zenarchopterus dispar* | + | + | + |
| 11 | *Glossogobius quiris* | + | + | + |
| 12 | *Pseudapocryptes lanceolatus* | + | + | + |
| 13 | *Eupleurogrammus glossodon* | + | + | + |
| 14 | *Pseudorhombus* sp. | + | ++ | +++ |
| 15 | *Pisodonophis boro* | + | + | + |
| 16 | *Tenualosa ilisha* | +++ | + | – |
| 17 | *Cynoglossus arel* | + | ++ | +++ |
| 18 | *Cynoglossus* sp. | + | + | +++ |
| 19 | *Leiognathus blochii* | – | + | + |
| 20 | *Leiognathus equalus* | + | + | + |
| 21 | *Hilsa* sp. | ++ | + | – |
| 22 | *Scatophagus argus* | + | + | + |
| 23 | *Liza parsia* | +++ | + | + |
| 24 | *Liza tade* | +++ | + | + |
| 25 | *Stolephorus commersonii* | + | + | + |
| 26 | *Stolephorus baganensis* | + | + | + |
| 27 | *Stolephorus kammalensis* | + | + | + |
| 28 | *Lutjanus johni* | + | + | + |
| 29 | *Setipinna taty* | + | + | + |
| 30 | *Lagocephalus lunaris* | – | + | + |
| 31 | *Escualosa thoracata* | – | + | + |
| 32 | *Epinephelus tauvina* | – | + | + |
| 33 | *Epinephelus coioides* | – | + | + |
| 34 | *Sphyraena* sp. | – | + | + |
| 35 | *Pomadasys* sp. | – | + | + |

(continued)

**Table 3.2** (continued)

| S. No | Species | Station 1 | Station 2 | Station 3 |
|---|---|---|---|---|
| 36 | *Sardinella longiceps* | – | + | + |
| 37 | *Periophthalmus* sp. | – | + | + |
| 38 | *Macrognathus* sp. | – | + | + |
| 39 | *Tetradon cutcutia* | – | + | + |
| 40 | *Bregmaceros meclellandii* | – | + | + |
| 41 | *Ichthyocampus carce* | – | + | + |
| 42 | *Stigmatogobius* sp. | – | + | + |
| 43 | *Channa* sp. | – | + | – |
| 44 | *Kurtus indicus* | – | + | – |
| 45 | *Harpodon nehereus* | – | + | – |
| 46 | *Moringua raitaborua* | – | + | – |
| 47 | *Suggrundus rodricensis* | – | + | – |

'–' means absent, '+' means present in small number, '++' means present in moderate level and '+++' means high abundance

In the maritime state of West Bengal preferably in central Indian Sundarbans sea level rise is about 3.00 mm/yr, which has increased the salinity value in the region (Table 3.3). This has resulted in more abundance of trash fishes like *Harpodon nehereus*, *Thryssa hamiltonii (Gray)*, *Thryssa baelama* etc. and less availability of commercially important fishes like *Tenualosa ilisha.*

**Table 3.3** Rising trend of salinity (in psu) in central Indian Sundarbans

| Year | Premonsoon | Monsoon | Postmonsoon |
|---|---|---|---|
| 1984 | 22.07 | 20.67 | 21.81 |
| 1985 | 23.59 | 20.74 | 22.08 |
| 1986 | 24.34 | 20.80 | 22.02 |
| 1987 | 24.22 | 20.14 | 22.67 |
| 1988 | 25.04 | 20.03 | 22.40 |
| 1989 | 25.69 | 20.35 | 23.08 |
| 1990 | 25.84 | 21.02 | 23.54 |
| 1991 | 25.01 | 21.67 | 23.57 |
| 1992 | 24.72 | 21.73 | 23.96 |
| 1993 | 25.60 | 21.86 | 23.77 |
| 1994 | 26.00 | 21.74 | 24.21 |
| 1995 | 26.04 | 21.99 | 24.99 |
| 1996 | 26.08 | 21.61 | 25.05 |

(continued)

**Table 3.3** (continued)

| Year | Premonsoon | Monsoon | Postmonsoon |
|---|---|---|---|
| 1997 | 26.83 | 22.01 | 24.92 |
| 1998 | 26.81 | 22.06 | 25.43 |
| 1999 | 26.75 | 20.64 | 25.41 |
| 2000 | 26.98 | 21.37 | 25.51 |
| 2001 | 27.04 | 22.60 | 25.96 |
| 2002 | 27.13 | 22.92 | 26.42 |
| 2003 | 27.45 | 22.98 | 26.48 |
| 2004 | 27.20 | 23.04 | 26.45 |
| 2005 | 27.18 | 23.16 | 27.56 |
| 2006 | 27.86 | 23.78 | 27.38 |
| 2007 | 27.69 | 23.83 | 27.69 |
| 2008 | 28.34 | 23.92 | 27.72 |
| 2009 | 31.68 | 22.87 | 28.30 |
| 2010 | 29.87 | 22.96 | 28.54 |
| 2011 | 30.13 | 23.37 | 28.89 |
| 2012 | 30.08 | 23.52 | 28.72 |
| 2013 | 30.46 | 23.69 | 28.92 |
| 2014 | 30.73 | 24.04 | 29.28 |
| 2015 | 30.81 | 23.89 | 29.80 |
| 2016 | 31.09 | 24.03 | 29.88 |
| 2017 | 30.80 | 25.03 | 29.30 |
| 2018 | 30.96 | 25.41 | 30.68 |

With this salinity data in the central part of Indian Sundarbans, we carried out a time series modeling to visualize the trend of salinity level using a nonlinear autoregressive model (NAR) treating seasonal salinity values as inputs and 32—year lead-time (Fig. 3.1) and observed significant rising trend in salinity with a base of 70% training, 15% validation and 15% testing (Fig. 3.2). This increased salinity due to reduced fresh water inflow in the central part of Indian Sundarbans coupled with sea level rise may significantly impact the spawning and larval development of Indian shad, *Tenualosa ilisha.*

Rising sea levels is a direct arm of climate change, which may inundate wetlands and other low-lying lands, erode beaches, intensify flooding, and increase the salinity of rivers, bays and estuaries. Salinity alteration due to seawater intrusion has been reported from various parts of the world. The rising trend of salinity over a period of 27 years has also been documented in the Matla River in the central part of Indian Sundarbans (Mitra et al. 2009). The biogeographic distribution of certain fish species is influenced by such alteration of aquatic salinity.

On the basis of this three decade data bank, we used Nonlinear Autoregressive Neural Network Model to forecast the salinity levels in central Indian Sundarbans

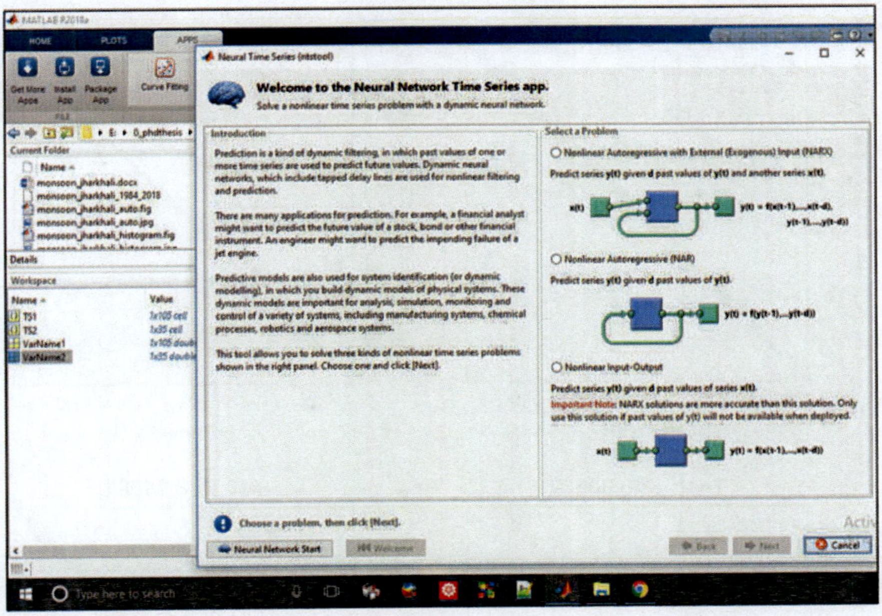


Fig. 3.1 Nonlinear Autoregression (NAR) model for evaluating the salinity trend

(with 32 years ahead time steps) for three seasons and observed a significant rise in all the seasons (Tables 3.4, 3.5 and 3.6; Figs. 3.3, 3.4 and 3.5).

The picture in the western Indian Sundarbans is totally reverse. There is a decrease of salinity in this region with the passage of time (Table 3.7), which is a congenial situation for the spawning of *Tenualosa ilisha*.

The salinity data in western Indian Sundarbans was also used to observe salinity trend using NAR considering seasonal salinity values as inputs and 32—year lead-time with a base of 70% training, 15% validation and 15% testing (Fig. 3.6). The output exhibits a significant decreasing trend of salinity, which is highly favourable for the spawning of *Tenualosa ilisha* during migration through this route of Hooghly-Bhagirathi River system from the inshore region of Bay of Bengal.

On the basis of this three decade data bank, we used Nonlinear Autoregressive Neural Network Model to forecast the salinity levels in western Indian Sundarbans (with 32 years ahead time steps) for three seasons and observed a significant decrease in salinity in all the seasons during 2050 (Tables 3.8, 3.9 and 3.10; Figs. 3.7, 3.8 and 3.9). It can, therefore, be forecasted that a congenial environment is ahead for the spawning and larval recruitment of *Tenualosa ilisha* in western sector of Indian Sundarbans.

An interesting observation in this context is the extension of the spawning ground of *Tenualosa ilisha* in more upstream zone after the construction of the Farakka barrage on the Gangetic stretch of India. The studies conducted during 1982–1992

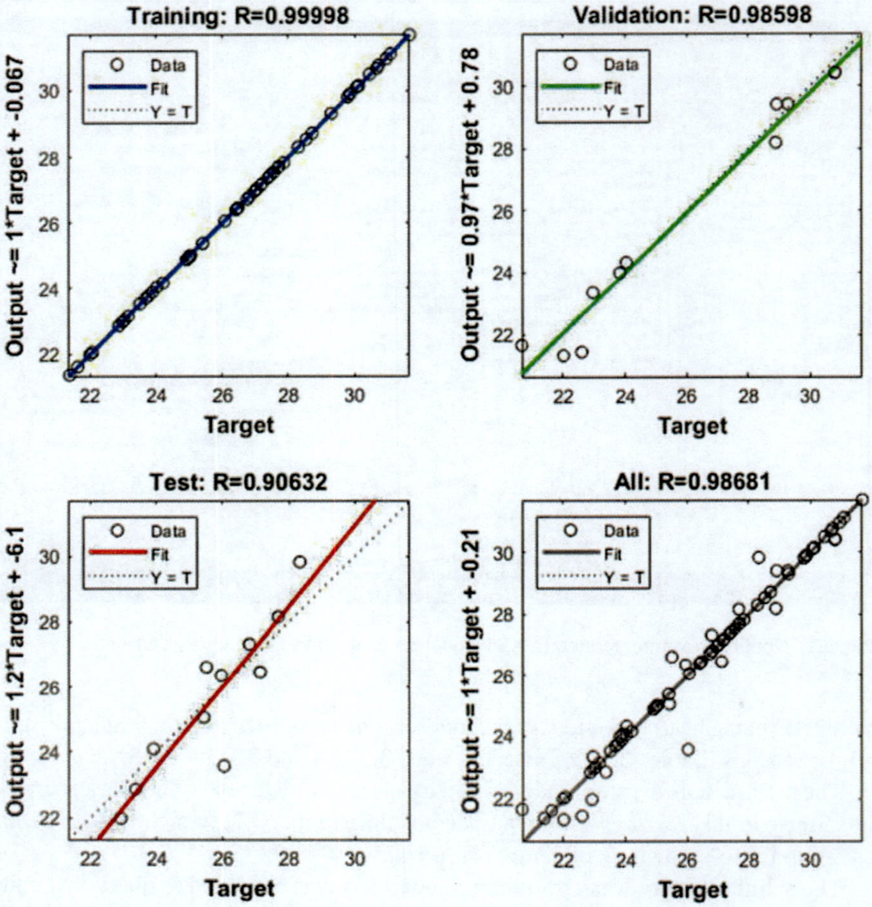


**Fig. 3.2** Outputs of Nonlinear Autoregression (NAR) model showing the trend of aquatic salinity in central Indian Sundarbans

clearly reveal that the ecology of Hooghly estuary (downstream zone of Ganga-Bhagirathi system) has undergone a major change due to construction of the Farakka barrage in 1975. The increased freshwater discharge in the western part of Indian Sundarbans has resulted in considerable decrease in salinity throughout the estuary. The fresh water zone has extended toward the mouth of the estuary and the marine and estuarine zone has been pushed almost towards the end of the lower stretch of the estuary (Sinha et al. 1996). This has resulted in the expansion of the spawning ground of *Tenualosa ilisha* in the post Farakka barrage period (De and Saigal 1989). Prior to 1975 (before the installation of the Farakka barrage), the spawning ground was restricted from Calcutta to Medgachi, but survey conducted during 1987 to 1989 showed that the spawning ground has expanded more in the upstream zone almost

**Table 3.4** Forecasting of salinity (in psu) in central Indian Sundarbans during premonsoon using Autoregressive Neural Network Model

| Year | Predicted salinity (in psu) in premonsoon |
|---|---|
| 2019 | 30.87 |
| 2020 | 30.89 |
| 2021 | 30.87 |
| 2022 | 30.87 |
| 2023 | 30.86 |
| 2024 | 30.86 |
| 2025 | 30.86 |
| 2026 | 30.86 |
| 2027 | 30.86 |
| 2028 | 30.86 |
| 2029 | 30.86 |
| 2030 | 30.86 |
| 2031 | 30.86 |
| 2032 | 30.86 |
| 2033 | 30.86 |
| 2034 | 30.86 |
| 2035 | 30.86 |
| 2036 | 30.86 |
| 2037 | 30.86 |
| 2038 | 30.86 |
| 2039 | 30.86 |
| 2040 | 30.86 |
| 2041 | 30.86 |
| 2042 | 30.86 |
| 2043 | 30.86 |
| 2044 | 30.86 |
| 2045 | 30.86 |
| 2046 | 30.86 |
| 2047 | 30.86 |
| 2048 | 30.86 |
| 2049 | 30.86 |
| 2050 | 30.86 |

towards Nabadwip. This case study clearly confirms that biogeographic distribution of fish (preferably migratory fish) and even their spawning ground coordinates are regulated by aquatic salinity (Mitra 2000).

**Table 3.5** Forecasting of salinity (in psu) in central Indian Sundarbans during monsoon using Autoregressive Neural Network Model

| Year | Predicted salinity (in psu) in monsoon |
|---|---|
| 2019 | 25.67 |
| 2020 | 25.93 |
| 2021 | 25.98 |
| 2022 | 27.05 |
| 2023 | 24.41 |
| 2024 | 30.28 |
| 2025 | 17.89 |
| 2026 | 23.60 |
| 2027 | 19.99 |
| 2028 | 21.15 |
| 2029 | 22.29 |
| 2030 | 22.62 |
| 2031 | 22.30 |
| 2032 | 21.81 |
| 2033 | 21.58 |
| 2034 | 21.73 |
| 2035 | 21.85 |
| 2036 | 21.79 |
| 2037 | 21.71 |
| 2038 | 21.74 |
| 2039 | 21.78 |
| 2040 | 21.77 |
| 2041 | 21.75 |
| 2042 | 21.75 |
| 2043 | 21.76 |
| 2044 | 21.76 |
| 2045 | 21.75 |
| 2046 | 21.75 |
| 2047 | 21.76 |
| 2048 | 21.76 |
| 2049 | 21.76 |
| 2050 | 21.76 |

### 3.2.2 *Alteration of Metabolic Activity and Growth*

The enzymatic activities of fishes are greatly regulated by ambient aquatic temperature. It has been documented that at higher temperature the structure and function of protein are significantly impaired in fishes. In case of fish it has been observed that

**Table 3.6** Forecasting of salinity (in psu) in central Indian Sundarbans during postmonsoon using Autoregressive Neural Network Model

| Year | Predicted salinity (in psu) in postmonsoon |
|---|---|
| 2019 | 29.38 |
| 2020 | 31.03 |
| 2021 | 29.31 |
| 2022 | 31.45 |
| 2023 | 29.12 |
| 2024 | 31.90 |
| 2025 | 28.83 |
| 2026 | 32.29 |
| 2027 | 28.58 |
| 2028 | 32.61 |
| 2029 | 28.43 |
| 2030 | 32.81 |
| 2031 | 28.38 |
| 2032 | 32.91 |
| 2033 | 28.36 |
| 2034 | 32.95 |
| 2035 | 28.35 |
| 2036 | 32.97 |
| 2037 | 28.35 |
| 2038 | 32.97 |
| 2039 | 28.35 |
| 2040 | 32.97 |
| 2041 | 28.35 |
| 2042 | 32.97 |
| 2043 | 28.35 |
| 2044 | 32.97 |
| 2045 | 28.35 |
| 2046 | 32.97 |
| 2047 | 28.35 |
| 2048 | 32.97 |
| 2049 | 28.35 |
| 2050 | 32.97 |

increase of aquatic temperature within thermal range for a particular species leads to increased growth rate provided that food is not a limiting factor. The feeding rates of fishes increase at higher temperatures due to which the biomass enhances. This is reflected through the Feed Conversion Ratio (FCR) of the fish, which exhibits a decrease with increase of temperature. We performed a pilot project in the Bali

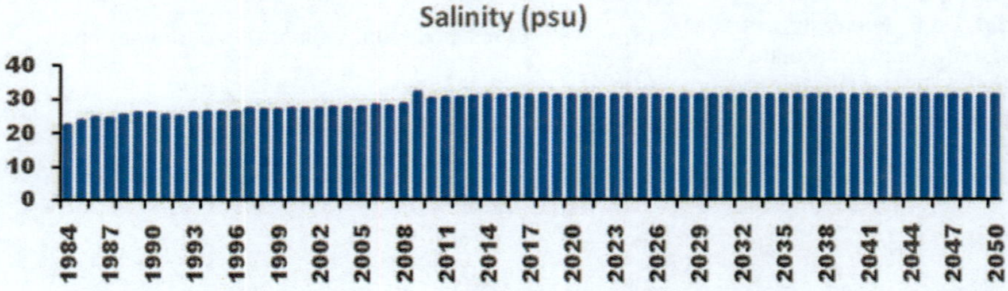


**Fig. 3.3** Predicted salinity (in psu) in central Indian Sundarbans during premonsoon using Nonlinear Autoregressive Neural Network Model; real time data from 1984 to 2018 has been used as training in the model

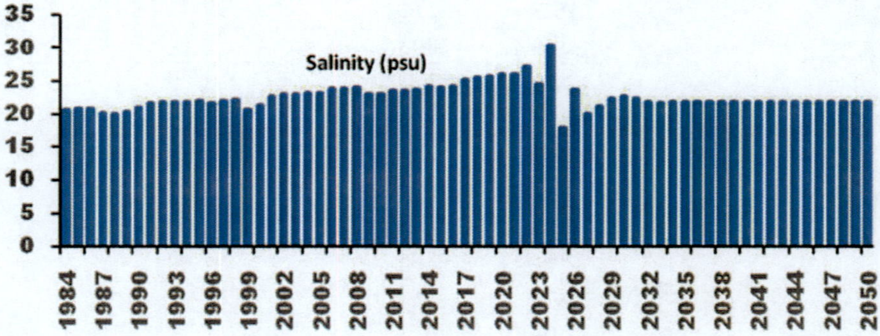


**Fig. 3.4** Predicted salinity (in psu) in central Indian Sundarbans during monsoon using Nonlinear Autoregressive Neural Network Model; real time data from 1984 to 2018 has been used as training in the model

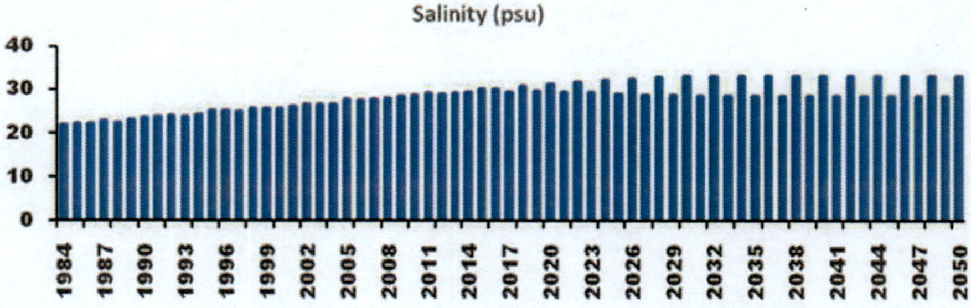


**Fig. 3.5** Predicted salinity (in psu) in central Indian Sundarbans during postmonsoon using Nonlinear Autoregressive Neural Network Model; real time data from 1984 to 2018 has been used as training in the model

**Table 3.7** Decreasing trend of salinity (in psu) in western Indian Sundarbans

| | Premonsoon | Monsoon | Postmonsoon |
|---|---|---|---|
| 1984 | 28.68 | 20.02 | 23.31 |
| 1985 | 28.34 | 19.67 | 21.03 |
| 1986 | 28.97 | 19.55 | 20.98 |
| 1987 | 28.00 | 19.42 | 21.02 |
| 1988 | 27.34 | 18.66 | 21.10 |
| 1989 | 26.57 | 18.44 | 21.24 |
| 1990 | 26.12 | 18.73 | 20.66 |
| 1991 | 26.07 | 18.53 | 20.54 |
| 1992 | 25.85 | 17.59 | 20.57 |
| 1993 | 25.81 | 17.97 | 20.43 |
| 1994 | 25.59 | 17.65 | 20.17 |
| 1995 | 25.34 | 16.95 | 20.29 |
| 1996 | 26.00 | 17.06 | 19.86 |
| 1997 | 26.05 | 17.13 | 19.91 |
| 1998 | 25.99 | 16.99 | 19.64 |
| 1999 | 25.57 | 16.87 | 19.54 |
| 2000 | 25.35 | 16.35 | 19.13 |
| 2001 | 25.20 | 16.13 | 19.23 |
| 2002 | 25.03 | 16.10 | 19.00 |
| 2003 | 24.90 | 16.41 | 18.84 |
| 2004 | 24.70 | 16.08 | 18.58 |
| 2005 | 24.35 | 16.19 | 18.18 |
| 2006 | 24.34 | 15.96 | 18.35 |
| 2007 | 24.20 | 15.78 | 18.11 |
| 2008 | 24.02 | 15.86 | 18.09 |
| 2009 | 28.07 | 17.04 | 18.02 |
| 2010 | 24.34 | 16.19 | 17.91 |
| 2011 | 24.00 | 15.94 | 17.78 |
| 2012 | 24.26 | 15.45 | 17.56 |
| 2013 | 23.23 | 15.13 | 18.44 |
| 2014 | 23.11 | 14.97 | 17.15 |
| 2015 | 22.91 | 12.78 | 19.40 |
| 2016 | 24.99 | 11.55 | 17.06 |
| 2017 | 23.09 | 10.57 | 18.33 |
| 2018 | 22.96 | 9.89 | 17.33 |

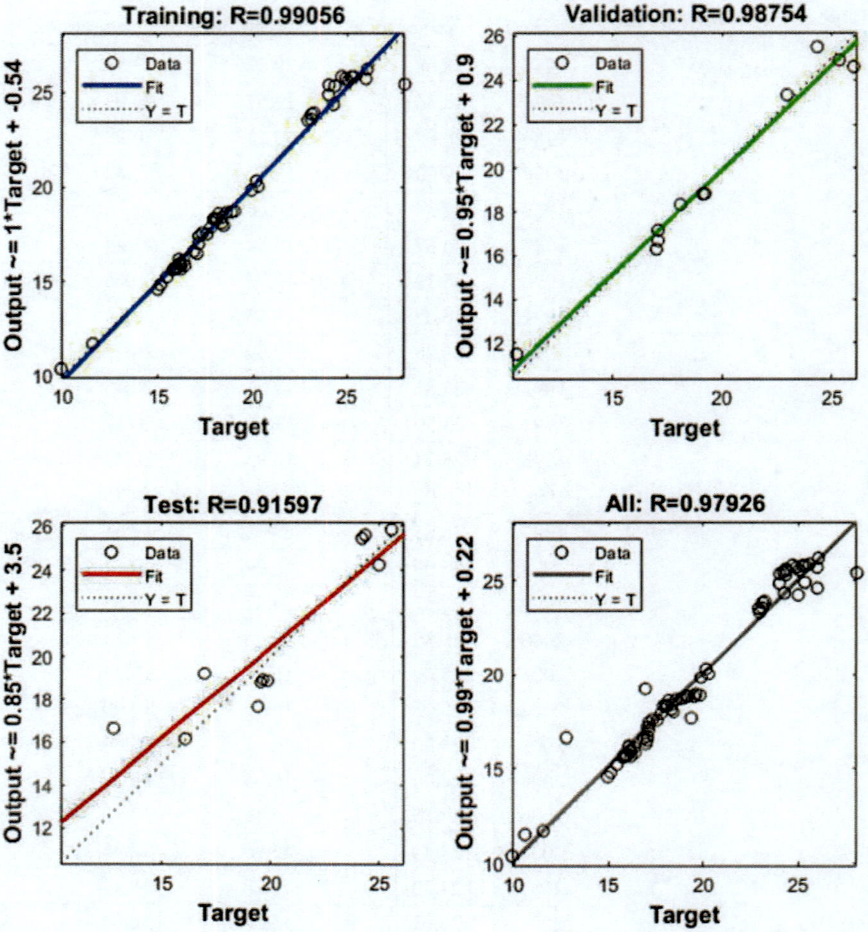


**Fig. 3.6** Outputs of Nonlinear Autoregression (NAR) model showing the trend of aquatic salinity in western Indian Sundarbans

island of Indian Sundarbans in a shrimp culture farm and monitored the FCR value of *Penaeus monodon* of three months age over a period of three decades (1990–2020). We observed a significant inverse relationship between ambient water temperature and FCR values (Table 3.11).

The decrease of FCR values with increase of water temperature (Fig. 3.10) is due to better conversion of feed into biomass of cultured shrimp with the rise of temperature. Thus, it can be concluded that at an increased temperature, within a specific thermal range the enzyme–substrate complex binds effectively, which poses a positive effect on the metabolic activities of the species leading to better growth.

**Table 3.8** Forecasting of salinity (in psu) in western Indian Sundarbans during premonsoon using Autoregressive Neural Network Model

| Year | Predicted salinity (in psu) in premonsoon |
|---|---|
| 2019 | 24.99 |
| 2020 | 23.12 |
| 2021 | 22.53 |
| 2022 | 25.31 |
| 2023 | 21.52 |
| 2024 | 23.12 |
| 2025 | 25.18 |
| 2026 | 23.15 |
| 2027 | 22.48 |
| 2028 | 25.34 |
| 2029 | 21.39 |
| 2030 | 23.28 |
| 2031 | 24.86 |
| 2032 | 23.98 |
| 2033 | 23.78 |
| 2034 | 24.27 |
| 2035 | 24.68 |
| 2036 | 24.43 |
| 2037 | 24.36 |
| 2038 | 24.44 |
| 2039 | 24.48 |
| 2040 | 24.45 |
| 2041 | 24.44 |
| 2042 | 24.45 |
| 2043 | 24.45 |
| 2044 | 24.45 |
| 2045 | 24.45 |
| 2046 | 24.45 |
| 2047 | 24.45 |
| 2048 | 24.45 |
| 2049 | 24.45 |
| 2050 | 24.45 |

### 3.2.3 *Alteration of Microbial Load in Fish Tissues*

Fish diseases are mostly seasonal and are regulated by environmental parameters preferably water temperature. Kawasaki (2001) provided specific evidence of the effects of global warming on world fisheries production, particularly on tuna and

**Table 3.9** Forecasting of salinity (in psu) in western Indian Sundarbans during monsoon using Autoregressive Neural Network Model

| Year | Predicted salinity (in psu) in monsoon |
|---|---|
| 2019 | 8.71 |
| 2020 | 8.51 |
| 2021 | 10.40 |
| 2022 | 11.32 |
| 2023 | 10.59 |
| 2024 | 8.24 |
| 2025 | 7.01 |
| 2026 | 9.62 |
| 2027 | 11.44 |
| 2028 | 11.21 |
| 2029 | 8.82 |
| 2030 | 6.91 |
| 2031 | 8.34 |
| 2032 | 11.37 |
| 2033 | 11.44 |
| 2034 | 9.26 |
| 2035 | 7.00 |
| 2036 | 7.65 |
| 2037 | 11.25 |
| 2038 | 11.46 |
| 2039 | 9.51 |
| 2040 | 7.11 |
| 2041 | 7.38 |
| 2042 | 11.14 |
| 2043 | 11.47 |
| 2044 | 9.70 |
| 2045 | 7.23 |
| 2046 | 7.22 |
| 2047 | 11.02 |
| 2048 | 11.47 |
| 2049 | 9.90 |
| 2050 | 7.36 |

sardines. The triggering of the fish disease due to warming of atmosphere was seen in American lobster. Massive, catastrophic summer-fall mortalities of lobsters in Long Island Sound began in August 1999, and the incidence continued to occur to a greater or lesser degree in subsequent summers. An extensive federally sponsored research program has identified summer warming of Long Island Sound bottom

**Table 3.10** Forecasting of salinity (in psu) in Western Indian Sundarbans during postmonsoon using Autoregressive Neural Network Model

| Year | Predicted salinity (in psu) in postmonsoon |
|---|---|
| 2019 | 18.99 |
| 2020 | 17.24 |
| 2021 | 19.43 |
| 2022 | 17.27 |
| 2023 | 18.65 |
| 2024 | 17.22 |
| 2025 | 19.41 |
| 2026 | 17.26 |
| 2027 | 18.68 |
| 2028 | 17.22 |
| 2029 | 19.43 |
| 2030 | 17.26 |
| 2031 | 18.62 |
| 2032 | 17.22 |
| 2033 | 19.40 |
| 2034 | 17.26 |
| 2035 | 18.73 |
| 2036 | 17.22 |
| 2037 | 19.46 |
| 2038 | 17.27 |
| 2039 | 18.54 |
| 2040 | 17.23 |
| 2041 | 19.33 |
| 2042 | 17.25 |
| 2043 | 18.90 |
| 2044 | 17.21 |
| 2045 | 19.46 |
| 2046 | 17.27 |
| 2047 | 18.53 |
| 2048 | 17.23 |
| 2049 | 19.31 |
| 2050 | 17.25 |

waters, coupled with hypoxia, as the most likely causes of the outbreak of disease. One of these diseases called "excretory calcinosis", discovered by scientists at Stony Brook University, is a gill tissue blood disorder resulting directly from warm water temperatures (Dove et al. 2004). Other lobster diseases also appear to result from the stress of high temperature and hypoxia. The result of these multiple stresses has

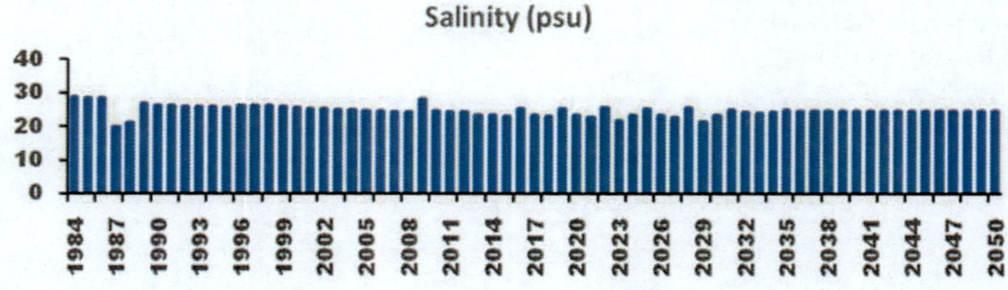


**Fig. 3.7** Predicted salinity (in psu) in western Indian Sundarbans during premonsoon using Nonlinear Autoregressive Neural Network Model; real time data from 1984 to 2018 has been used as training

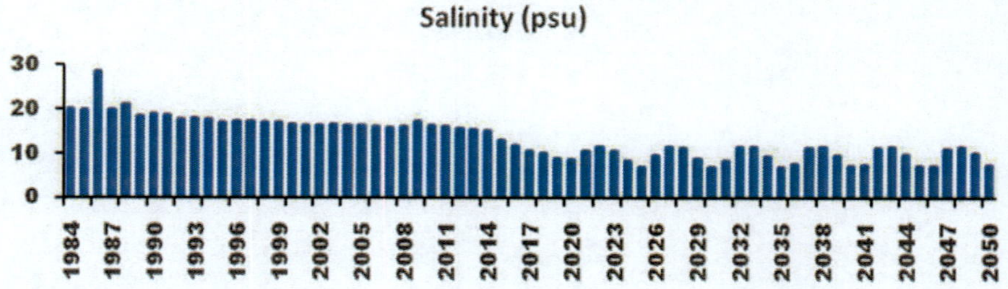


**Fig. 3.8** Predicted salinity (in psu) in western Indian Sundarbans during monsoon using Nonlinear Autoregressive Neural Network Model; real time data from 1984 to 2018 has been used as training in the model

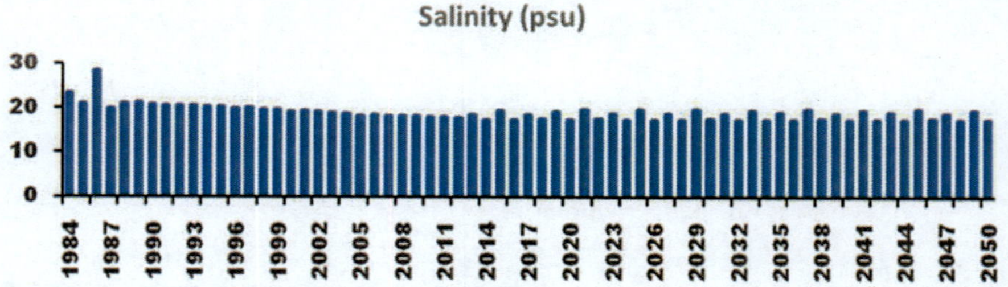


**Fig. 3.9** Predicted salinity (in psu) in western Indian Sundarbans during postmonsoon using Nonlinear Autoregressive Neural Network Model; real time data from 1984 to 2018 has been used as training in the model

been a 75% reduction in total landings and 85% reduction in the overall abundance of the population. These diseases now appear to be moving northward.

An experiment conducted by the authors at the roof-top tank of Techno India University, West Bengal (India) exhibited higher levels of total coliform and fecal coliform bacteria in the muscle, gill and hepatopancreas of *Macrobrachium rosenburgii* in 2019 compared to 2018 in all the three seasons (premonsoon, monsoon, postmonsoon). The increased microbial load during 2019 may be attributed to higher

**Table 3.11** Surface water temperature and FCR values of shrimps cultured at Bali island of Indian Sundarbans

| Year | Water temperature (°C) | FCR value |
|---|---|---|
| 1990 | 32.7±0.5 | 2.05 |
| 1991 | 33.1±0.5 | 1.98 |
| 1992 | 32.9±0.3 | 2.02 |
| 1993 | 33.0±0.4 | 1.96 |
| 1994 | 33.3±0.5 | 1.90 |
| 1995 | 33.0±0.4 | 2.00 |
| 1996 | 32.9±0.4 | 1.88 |
| 1997 | 33.0±0.6 | 1.85 |
| 1998 | 33.2±0.6 | 1.79 |
| 1999 | 33.5±0.6 | 1.74 |
| 2000 | 33.6±0.6 | 1.87 |
| 2001 | 33.8±0.6 | 1.69 |
| 2002 | 33.6±0.5 | 1.68 |
| 2003 | 33.4±0.5 | 1.71 |
| 2004 | 33.7±0.7 | 1.64 |
| 2005 | 33.5±0.7 | 1.52 |
| 2006 | 34.1±0.7 | 1.49 |
| 2007 | 34.0±0.7 | 1.46 |
| 2008 | 34.3±0.6 | 1.44 |
| 2009 | 36.0±0.6 | 1.35 |
| 2010 | 34.4±0.6 | 1.33 |
| 2011 | 34.6±0.6 | 1.41 |
| 2012 | 34.7±0.6 | 1.28 |
| 2013 | 34.8±0.5 | 1.30 |
| 2014 | 34.8±0.7 | 1.33 |
| 2015 | 35.1±0.7 | 1.29 |
| 2016 | 35.3±0.7 | 1.26 |
| 2017 | 35.8±0.7 | 1.28 |
| 2018 | 36.1±0.7 | 1.22 |
| 2019 | 36.5±0.7 | 1.27 |
| 2020 | 36.8±0.7 | 1.21 |

temperature witnessed in 2019 (average yearly temperature = 38.8 °C) compared to 2018 (average yearly temperature = 36.6 °C) in the city of Kolkata (Figs. 3.11, 3.12 and 3.13).

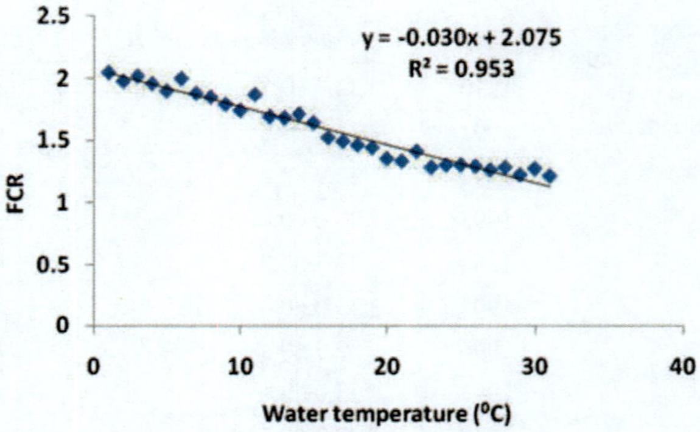


**Fig. 3.10** Decrease of FCR values with increase of water temperature

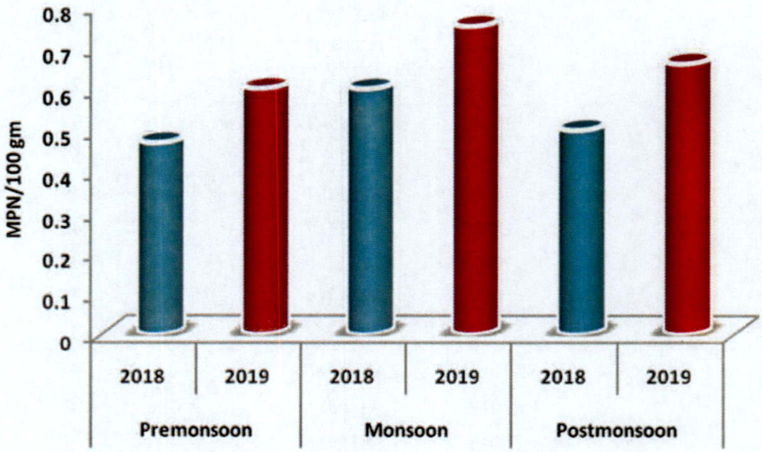


**Fig. 3.11** Total coliform (MPN/100gm) in the muscle of *Macrobrachium rosenburgii* in the roof-top tank of Techno India University, West Bengal (India)

## 3.3 Acidification—Effect on Fish Community of Indian Sundarbans

The atmospheric $CO_2$ has accelerated with a high magnitude since the pre-industrial scenario and has exceeded 400 ppm. Approximately 30% of the excess $CO_2$ in the atmosphere has been absorbed by the oceans (Sabine et al. 2004), which, owing to their expansive surface area, experience a parallel increase in the partial pressure of $CO_2$ ($PCO_2$) with the atmosphere (Doney 2010). If the current trajectory is maintained, increases in atmospheric $CO_2$ levels are predicted to exceed 900 ppm by 2100 (Meinshausen et al. 2011). Increasing $PCO_2$ of marine systems is often accompanied by a decrease in pH of the water, termed ocean acidification (OA) (Doney 2010).

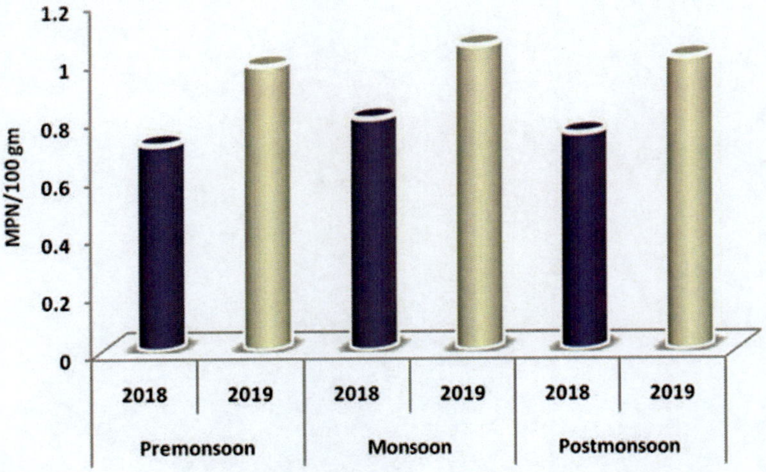


**Fig. 3.12** Total coliform (MPN/100gm) in the gills of *Macrobrachium rosenburgii* in the roof-top tank of Techno India University, West Bengal (India)

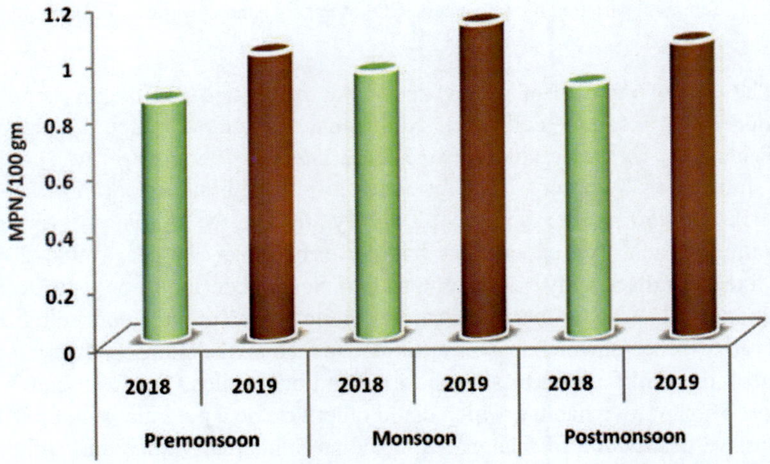


**Fig. 3.13** Total coliform (MPN/100gm) in the hepatopancreas of *Macrobrachium rosenburgii* in the roof-top tank of Techno India University, West Bengal (India)

The scenario in mangrove dominated Indian Sundarban estuaries is no exception to this global trend.

We carried out a seasonal study on atmospheric carbon dioxide concentration at Sagar Island at 18 m above the ground on the campus of Youth Hostel at Sagar South (21° 38′ 51.55″ N and 88° 02′ 20.97″ E), with a non-dispersive infrared gas analyzer (LI6262, LI-COR Inc., USA). We installed and air intake at 1 m above the roof of the building. A diaphragm pump was used to draw air through a ¼ inch Teflon tube at a rate of 10 L per minutes with most of the air being vented

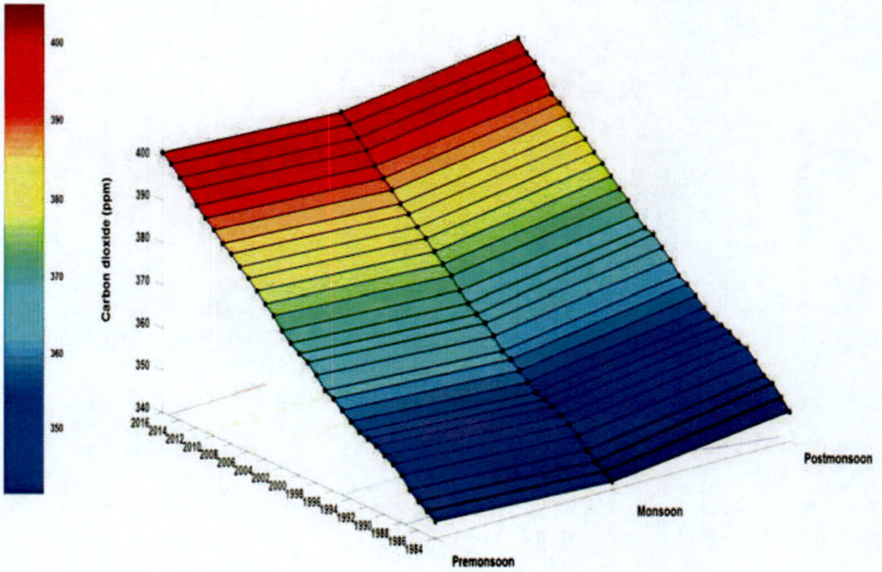


**Fig. 3.14** Temporal variation of atmospheric $CO_2$ at Sagar Island in western Indian Sundarbans

off. The step of removal of water vapour was performed and the air sample was introduced into a sample cell of the NDIR analyzer, and the output voltage of the NDIR analyzer was integrated at five-minute interval. The analyzer was calibrated every four hours by successively introducing four calibrated working gases (340, 380, 410 and 450 ppm carbon dioxide in dry air) into the NDIR analyzer cell for five minutes each. The experiment was performed in every year (1984–2016) in three fixed months namely May (premonsoon), September (monsoon) and December (postmonsoon). A linear increase in atmospheric carbon dioxide concentration was observed (Fig. 3.14), which may be attributed to factors like intense industrialization, urbanization, proliferation of shrimp farms and brick kilns (Fig. 3.15 and 3.16) at the cost of mangrove habitat, which could otherwise be a potential sink of $CO_2$.

Massive destruction of mangroves in Indian Sundarbans since mid eighties has squeezed the reservoir of mangrove storage due to which Estuarine Acidification (EA) has gradually crept in the lower Gangetic water. Sundarban, being a rich habitat for marine and estuarine fishes cannot escape the situation. Impacts of acidification on organisms with calcareous parts are highly identifiable, but in most fishes with efficient adaptive mechanism the impact is mostly masked. We have highlighted here few probable impacts of EA on the fish community.

**Fig. 3.15** Shrimp farms in Indian Sundarbans at the cost of mangroves

**Fig. 3.16** Brick kilns in Indian Sundarbans have come up on the mangrove habitat

### 3.3.1 Effect on Calcareous Structures

Fish are not ideally considered as calcifying organisms, but they produce $CaCO_3$ precipitates in the intestinal lumen and otoliths in the inner ear. Otoliths are $CaCO_3$ concretions that serve the purpose of sound detection and gravity sensing (Tohse and Mugiya 2004; Tohse et al. 2001), whereas intestinal $CaCO_3$ precipitates are produced to provide service like water absorption in the intestine and subsequent osmoregulation. It has been documented that there is an increase in the size and density of at elevated level of $CO_2$. A study on sea bass put forward the hypothesis that reduced aragonite saturation states in water with elevated $CO_2$ would impair otolith growth and demonstrated increased rather than decreased otolith size in larval fish exposed to environmentally realistic $CO_2$ levels of 993 μAtm (Checkley et al. 2009).

### 3.3.2 Effect on Growth and Condition Index

The growth and condition index of fishes are highly species- specific and life cycle stage-specific. At elevated level of $CO_2$ no significant change in condition factor was observed, but a reduction in condition factor was demonstrated in juvenile Atlantic cod (Moran and Støttrup 2011). We made a thorough study on the condition index of few dominant adult finfish species in both Hooghly and Matla estuary and observed a decrease in condition index with the decrease in aquatic pH. Individual length and weight of the selected species were measured to evaluate the condition factor[3] as per the following expression:

$$K = \frac{\overline{w}}{\left(\overline{TL}\right)^3} \times 10^3$$

where $K$ is the condition factor, $\overline{w}$ is the average weight (g) and $\overline{TL}$ is the average total length (cm). Condition factor is an indication of the well being of an organism and is based on the hypothesis that heavier fish of a given length are in a better condition than light weighed fish of the same species. It has been used as an index of growth and feeding intensity, decreases with increase in length and also influences the reproductive cycle in fish. On this background we have used this parameter/index as proxy to detect the impact of lowering of pH in the two major estuaries of Indian Sundarbans namely Hooghly and Matla.

Our first order analysis clearly reflects a pronounced variation in condition factor of the selected species between the two sectors and also between the two time periods (Table 3.12). Significantly higher condition factor of commercially important finfish was observed in the species collected from the Hooghly estuary, where the rate of pH decrease is less (0.0016) compared to Matla estuary (0.0025). We interpret from

**Table 3.12** Condition factor of the selected fin fish species in the Hooghly and Matla estuaries of Indian Sundarbans

| Commercially important fin fish | Western sector (Hooghly estuary) | | Central sector (Matla estuary) | |
|---|---|---|---|---|
| | 1984 | 2019 | 1984 | 2019 |
| | pH = 8.30 | pH = 8.24 | pH = 8.34 | pH = 8.25 |
|  *Tenualosa ilisha* (Family: Clupeidae) | 0.934 | 0.652 | 0 | 0 |
|  *Pama pama* (Family: Sciaenidae) | 1.102 | 0.429 | 0.901 | 0.328 |
|  *Pampus* spp. (Family: Stromateidae) | 0.932 | 0.544 | 0.872 | 0.431 |

(continued)

**Table 3.12** (continued)

| Commercially important fin fish | Western sector (Hooghly estuary) | | Central sector (Matla estuary) | |
|---|---|---|---|---|
| | 1984 | 2019 | 1984 | 2019 |
|  *Ilisha elongata* (Family: Pristigasteridae) | 0.878 | 0.613 | 0 | 0 |
|  *Lates calcarifer* (Family: Centropomidae) | 1.401 | 0.560 | 1.050 | 0.477 |

(continued)

**Table 3.12** (continued)

| Commercially important fin fish | Western sector (Hooghly estuary) | | Central sector (Matla estuary) | |
|---|---|---|---|---|
| | 1984 | 2019 | 1984 | 2019 |
|  *Pangasius pangasius* (Family: Pangasiidae) | 0.938 | 0.475 | 0.766 | 0.405 |
|  *Liza parsia* (Family: Mugilidae) | 0.957 | 0.516 | 0.623 | 0.385 |
|  *Liza tade* (Family: Mugilidae) | 1.008 | 0.933 | 0.918 | 0.864 |

(continued)

**Table 3.12** (continued)

| Commercially important fin fish | Western sector (Hooghly estuary) | | Central sector (Matla estuary) | |
|---|---|---|---|---|
| | 1984 | 2019 | 1984 | 2019 |
|  *Tenualosa toli* (Family: Clupeidae) | 1.326 | 0.733 | 1.004 | 0.667 |
|  *Polynemus paradiseus* (Family: Polynemidae) | 0.994 | 0.705 | 0.718 | 0.540 |

(continued)

**Table 3.12** (continued)

| Commercially important fin fish | Western sector (Hooghly estuary) | | Central sector (Matla estuary) | |
|---|---|---|---|---|
| | 1984 | 2019 | 1984 | 2019 |
|  *Otolithoides biauritus* (Family: Sciaenidae) | 1.105 | 0.917 | 0.889 | 0.520 |
|  *Tachysurus jella* (Family: Ariidae) | 0.982 | 0.764 | 0.653 | 0.388 |
|  *Sciaena biauritus* (Family: Sciaenidae) | 1.117 | 0.955 | 1.006 | 0.553 |

(continued)

**Table 3.12** (continued)

| Commercially important fin fish | Western sector (Hooghly estuary) | | Central sector (Matla estuary) | |
|---|---|---|---|---|
| | 1984 | 2019 | 1984 | 2019 |
|  *Eleutheronema tetradactylum* (Family: Polynemidae) | 1.006 | 0.489 | 0.919 | 0.475 |

*Note* The value 0 (zero) represents the non availability of the species in the estuarine water this data sets that stability in aquatic pH is a major driver in providing a congenial environment of fish species.

### 3.3.3 Effect on Bioaccumulation

Acidification has been observed to increase the bioaccumulation in the fish tissue. Our data bank since 1980 reflects an increasing trend of dissolved Cu in western Indian Sundarbans (Table 3.13), which is an indication of water quality variation. It is very difficult to relate the oscillation of dissolved Cu with climate change due to noise created by recent industrialization and urbanization in the Haldia industrial complex in the upstream zone. We also lack data on the effect of temperature increase on Cu toxicity. It has been documented that lowering of pH (which is a consequence of $CO_2$ increase) favours transference of Cu from the sediment (bed material) to aquatic phase (Mitra et al. 1994; Mitra 1998a, 1998b) and thus an indirect, but feeble string of Cu bioaccumulation and elevation of atmospheric $CO_2$ can be sketched.

Statistical analyses of our 18 years data set on aquatic pH, dissolved Cu and biologically available Cu in sediment of western Indian Sundarbans (Table 3.13) confirm the process of transference of Cu from the sediment bed to the upper aquatic

**Table 3.13** Concentrations of dissolved Cu, biologically available Cu in sediment and surface water pH in western Indian Sundarbans

| Year | Dissolved Cu (ppb) | Biologically available Cu (ppm) | pH |
|---|---|---|---|
| 2000 | 38.73 | 24.66 | 8.33 |
| 2001 | 41.84 | 20 | 8.33 |
| 2002 | 50.93 | 15.81 | 8.32 |
| 2003 | 49.65 | 18.24 | 8.32 |
| 2004 | 49.94 | 20.66 | 8.32 |
| 2005 | 50.41 | 16.96 | 8.31 |
| 2006 | 55.07 | 16.65 | 8.32 |
| 2007 | 55.87 | 14.7 | 8.31 |
| 2008 | 54.85 | 15.19 | 8.31 |
| 2009 | 62.96 | 12.97 | 8.31 |
| 2010 | 66.02 | 13.41 | 8.3 |
| 2011 | 65.95 | 14.29 | 8.3 |
| 2012 | 61.12 | 15.72 | 8.3 |
| 2013 | 69.73 | 11.98 | 8.29 |
| 2014 | 66.93 | 13.83 | 8.29 |
| 2015 | 57.4 | 16.4 | 8.29 |
| 2016 | 66.96 | 14.69 | 8.28 |
| 2017 | 69.94 | 11.87 | 8.28 |

phase ($\mathbf{r}_{\text{dissolved Cu} \times \text{sediment Cu}} = -0.903, p < 0.01$; $\mathbf{r}_{\text{dissolved Cu} \times \text{pH}} = -0.887, p < 0.01$ and $\mathbf{r}_{\text{sediment Cu} \times \text{pH}} = 0.747, p < 0.01$) in a condition of low pH value. Considering dissolved Cu as variable 1 (VAR_1), biologically available Cu as variable 2 (VAR_2) and pH as variable 3 (VAR_3), the scatter plot generated through SYSTAT reflects that higher pH favours the precipitation of Cu from water to sediment, whereas lower pH increases the concentrations of dissolved Cu (Fig. 3.17). The vulnerability of the increasing trend of dissolved heavy metals cannot be ignored as the process is related to bioaccumulation and biomagnification leading to human health hazard.

We have also observed increasing trend of heavy metals since 2000 (Table 3.14a–c), which is inversely proportional to trend of pH in the same period of time. A long term research needs to be undertaken to investigate the interrelationship between the heavy metal level in fish tissue and indicators of climate change (like water temperature or pH). Simulation experiment may be an effective approach in this context.

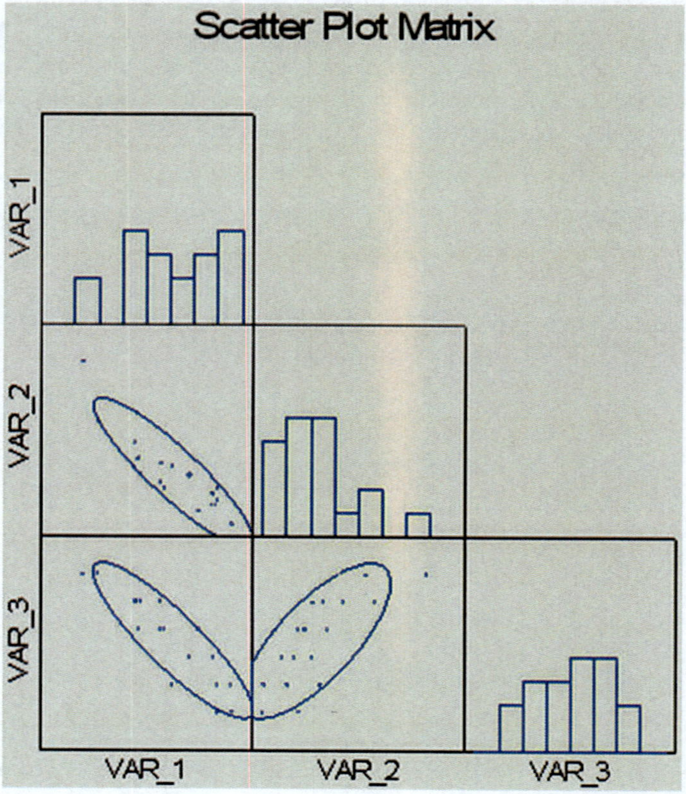


**Fig. 3.17** Scatter plot matrix showing the inter-relationship between dissolved Cu (VAR_1), biologically available Cu in sediment (VAR_2) and surface water pH (VAR_3)

**Table 3.14** **a** Zn concentrations (in ppm dry wt.) in fin fish muscles collected from Hooghly estuary, **b** Cu concentrations (in ppm dry wt.) in fin fish muscles collected from Hooghly estuary, **c** Pb concentrations (in ppm dry wt.) in fin fish muscles collected from Hooghly estuary

| **(a)** Zn concentrations (in ppm dry wt.) in fin fish muscles collected from Hooghly estuary | | | | | |
|---|---|---|---|---|---|
| Species | 2015 | 2016 | 2017 | 2018 | 2019 |
| *Coilia* sp. | 21.12 | 28.00 | 30.59 | 29.71 | 27.11 |
| *Thryssa hamiltonii* | 17.61 | 25.77 | 26.96 | 23.71 | 32.5 |
| *Rhinomugil corsula* | 19.23 | 29.23 | 29.12 | 38.73 | 38.1 |
| *Mugil cephalus* | 25.21 | 27.88 | 29.98 | 41.94 | 44.88 |
| *Sillago sihama* | 12.85 | 19.41 | 18.86 | 24.32 | 21.12 |
| *Pisodonophis boro* | 28.7 | 39.31 | 41 | 57.79 | 54.99 |
| *Tenualosa ilisha* | 21.9 | 29 | 28.47 | 33.6 | 34.08 |
| *Liza parsia* | 29.34 | 35.5 | 38.13 | 42.93 | 39.2 |
| *Liza tade* | 33.11 | 46.21 | 49.22 | 52.89 | 48.65 |
| *Stolephorus commersonii* | 14.23 | 20.35 | 12.66 | 18.24 | 10.84 |
| **(b)** Cu concentrations (in ppm dry wt.) in fin fish muscles collected from Hooghly estuary | | | | | |
| Species | 2015 | 2016 | 2017 | 2018 | 2019 |
| *Coilia* sp. | 3.19 | 6.34 | 6.68 | 7.78 | 9.98 |
| *Thryssa hamiltonii* | 6.68 | 10.27 | 10.21 | 7.13 | 11.33 |
| *Rhinomugil corsula* | 5.82 | 7.25 | 6.81 | 8.63 | 9.44 |
| *Mugil cephalus* | 10.49 | 15.04 | 19.45 | 16.06 | 20.59 |
| *Sillago sihama* | 3.08 | 6.08 | 4.95 | 6.40 | 5.04 |
| *Pisodonophis boro* | 13.60 | 18.09 | 17.47 | 18.64 | 21.21 |
| *Tenualosa ilisha* | 4.84 | 7.44 | 7.83 | 7.62 | 9.38 |
| *Liza parsia* | 5.42 | 9.26 | 8.74 | 7.98 | 10.0 |
| *Liza tade* | 6.49 | 10.82 | 10.80 | 13.41 | 9.10 |
| *Stolephorus commersonii* | 3.83 | 0 | 5.45 | 4.46 | 5.0 |
| **(c)** Pb concentrations (in ppm dry wt.) in fin fish muscles collected from Hooghly estuary | | | | | |
| Species | 2015 | 2016 | 2017 | 2018 | 2019 |
| *Coilia* sp. | 1.05 | 0.32 | 2.21 | 3.20 | 2.94 |
| *Thryssa hamiltonii* | 3.05 | 3.35 | 3.05 | 2.69 | 4.57 |
| *Rhinomugil corsula* | 4.22 | 4.63 | 4.98 | 2.53 | 2.28 |
| *Mugil cephalus* | 7.28 | 7.53 | 5.08 | 5.77 | 7.36 |
| *Sillago sihama* | 2.1 | 1.43 | 1.50 | 2.54 | 2.34 |
| *Pisodonophis boro* | 4.53 | 3.32 | 5.97 | 7.08 | 4.80 |
| *Tenualosa ilisha* | 4.17 | 3.60 | 2.37 | 4.65 | 4.65 |
| *Liza parsia* | 2.06 | 2.67 | 2.08 | 2.53 | 1.97 |
| *Liza tade* | 2.83 | 2.23 | 3.33 | 2.67 | 1.96 |
| *Stolephorus commersonii* | 2.04 | 1.38 | 1.17 | 1.84 | 1.60 |

**Table 3.15** Egg diameter of *Mystus gulio* in relation to pH of ambient aquatic phase

| Year | Sagar Island | | Bali Island | |
|---|---|---|---|---|
| | Mean Egg diameter (in mm); Sample size = 1875 | Mean pH (mean of 50 readings covering three seasons) | Mean Egg diameter (in mm); Sample size = 1655 | Mean pH (mean of 50 readings covering three seasons) |
| 2010 | 1.5 | 8.30±0.03 | 1.7 | 8.31±0.04 |
| 2011 | 1.3 | 8.30±0.03 | 1.8 | 8.31±0.04 |
| 2012 | 1.4 | 8.28±0.02 | 1.3 | 8.32±0.04 |
| 2013 | 1.2 | 8.27±0.02 | 1.4 | 8.32±0.04 |
| 2014 | 0.8 | 8.27±0.02 | 1.6 | 8.30±0.03 |
| 2015 | 0.9 | 8.28±0.02 | 1.2 | 8.30±0.03 |
| 2016 | 1.1 | 8.27±0.02 | 1.0 | 8.29±0.03 |
| 2017 | 0.7 | 8.25±0.03 | 0.9 | 8.29±0.03 |
| 2018 | 0.1 | 8.23±0.02 | 1.0 | 8.30±0.03 |
| 2019 | 0.6 | 8.21±0.03 | 0.8 | 8.28±0.02 |

### 3.3.4 Effect on Spawning

Researchers have documented that acidification of the ocean, seas and estuarine waters are likely to interfere in the spawning process of fish. It is known that both the male and the female gonads exhibit cyclic morphological and histological changes before reaching full maturity and becoming ripe. The release of gametes from the body into ambient aquatic phase is called spawning. During spawning the egg diameter attains maximum value. We have documented the egg diameter of *Mystus gulio*, the estuarine cat fish in the waters adjacent to Sagar island and Bali island for a decade of 2010–2019 and observed a decrease in egg diameter in both the stations (Table 3.15). Our result clearly confirms the adverse impact of estuarine acidification on the egg diameter which may pose a negative effect on the spawning process of the species. Detailed literature on this subject is still scanty and needs to be studied to fill the gap area.

## 3.4 Take Home Messages

A. The fish fauna of the estuarine waters around Sundarbans has been classified into residents and transients (migrants). The species whose individuals of different sizes are present during all the months of the year in any zone of the estuary are referred to as resident species. The important resident species of fish are *Mugil parsia, Mugil tade, Polynemus paradiseus, Polydactylus indicus, Otolithoides biauritus, Lates calcarifer, Hilsa toli, Arius jella, Harpodon nehereus, Setipinna taty, Ilisha elongata, Setipinna phasa, Coilia ramcarati, Pama pama* and

*Sillaginopsis panijus.* The transient or migratory fishes enter and stay in the Bay of Bengal associated estuaries for a short period. Depending on nature of migration, the migrants may be divided into three categories like (1) Marine forms that migrate upstream and spawn in freshwater areas of the estuary like *Tenualosa ilisha, Polynemus paradiseus, Sillaginopsis panijus and Pama pama* (2) Freshwater species, which spawn in saline area of the estuary like *Pangasius pangasius* (3) Marine species, that spawns in less saline water of the estuary like *Arius jella, Osteogeneious militaris* and *Polydactylus indicus.*

B. Climate change is a bitter truth associated with the present century and has both direct and indirect impacts on fish stocks which are exploited commercially. Direct effects act on physiology and behaviour of fishes and alter their growth, reproduction, mortality and distribution. Indirect effects encompass events like alteration of aquatic productivity, biotic community structure and composition of the marine and estuarine ecosystems on which fishes depend for food and survival. Changes in primary and secondary production will obviously have a major effect on fisheries production, but it is not possible in the current state of knowledge to make accurate quantitative predictions of changes in global marine primary production solely due to climate.

C. Ocean acidification due to increased $CO_2$ level is a well accepted concept. Most of our knowledge on the direct effects of ocean acidification on marine organisms focuses on species known as "marine calcifiers" (e.g., corals, molluscs etc.) that build skeletons or shells made of calcium carbonate. Many of these species will suffer impaired ability to build skeletons as pH decreases. We know less about the direct impacts of acidification on harvested species like fishes and squids. In these species, the response to acidification is likely to involve physiological diseases including acidosis of tissue and body fluids leading to impaired metabolic function. Egg and larval stages are likely to be much more susceptible than adults, suggesting that reduced reproductive success will be among the first symptoms to appear. The indirect effects of acidification on fisheries will include loss of reef habitat and breeding ground constructed by marine calcifiers.

D. Massive destruction of mangroves in Indian Sundarbans since mid eighties has squeezed the reservoir of mangrove storage due to which Estuarine Acidification (EA) has gradually crept in the lower Gangetic water. Sundarban, being a rich habitat for marine and estuarine fishes cannot escape the situation. Impacts of acidification on organisms with calcareous parts are highly identifiable, but in most fishes with efficient adaptive mechanism the impact is mostly masked. Few identifiable indicators to assess the impact of EA on fish community of Indian Sundarbans are otolith growth, condition index, bioaccumulation in fish tissue and spawning related parameters.

## Annexure 3: Compartmentation of Heavy Metals due to Acidification in Matla Estuary of Indian Sundarbans

### Background of the Study

Over the last century, the atmospheric concentration of carbon dioxide has risen at a rate 100 times faster than any change observed during the past 650,000 years (Siegenthaler et al. 2005). There is broad consensus that this ongoing change is a direct result of human activity, principally by fossil fuel burning, cement production and changing land use (Hansen et al. 2007). Atmospheric levels of carbon dioxide have consequently increased from pre-industrial levels of 280 ppm to a concentration of approximately 380 ppm (Feely et al. 2004). Almost 50% of all anthropogenic carbon dioxide emitted to the atmosphere has diffused passively into the ocean, thereby significantly decreasing the rate of global warming (Sabine et al. 2004). Concentrations of atmospheric carbon dioxide are rising at a rate of 3.3% per year and will continue this rising trend (Canadell et al. 2007). Hydrological models predict that, based on proposed future emissions of carbon dioxide, the average oceanic pH will decline by 0.3–0.5 by the year 2100 and by 0.7 within the next 300 years (Caldeira and Wickett 2003). Leakage from carbon dioxide seabed storage would create locally faster and stronger acidification than that induced by atmospheric carbon dioxide (Hawkins 2004).

The process of acidification largely regulates the concentration of conservative pollutants (primarily heavy metals) in the water and sediments. It has been documented by several researchers that aquatic pH controls the process of dissolution / precipitation and there by regulates the level of heavy metals in the aquatic phase and the underlying sediment compartments (Mitra et al. 1992; Mitra and Choudhury 1993; Byrne 2002). Depending on the aquatic pH, bioaccumulation and subsequent toxic effects occur in plankton, nekton or benthic community.

On the basis of the global trend of acidification, we try to focus on the issue at the local level in the Matla River situated in the central Indian Sundarbans of the lower Gangetic delta region and evaluat e the role of the phenomenon on the compartmentation of selected heavy metals (Zn, Cu and Pb), that are dominant in the present geographical locale (Mitra 1998a, 1998b; Mitra et al. 2011).

### Materials and Methods

### Study Area

The lower Gangetic delta region, at the apex of Bay of Bengal sustains the famous mangrove dominated ecosystem, the Sundarbans. The deltaic complex has an area of 9630 sq. Km and houses 102 islands. The western sector of the deltaic lobe receives the snowmelt water of mighty Himalayan glaciers after being regulated through several barrages on the way. The central sector on the other hand, is fully deprived from such supply due to heavy siltation and clogging of the Bidyadhari channel since the late fifteenth century (Chaudhuri and Choudhury 1994). The lower part of the deltaic complex has multifarious industries such as paper, textiles, chemicals, pharmaceuticals, plastic, shellac, food, leather, jute, tyres and cycle rims (UNEP 1982). These units are point sources of heavy metals in the estuarine water (Mitra et al. 1992). The central part of the delta complex receives the water of the Matla River. Industries are almost absent along the bank of the river, but the water is contaminated with sewage from the highly urbanized city of Kolkata. Reports of considerable concentrations of Zn, Cu and Pb are also available in this region, which originates mainly from the antifouling paints used for conditioning the fishing vessels and trawlers (Mitra et al. 1992, 2011; Mitra and Choudhury 1993; Mitra 1998a, 1998b). The present study was conducted 1.5 km off Bonnie camp (21° 49′42.9″ N/88° 37′13.7″ E) in the Ajmalmari river (local name) that receives the water from the Matla River.

A total of 30 samples (for both water and sediment) were collected for each parameter in May (summer season in the present geographical locale) of every year (1984–2013) during high tide condition and the average results of every parameter were considered for statistical analyses.

**Measurement of Aquatic pH**

pH of the surface water in the selected sampling station was measured with a portable pH meter (sensitivity = ± 0.02).

**Analysis of Dissolved Zn, Cu and Pb**

Surface water samples were collected from the sampling station using 10-l Teflon-lined Go-Flo bottles, fitted with Teflon taps and deployed on a rosette or on Kevlar line, with additional surface sampling carried out by hand. Shortly after collection, samples were filtered through Nuclepore filters (0.4 μm pore diameter) and aliquots of the filters were acidified with sub-boiling distilled nitric acid to a pH of about 2 and stored in cleaned low-density polyethylene bottles. Dissolved heavy metals were separated and pre-concentrated from the seawater using dithiocarbamate complexation and subsequent extraction into Freon TF, followed by back extraction into $HNO_3$ as per the procedure of Danielsson et al (1978). Extracts were analyzed for Zn, Cu and Pb by Atomic Absorption Spectrophotometer (Perkin Elmer: Model 3030). The accuracy of the dissolved heavy metal determinations is indicated by good agreement between our values and reported for certified reference seawater materials (CASS 2) (Table 3.16).

**Analysis of Biologically Available Zn, Cu and Pb**

Sediment samples from surface (1 cm depth) were collected by scrapping using a pre-cleaned and acid washed plastic scale and immediately kept in clean polythene

**Table 3.16** Analysis of reference material for near shore seawater (CASS 2)

| Element | Certified value ($\mu g\ l^{-1}$) | Laboratory results ($\mu g\ l^{-1}$) |
|---|---|---|
| Zn | 1.97±0.12 | 2.01±0.14 |
| Cu | 0.675±0.039 | 0.786±0.058 |
| Pb | 0.019±0.006 | 0.029±0.009 |

**Table 3.17** Analysis of reference material (NIES Sargasso sample) for sediments obtained from the National Institute of Environmental Studies, Japan

| Element | Certified value ($\mu g\ g^{-1}$) | Laboratory results ($\mu g\ g^{-1}$) |
|---|---|---|
| Zn | 28.6 | 26.2 |
| Cu | 14.9 | 13.7 |
| Pb | 2.4 | 2.9 |

bags, which were sealed. The samples were washed with metal free double distilled water and dried in an oven at 105 °C for 5–6 h, freed from visible shells or shell fragments, ground to powder in a mortar and stored in acid washed polythene bags. Analyses of biologically available metals were done after re-drying the samples, from which 1 gm was taken and digested with 0.5 (N) HCl as per the standard procedure outlined by Malo (1977). The resulting solutions were then stored in polythene containers for analysis. The solutions were finally aspirated in the flame Atomic Absorption Spectrophotometer (Perkin Elmer: Model 3030) for the determination of metal concentrations. No detectable trace metals were found in the reagent blank. Analysis of the NIES Sargasso sample was carried out to assure the quality of the data (Table 3.17).

**Statistical Analysis**
Inter-relationships between aquatic pH, selected dissolved heavy metals and biologically available heavy metals in sediment were determined through correlation coefficient values, scatter plots and allometric equations for all possible combinations. All statistical calculations were performed with SPSS 9.0 for Windows.

**Results**

1. **Surface water pH**

The surface water pH exhibited variation within a small range. Highest value was recorded during 1985 (8.32) and lowest value (8.29) was recorded during 2013 (Fig. 3.18). The gradual lowering of pH (0.001/yr) clearly confirms the phenomenon of acidification in the Matla estuarine system.

2. **Dissolved metal**

The order of dissolved heavy metals in the estuarine water is Zn > Cu > Pb. Dissolved Zn ranged from 123.66 ppb during (1984) to 265.00 ppb during (2013) (Fig. 3.19). Dissolved Cu ranged from 44.75 ppb during (1984) to 97.14 ppb during (2013) (Fig. 3.20). During the study period the value of dissolved Pb ranged from 7.39 ppb (1984) to 17.39 ppb (2013). All the selected dissolved metals thus exhibit an increasing trend.

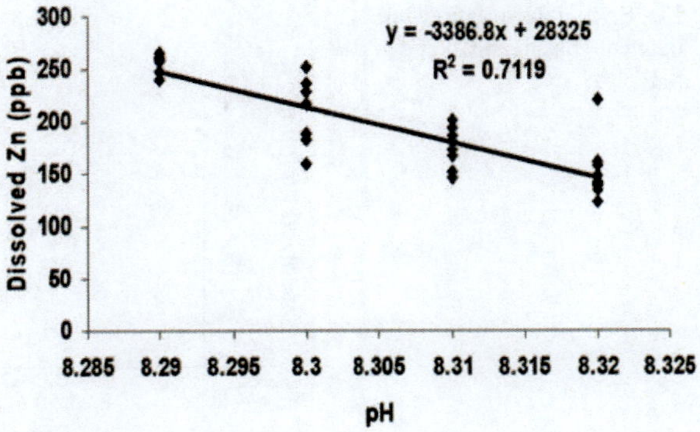


**Fig. 3.18** Inter-relationship between dissolved Zn and pH

### 3. Sediment metal

In sediment compartment, the biologically available metals exhibited a decreasing trend. In case of Zn, the value decreased from 64.71 ppm (during 1984) to 29.66 ppm (during 2013). In case of Cu, the value ranged from 7.31 ppm (during 2013) to 23.61 ppm (during 1984). In case of Pb, the lowest value was observed during 2013 (1.08 ppm) and the highest value was recorded during 1984 (10.78 ppm). It is also noted that the order of biologically available heavy metals in sediment is similar to that of dissolved heavy metals (Zn > Cu > Pb).

**Discussion**

Ocean acidification is predicted to occur under current IPCC $CO_2$ emissions scenarios (Houghton et al. 2001). Around 50% of the emissions of carbon dioxide are being absorbed by the oceans, increasing the $pCO_2$ with a concomitant decrease in the surface pH by 0.3–0.4 units by the end of the century (Feely et al. 2004; Caldeira and Wickett 2005; Orr et al. 2005). At present, the carbon dioxide interaction with seawater reduces the carbonate ion concentration, which is thought to regulate the calcification of extracellularly calcifying organisms (Spero et al. 1997; Marubini et al. 2008). This process is governed by the $CaCO_3$ saturation state ($\Omega$ $(CO_3{}^{2-})$ $(Ca^{2+})/K^*_{sp}$), where $K^*_{sp}$ is the apparent stoichiometric solubility product. For values of $\Omega < 1$ (undersaturated), seawater is corrosive, and dissolution may proceed (Isaji 1995). The process triggers the rate of dissolution of precipitated heavy metals from the sediment compartment to the water column and alters the speciation of selected heavy metals in the study area. The scatter plots (Figs. 3.18, 3.19, 3.20, 3.21, 3.22 and 3.23) explain the significant negative and positive relationships of aquatic pH with dissolved and biologically available heavy metals in surface sediments respectively. The $R^2$ values stated in Figs. 3.18, 3.19, 3.20, 3.21, 3.22 and 3.23 indicate strong inverse relationships between aquatic pH and dissolved heavy metals. However, with biologically available heavy metals, the positive relationship is highly significant for Zn, and moderately significant for Cu and Pb.

The present findings may serve as baseline report to link the climate change induced aquatic pH variation with pollution level of the estuarine and coastal waters. The results of the study also suggest that heavy metal concentrations in coastal and

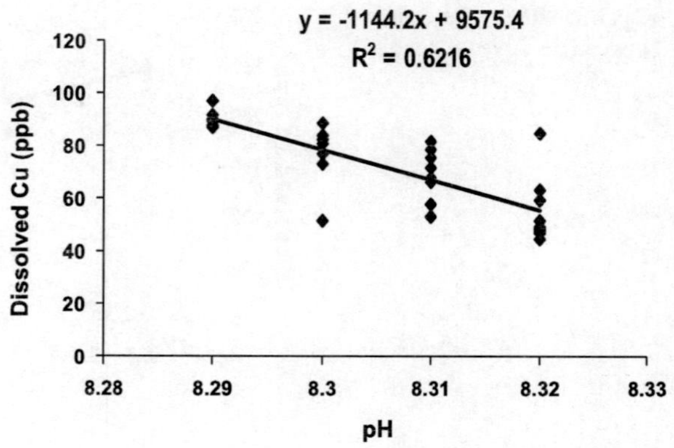


**Fig. 3.19** Inter-relationship between dissolved Cu and pH

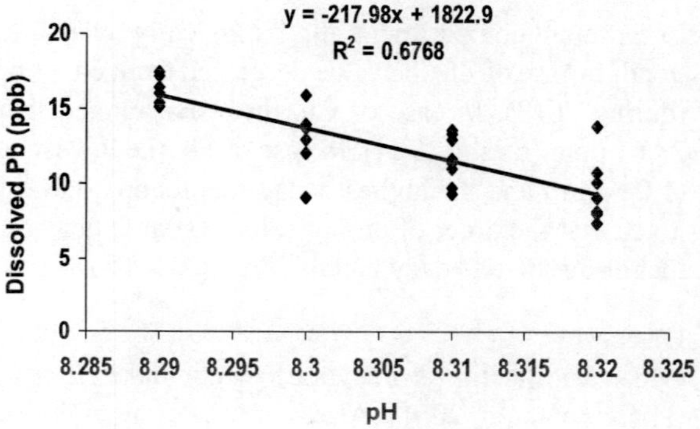


**Fig. 3.20** Inter-relationship between dissolved Pb and pH

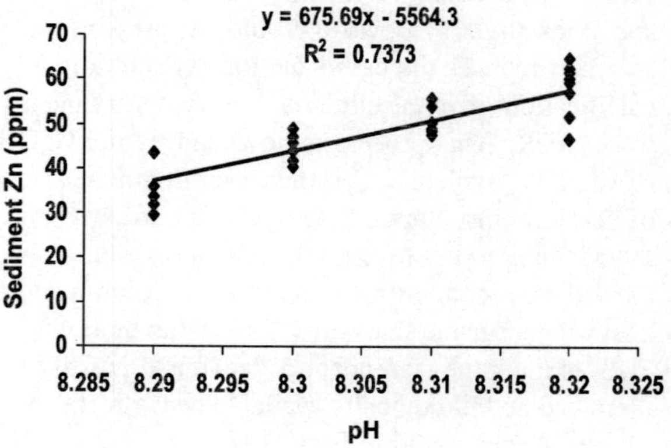


**Fig. 3.21** Inter-relationship between sediment Zn and pH

estuarine waters need to be monitored considering the phenomenon of acidification, which is a major driver in altering the species of heavy metals in the coastal and estuarine ecosystem. The subject needs further studies and researches as acidification may trigger the process of bioaccumulation of heavy metals in edible estuarine

Fig. 3.22 Inter-relationship between sediment Cu and pH

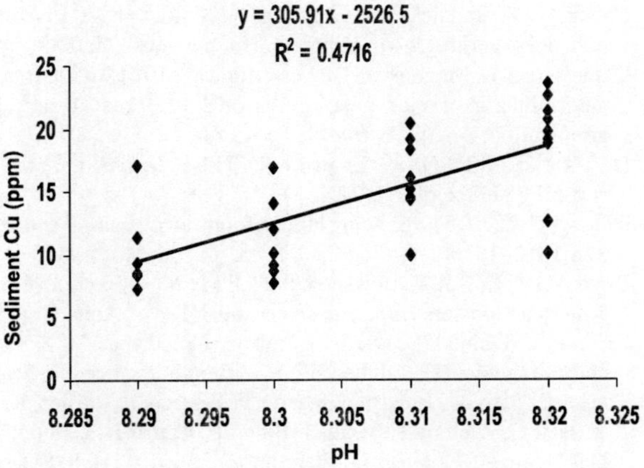


Fig. 3.23 Inter-relationship between sediment Pb and pH

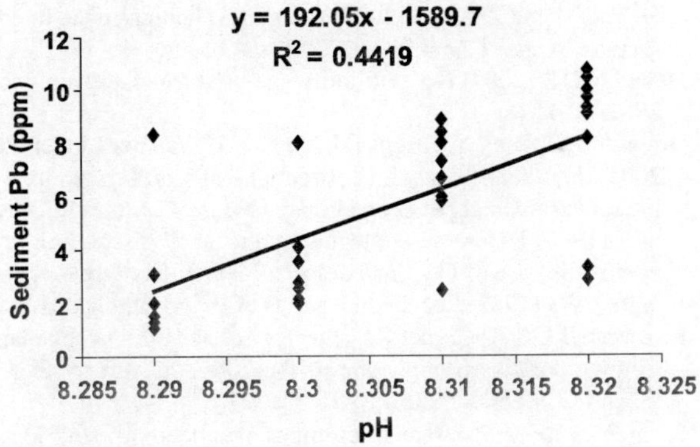


species (like fishes, crabs, oysters, shrimps etc.) and pose subsequent vulnerability to human health.

## References

Byrne RH (2002) Inorganic speciation of dissolved elements in seawater: the influence of pH on concentration ratios. Geochem Trans 3:11–16

Caldeira K, Wickett ME (2003) Anthropogenic carbon and ocean pH. Nature 425:365

Caldeira K, Wickett ME (2005) Ocean model predictions of chemistry changes from carbon dioxide emissions to the atmosphere and ocean. J Geophy Res 110:C09S04

Canadell JG, Le Quéré C, Raupach M, Field CB, Buitenhuis ET, Ciais P, Conway TJ, Gillet NP, Hougton RA, Marland G (2007) Contributions to accelerating atmospheric $CO_2$ growth from economic activity, carbon intensity, and efficiency of natural sinks. Proc Nat Acad Sci USA 104:18866–18870

Chaudhuri AB, Choudhury A (1994) Mangroves of the Sundarbans. IUCN, India

Checkley DM Jr, Dickson AG, Takahashi M, Radich JA, Eisenkolb N, Asch R (2009) Elevated $CO_2$ enhances otolith growth in young fish. Science 324:1683

Danielsson LG, Magnusson B, Westerlund S (1978) An improved metal extraction procedure for the determination of trace metals in seawater by atomic absorption spectrometry with electrothermal atomization. Analy Chem Acta 98:45–57

De TK, Saigal BN (1989) Spawning of Hilsa, *Tenualosa ilisha* (Hamilton) in the Hooghly estuary. J of Inl Fish Soc of Ind. 21:46–48

Doney SC (2010) The growing human footprint on coastal and open-ocean biogeochemistry. Science 328:1512–1516

Dove ADM, LoBue C, Bowser P, Powell M (2004) Excretory calcinosis: a new fatal disease of wild American lobsters *Homarus americanus*. Dis of Aqua Orga 58(2–3):215–221

Feely RA, Sabine CL, Lee K, Berelson W, Kleypas JA, Fabry VJ, Millero FJ (2004) Impact of anthropogenic $CO_2$ on the $CaCO_3$ system in the oceans. Science 281:237–240

Hansen J, Sato M, Ruedy R, Kharecha P, Lacis A, Miller, R, Nazarenko L, Lo K, Schmidt GA, Russell G, Aleinov I, Bauer S, Baum E, Cairns B, Canuto V, Chandler M, Cheng Y, Cohen A, Del Genio A, Faluvegi G, Fleming E, Friend A, Hall T, Jackman C, Jonas J, Kelley M, Kiang NY, Koch D, Labow G, Lerner J, Menon S, Novakov T, Oinas V, Perlwitz J, Perlwitz J, Rind, Romanou A, Schmunk R, Shindell D, Stone P, Sun S, Streets D, Tausnev N, Thresher D, Unger N. Yao M and Zhang S (2007) Dangerous human-made interference with climate: a GISS model E study. Atmos Chem Phys 7: 2287–2312

Hawkins DG (2004) No exit: thinking about leakage from geologic carbon storages sites. Energy 29:1571–1578

Houghton JT, Ding Y, Griggs DJ, Noguer M, Van der Linden PG, Xiaosu D (2001) Climate change 2001: the scientific basis: contributions of working group i to the third assessment report of the intergovernmental panel on climate change. Cambridge University Press, Cambridge, UK

Isaji S (1995) Defensive strategies against shell dissolution in bivalves inhabiting acidic environments: the case of Geloina (Corbiculidae) in mangrove swamps. Veliger 38:235–246

Jhingran VG (1982) Fish and Fisheries of India. Hindusthan Publishing Corporation (India), Delhi

Kawasaki T (2001) Climate change, regime shift and -stock management, presented to International Commemorative Symposium of the 70th Anniversary of the Jap Soc of Fish Sci, Yokohama, Japan~ October 1–5 2001. p 6

Khan RA (2003) Fish faunal resources of sunderban estuarine system with special reference to the biology of some commercially important species. Zoological Survey of India, Occ. Paper No. 209

Malo BA (1977) Partial extraction of metals from aquatic sediments. Environ Sci Technol 11:277–288

Mandal AK, Nandi NC (1989) Fauna of sundarban mangrove ecosystem, West Bengal. Fauna of Conservation Areas, Zoological Survey of India, India

Marubini F, Ferrier-Pages C, Furla P, Allemand D (2008) Coral calcification responds to seawater acidification: a working hypothesis toward a physiological mechanism. Coral Reefs 27:491–499

Meinshausen M, Smith SJ, Calvin K, Daniel JS, Kainuma MLT, Lamarque J, Matsumoto K, Montzka SA, Raper SCB, Riahi K et al (2011) The RCP greenhouse gas concentrations and their extensions from 1765 to 2300. Clim Change 109:213–241

Mitra A (1998a) Status of coastal pollution in West Bengal with special reference to heavy metals. J Ind Ocen Stud 5(2):135–138

Mitra A (1998b) Status of coastal pollution in West Bengal with special reference to heavy metals. J Ind Ocea Stud 5(2):135–138

Mitra A, Choudhury A (1993) Seasonal variations in metal content in the gastropod *Nerita articulata* (*Gould*). Ind J Environ Health NEERI 35(1):31–35

Mitra A, Choudhury A, Zamaddar YA (1992) Effects of heavy metals on benthic molluscan communities in Hooghly estuary. Proc Zool Soc 45:481–496

Mitra A, Banerjee K, Sinha S (2011) Shrimp tissue quality in the lower Gangetic delta at the apex of Bay of Bengal. Toxicol Environ Chem 93(3):565–574

Mitra A (2000) The Northeast coast of the Bay of Bengal and deltaic Sundarbans. In: Seas at the Millennium—An environmental evaluation, Chapter 62 (Editor: Charles Sheppard, University of Warwick, Coventry, UK), Elsevier Science 143–157

Mitra A, Trivedi S, Choudhury A (1994) Inter-relationship between trace metal pollution and physico-chemical variables in the frame work of Hooghly estuary. Indian Ports 27–35

Mitra A, Gangopadhyay A, Dube A, Andre CKS, Banerjee K (2009) Observed changes in water mass properties in the Indian Sundarbans (Northwestern Bay of Bengal) during 1980–2007. Curr Sci 1445–1452

Moran D, Støttrup JG (2011) The effect of carbon dioxide on growth of juvenile Atlantic cod Gadus morhua L. Aquat Toxicol (amst) 102:24–30

Orr JC, Fabry VJ, Olivier A, Bopp LD, Scott C, Feely Richard A, Anand G, Nicolas G, Akio I, Fortunat J, Key Robert M, Keith L, Ernst M-R, Richard M, Patrick M, Anne M, Najjar Raymond G, Gian-Kasper P, Rodgers Keith B, Sabine Christopher L, Sarmiento Jorge L, Reiner S, Slater Richard D, Totterdell Ian J, Marie-France W, Yasuhiro Y, Yool A (2005) Anthropogenic ocean acidification over the twenty-first century and its impact on calcifying organisms. Nature 437:681–686

Pillay TVR (ed) (1967) FAO fishery report, (44), 1 (2): 55 p. In: Proceedings of the FAO world symposium on warm-water Pond Fish Culture, Rome, 18–25 May 1966

Sabine CL, Feely RA, Gruber N, Key RM, Lee K, Bullister JL, Wanninkhof R, Wong CS, Wallace DWR, Tilbrook B, Millero FJ, Peng PH, Kozyr A, Ono T, Rios AF (2004) The oceanic sink for anthropogenic $CO_2$. Science 305:367–371

Siegenthaler U, Stocker TF, Monnin E, Luethi D, Schwander J, Stauffer B, Raynaud D, Barnola JM, Fisher H, Masson-Delmotte V, Jouzel J (2005) Stable carbon cycle–climate relationship during the last Pleistocene. Science 310:1313–1317

Sinha M, Mukhopadhyay MK, Mitra PM, Bagchi MM, Karmakar HC (1996) Impact of Farakka Barrage on the hydrology and Fishery of Hooghly Estuary. Estuaries 19(3):710–722

Spero HJ, Bijma J, Lea DW, Bemis BE (1997) Effect of seawater carbonate ion concentration on foraminiferal carbon and oxygen isotopes. Nature 390:497–500

Tohse H, Mugiya Y (2004) Éffects of enzyme and anion transport inhibitors on in vitro incorporation of organic carbon and calium into endolymph and otoliths in salmon *Oncorhynchus masou.* Comp Biochem Physiol A 128:177–184

Tohse H, Ando H, Mugiya Y (2001) Biochemical properties and immunohistochemical localization of carbonic anhydrase in the sacculus of the inner ear in the salmon *Oncorhynchus masou.* Comp Biochem Physiol A Mol Integr Physiol 137:87–94

UNEP (1982) Pollution and the marine environment in the Indian Ocean. Geneva, Switzerland: UNEP Regional Seas Programme Activity Centre

## *Internet Reference*

www.nltr.org

# Chapter 4
# Impact of Acidification on Molluscs

**Abstract** The lowering of pH of the coastal and estuarine waters lead to a shortage of carbonate, which is the primary building block to build the shells of corals and molluscs. The main groups of animals that are highly vulnerable to acidification, apart from corals are oysters, clams, crabs and lobsters. The present chapter depicts the vulnerability of these shelled organisms to acidification with special focus on edible oyster (*Saccostrea* spp.) in which thinning of shell is observed with the lowering of ambient aquatic pH. The ground zero data obtained directly from the field study in Indian Sundarbans shows a direct relationship between aquatic pH and thickness of the oyster shell.

## 4.1 Estuarine Molluscan Community: An Overview

Mollusca constitutes a very large phylum in the domain of invertebrate animals with around 1,25,000 living species. This phylum encompasses of about 23% of all marine organisms and has six major classes (Table 4.1).

Members of this phylum have soft body with hard calcareous shell. However, in some groups of terrestrial slugs the shell is greatly reduced and sometimes internal which is basically functionless. Marine shelled molluscs have a complex life cycle in which the larval stage is pelagic and the adult stage is benthic in nature. Molluscs provide several ecosystem services which can be categorized into provisioning, regulating, supporting and cultural services (Table 4.2).

The major species of gastropods found in coastal and estuarine waters are listed in Table 4.3

Bivalves are amongst the important classes of phylum mollusca. They constitute a large class of molluscs, also known as pelecypods and possess hard calcareous shell made of two parts or ‘valves’. Few bivalves are carnivorous in nature, eating much larger prey than the tiny microalgae eaten by other bivalves. The best known examples of bivalves are clams, mussels, scallops and oysters. Gigantic shaped bivalves weighing about 5 kg are also seen in the marine and coastal ecosystems. Some common bivalves of the coastal and estuarine waters are listed in Table 4.4.



A. Mitra and S. Zaman, *Estuarine Acidification*,
https://doi.org/10.1007/978-3-030-84792-0_4

**Table 4.1** Major classes of phylum mollusca with examples

| Class | Example |
|---|---|
| Bivalvia | *Saccostrea cucullata, Crassostrea cuttackensis, C. madrasensis, Macoma birmanica* |
| Gastropoda | *Telescopium telescopium, Cerithedia obtusa, C. cingulata* |
| Cephalopoda | *Loligo indica, Octopus rugosus, Octopus alatus* |
| Polyplacophora | *Cryptochiton stelleri, Chaetopleura apiculata* |
| Scaphopoda | *Dentalium laqueatum* |
| Monoplacophora | *Neopilina galathea* |

**Table 4.2** Ecosystem services of molluscs

| Ecosystem service type | Description |
|---|---|
| Provisioning | Molluscan meat/flesh is a source of food, shells of molluscs are used as decorative items, calcium of molluscan shell is used in pharmaceutical industries |
| Regulating | Bioremediation, carbon sequestration |
| Supporting | Oyster bed serves as substratum of several marine and estuarine organisms like sea anemone, barnacles, seaweeds etc. |
| Cultural | Conch is used in several religious rituals |

Apart from class gastropoda and bivalvia, phylum mollusca also encompasses class cephalopoda, which has representatives in coastal and marine ecosystems. They are characterized by a completely merged head and foot, with a ring of arms and/or tentacles surrounding the head. Few estuarine cephalopods are listed in Table 4.5.

The mangrove ecosystem is highly dynamic as water covers almost one-third of the tree during high tide (Fig. 4.1), but it is also noted for its vast intertidal mudflats that get exposed during low tide (Fig. 4.2).

In mangrove forest, molluscs occupy almost all the levels in the food web such as predators, herbivores, detritus and filter feeders, and thus they play an important role in maintaining the function and productivity of mangroves. Gastropods and bivalves are the two major classes of phylum mollusca occupying mangrove forest areas. These macrobenthic molluscs can be broadly grouped under three categories on the basis of their niche preference namely (i) epifauna—those living on mud or surface area of the land, (ii) infauna—those burying themselves in the substratum, and (iii) arboreal—mostly living on the vegetation.

Examples of epifauna are *Telescopium telescopium, Cerithedia cingulata* etc. (Fig. 4.3).

*Macoma birmanica* is an infauna widely found in the intertidal mudflat of Sundarbans, which is under class bivalvia and possesses inhalant and exhalent siphons for in-taking of food materials and expulsion of wastes respectively (Fig. 4.4). *Cerithedia obtusa* is a very common arboreal gastropod that is found grazing on the algal mat on mangrove stem (Fig. 4.5).

**Table 4.3** Major gastropod species found in coastal and estuarine waters

| S. No. | Gastropod species |
|---|---|
| 1 | *Amaea acuminate* |
| 2 | *Assiminea beddomeana* |
| 3 | *Assiminea brevicula* |
| 4 | *Assiminea francesiae* |
| 5 | *Assiminea hungerfordiana* |
| 6 | *Assiminea microsculpta* |
| 7 | *Assiminea nitida* |
| 8 | *Assiminea theobaldiana* |
| 9 | *Assiminea woodmasoniana* |
| 10 | *Auricula translucens* |
| 11 | *Batillaria sordid* |
| 12 | *Bulla ampulla* |
| 13 | *Canarium erythrinum* |
| 14 | *Cassidula aurisfelis* |
| 15 | *Cassidula bensoni* |
| 16 | *Cassidula mustelina* |
| 17 | *Cassidula nucleus* |
| 18 | *Cerithidea alata* |
| 19 | *Cerithidea cingulata* |
| 20 | *Cerithidea decollate* |
| 21 | *Cerithidea djadjariensis* |
| 22 | *Cerithidea fluviatilis* |
| 23 | *Cerithidea obtuse* |
| 24 | *Cerithidea quadrata* |
| 25 | *Cerithidea weversi* |
| 26 | *Cerithium citrinum* |
| 27 | *Cerithium columna* |
| 28 | *Cerithium coralium* |
| 29 | *Cerithium scabridum* |
| 30 | *Cerithium trailli* |
| 31 | *Cerithium zonatum* |
| 32 | *Chicoreus brunneus* |
| 33 | *Cithon oualaniensis* |
| 34 | *Clathrella clathrata* |
| 35 | *Clithon bicolor* |
| 36 | *Clithon corona* |
| 37 | *Cuthona annandalei* |

(continued)

**Table 4.3** (continued)

| S. No. | Gastropod species |
|---|---|
| 38 | *Cyclostrema (Tubiola) innocens* |
| 39 | *Dolomena variabilis* |
| 40 | *Dostia crepidularia* |
| 41 | *Drupella margariticola* |
| 42 | *Ellobium aurisjudae* |
| 43 | *Ellobium gangeticum* |
| 44 | *Elysia bangtawaensis* |
| 45 | *Engina alveolata* |
| 46 | *Ergalatax contracta* |
| 47 | *Ergalatax heptagonalis* |
| 48 | *Gangetica milicea* |
| 49 | *Haminoea crocata* |
| 50 | *Hemifusus pugilinus* |
| 51 | *Indoplanorbis exustus* |
| 52 | *Lambis (Lambis) lambis* |
| 53 | *Littoraria carinifera* |
| 54 | *Littoraria melanostoma* |
| 55 | *Littoraria scabra* |
| 56 | *Littoraria undulate* |
| 57 | *Littorina intermedia* |
| 58 | *Lunella cinerea* |
| 59 | *Mainwaringia paludomoidea* |
| 60 | *Mauritia Arabica* |
| 61 | *Melampus caffer* |
| 62 | *Melampus castaneus* |
| 63 | *Melampus ceylonicus* |
| 64 | *Melampus coffea* |
| 65 | *Melampus pulchella* |
| 66 | *Melampus singaporensis* |
| 67 | *Melampus striatus* |
| 68 | *Melanoides tuberculata* |
| 69 | *Mitra (Strigatella) paupercula* |
| 70 | *Mitra (Strigatella) scutulata* |
| 71 | *Monodonta (Monodonta) labio* |
| 72 | *Morula (Morula) anaxeres* |
| 73 | *Muricopsis bombayanus* |
| 74 | *Naquetia capucina* |

(continued)

**Table 4.3** (continued)

| S. No. | Gastropod species |
|---|---|
| 75 | *Nassarius foveolatus* |
| 76 | *Nassarius globosus* |
| 77 | *Nassarius immerse* |
| 78 | *Nassarius olivaceus* |
| 79 | *Nassarius orissaensis* |
| 80 | *Nassarius stolatus* |
| 81 | *Nassarius subconstrictus* |
| 82 | *Nerita albicilla* |
| 83 | *Nerita articulate* |
| 84 | *Nerita chamaeleon* |
| 85 | *Nerita chameleon* |
| 86 | *Nerita crepidularia* |
| 87 | *Nerita insculpta* |
| 88 | *Nerita planospira* |
| 89 | *Nerita polita* |
| 90 | *Nerita semirugosa* |
| 91 | *Nerita squamulata* |
| 92 | *Nerita undulate* |
| 93 | *Neritina depressa* |
| 94 | *Neritina smithi* |
| 95 | *Neritina violacea* |
| 96 | *Nodilittorina* |
| 97 | *Notocochlis qualteriana* |
| 98 | *Notocochlis tigerina* |
| 99 | *Onchidium tenerum* |
| 100 | *Onchidium tigrinum* |
| 101 | *Onchidium typhae* |
| 102 | *Pascula ochrostoma* |
| 103 | *Peronia verruculata* |
| 104 | *Pila virens* |
| 105 | *Pila globosa* |
| 106 | *Planaxis sulcatus* |
| 107 | *Polinices tumidus* |
| 108 | *Potamacmaea fluviatilis* |
| 109 | *Potamides cingulatus* |
| 110 | *Pseudanachis duclosiana* |
| 111 | *Pseudonerita obtuse* |

(continued)

**Table 4.3** (continued)

| S. No. | Gastropod species |
|---|---|
| 112 | *Pseudonerita sulculosa* |
| 113 | *Pugilina cochlidium* |
| 114 | *Purpura bufo* |
| 115 | *Purpura persica* |
| 116 | *Pythia plicata* |
| 117 | *Rhinoclavis (Rhinoclavis) sinensis* |
| 118 | *Rhinoclavis aspera* |
| 119 | *Rhinoclavis vertagus* |
| 120 | *Salinator burmana* |
| 121 | *Septaria caerulescens* |
| 122 | *Stenothyra blanfordiana* |
| 123 | *Stenothyra deltae* |
| 124 | *Syncera brevicula* |
| 125 | *Telescopium telescopium* |
| 126 | *Terebralia palustris* |
| 127 | *Thais (Thalessa) virgata* |
| 128 | *Thaisella blanfordi* |
| 129 | *Thaisella lacera* |
| 130 | *Thaisella tissoti* |
| 131 | *Thiara scabra* |
| 132 | *Umbonium vestiarium* |
| 133 | *Vanikoro cancellata* |

Molluscs are known for their unique property to serve as indicator species. Macrobenthic molluscs can tolerate a wide range of environmental oscillation although to some molluscs like oyster, low salinity (below 10 psu) often poses stress on their survival. We have made an extensive study on the macrobenthic molluscan diversity in the estuaries of east coast of Indian sub-continent during May 2014–April 2015. The intertidal mudflats in four maritime states of India namely West Bengal (Shankarpur coast), Odisha (Bahuda estuary), Andhra Pradesh (East Godavari estuary) and Puducherry (Kalapet coast) were our selected sites. A total of 17 dominant species were observed under the molluscan community encompassing two major classes namely bivalvia and gastropoda. Population density by quadrat method was enumerated in all the sites to evaluate the magnitude of environmental stress (Figs. 4.6, 4.7, 4.8, 4.9, 4.10, 4.11, 4.12, and 4.13).

From the data of population density of each sites, we evaluated the Shannon—Weiner species diversity index as per the standard expression.

**Table 4.4** Major bivalve species found in the coastal and estuarine waters

| S. No. | Bivalve species |
|---|---|
| 1 | *Anadara granosa* |
| 2 | *Anadara rhombea* |
| 3 | *Bactronophorus thoracites* |
| 4 | *Bankia bipennata* |
| 5 | *Bankia campanellata* |
| 6 | *Bankia carinata* |
| 7 | *Bankia nordi* |
| 8 | *Bankia rochi* |
| 9 | *Barbatia candida* |
| 10 | *Barnea candida* |
| 11 | *Cardites bicolor* |
| 12 | *Codakia tigerina* |
| 13 | *Crassostrea cuttackensis* |
| 14 | *Crassostrea gryphoides* |
| 15 | *Crassostrea madrassensis* |
| 16 | *Crassostrea palmipes* |
| 17 | *Dicyathifer manni* |
| 18 | *Donax cuneatus* |
| 19 | *Donax incarnatus* |
| 20 | *Donax lubricus* |
| 21 | *Dosinia (Asa) tumida* |
| 22 | *Enigmonia aenigmatica* |
| 23 | *Gafrarium pectinatum* |
| 24 | *Geloina galatheae* |
| 25 | *Geloina siamica* |
| 26 | *Glauconome cerea* |
| 27 | *Glauconome chinensis* |
| 28 | *Isognomon ephippium* |
| 29 | *Katelysia opima* |
| 30 | *Laternula truncata* |
| 31 | *Lyrodus massa* |
| 32 | *Lyrodus pedicellatus* |
| 33 | *Macoma birmanica* |
| 34 | *Macoma qubernaculum* |
| 35 | *Mactra cuneata* |
| 36 | *Marcia opima* |
| 37 | *Martesia striata* |

(continued)

**Table 4.4** (continued)

| S. No. | Bivalve species |
|---|---|
| 38 | *Meretrix attenuata* |
| 39 | *Meretrix casta* |
| 40 | *Meretrix meretrix* |
| 41 | *Modiolus americanus* |
| 42 | *Modiolus modulaides* |
| 43 | *Modiolus striatulus* |
| 44 | *Modiolus traillii* |
| 45 | *Modiolus undulatus* |
| 46 | *Nausitora dunlopei* |
| 47 | *Nausitora hedleyi* |
| 48 | *Nausitora hedleyi* |
| 49 | *Neotrapezium sublaevigatum* |
| 50 | *Nototeredo edax* |
| 51 | *Nuculana mauritiana* |
| 52 | *Paphia malabarica* |
| 53 | *Paphia undulata* |
| 54 | *Pelecyora trigona* |
| 55 | *Perna perna* |
| 56 | *Perna viridis* |
| 57 | *Pharella javanicus* |
| 58 | *Placenta placenta* |
| 59 | *Placuna placenta* |
| 60 | *Polymesoda bengalensis* |
| 61 | *Potamocorbula abbreviata* |
| 62 | *Saccostrea cucullata* |
| 63 | *Siliqua albida* |
| 64 | *Solen annandalei* |
| 65 | *Solen brevis* |
| 66 | *Solen kempi* |
| 67 | *Sphenia pedata* |
| 68 | *Strigilla splendida* |
| 69 | *Tanysiphon rivalis* |
| 70 | *Tegillarca granosa* |
| 71 | *Tellina ala* |
| 72 | *Tellina bruguieri* |
| 73 | *Tellina iridescens* |
| 74 | *Teredo furcifera* |

(continued)

**Table 4.4** (continued)

| S. No. | Bivalve species |
|---|---|
| 75 | *Theora opalina* |
| 76 | *Uperotus rehderi* |
| 77 | *Villorita cyprinoides* |

**Table 4.5** Major cephalopods found in the coastal and estuarine waters

| S. No. | Species |
|---|---|
| 1 | *Loligo indica* |
| 2 | *Octopus rugosus* |
| 3 | *Sepia aculeata* |
| 4 | *Sepia inermis* |

**Fig. 4.1** Mangrove forest almost under water during high tide

$$\text{Species diversity index } (H) = -\sum_{i=1}^{-s} Pi \log_e Pi$$

$$or, \quad (H) = -\sum_{i=1}^{-s} ni/N \text{ loge } ni/N$$

**Fig. 4.2** Exposed inter-tidal mudflats in mangrove forest during low tide

where,

$Pi$ = Importance probability for each species

$ni$ = Importance value for each species

N = Total of importance values

S = Total number of species.

The Shannon—Weiner species diversity index, which is a mirror of degree of stress/congeniality ranged from 2.180 to 2.226 (in Shankarpur coast of West Bengal), 2.693 to 2.765 (in Bahuda estuary of Odisha), 2.176 to 2.474 in (East Godavari estuary of Andhra Pradesh) and 2.138 to 2.578 (in Kalapet coast of Puducherry). Considering the mean values of Shannon-Weiner species diversity index, it is observed that the value is highest in Bahuda estuary (2.787), followed by East Godavari estuary (2.638), Kalapet coast (2.599) and Shankarpur coast (2.278). The picture of lowest diversity index of macrobenthic molluscs in Shankarpur coast despite of considerable population density is the outcome of uneven distribution of population amongst the species, which is an indication of high degree of environmental stress in and around the selected sampling site (Figs. 4.14, 4.15, 4.16, and 4.17). The high degree of environmental stress may be attributed to lowering of pH as witnessed by us in a nearby station known as Digha (*vide* Table 4.6 for the long term data set), which is a popular tourist spot in the maritime state of West Bengal (21° 0.62′ 22″ N, 87° 50′ 66″ E).

**Fig. 4.3** *Telescopium telescopium*—a common epifauna (under class gastropoda) in mangrove ecosystem of Indian Sundarbans

The present study has great implications as macrobenthic molluscan community is noted for its notable ecosystem services mainly in the verticals of bioremediation, biomonitoring (as indicator species), carbon storage, livelihood etc.

## 4.2 Effect of Acidification on Oyster Community

We have carried out an extensive literature survey to find out the researches related to the effects of acidification on marine and estuarine molluscan species and documented that the first study on this subject was performed by Loosanoff and Tommers (1947). In the present geographical locale impact of lowering of pH on molluscan community was documented by (Mitra and Zaman 2015). A thorough literature survey on the impact of acidification on shelled molluscs reveal several adverse impacts on the community as listed in Table 4.6.

Oyster is widely distributed in the coastal and estuarine waters of the maritime state of West Bengal. A long term study carried out for 35 years on the condition index (CI) of *Saccostrea cucullata* sampled from Digha sea beach in the maritime state of West Bengal revealed significant direct relationship between aquatic pH and condition index of the species (Table 4.7).

**Fig. 4.4** *Macoma birmanica*—a common infauna (under class bivalvia) in the mangrove ecosystem of Indian Sundarbans

On the basis of more than three decade data bank (Table 4.8), we used Non-linear Autoregressive Neural Network Model to forecast the aquatic pH levels in Digha (with 30 years ahead time steps) and observed a significant decrease of pH in the coastal water of Digha as time progresses (Table 4.8 and Fig. 4.18).

In this study we have used condition index as proxy to evaluate the role of pH on the oyster health. More value of this index is an indication of congenial environmental variables for the growth and survival of the species. We observed a significant decrease of CI with the passage of time (Table 4.9) and using Nonlinear Autoregressive Neural Network Model to forecast the CI, we observe the adverse impact of coastal acidification on the health of the oyster species, *Saccostrea cucullata* from the output data. The CI value has decreased with time and has touched 4.93 in 2050 (Fig. 4.19).

Another important impact of acidification on molluscan species is the enhancement of heavy metal level in the body tissues of bivalves and gastropods. We have made a time series analysis on the heavy metal levels in selective molluscan species and observed that the values increased in the species inhabiting Hooghly estuary, but decreased in the species sampled from Matla estuary (Tables 4.10, 4.11, 4.12, 4.13, 4.14, and 4.15). This variation may be attributed to a number of factors like distribution, mobility, biological availability of chemical elements (*i.e.* it's chemical or physical association), pH, redox potential and availability of reactive species such

**Fig. 4.5** *Cerithedia obtusa*—an arboreal gastropod in the mangrove ecosystem of Indian Sundarbans

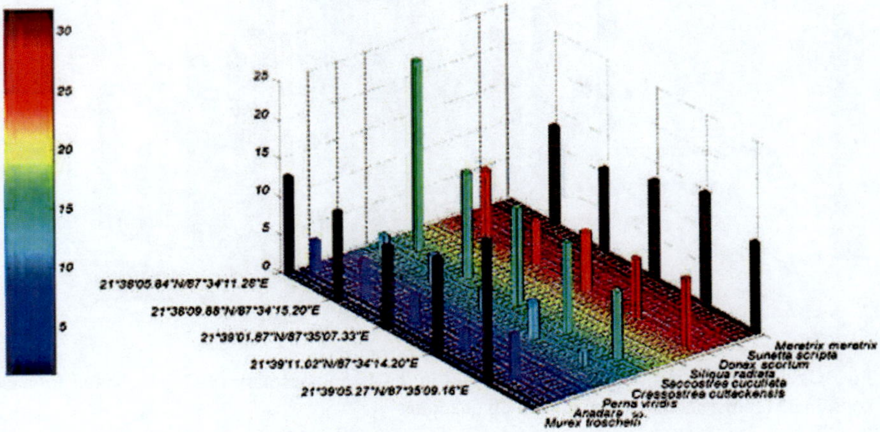


**Fig. 4.6** Population density of bivalve species in the Shankarpur coast of West Bengal; value for each coordinate represents the mean value of 10 quadrates

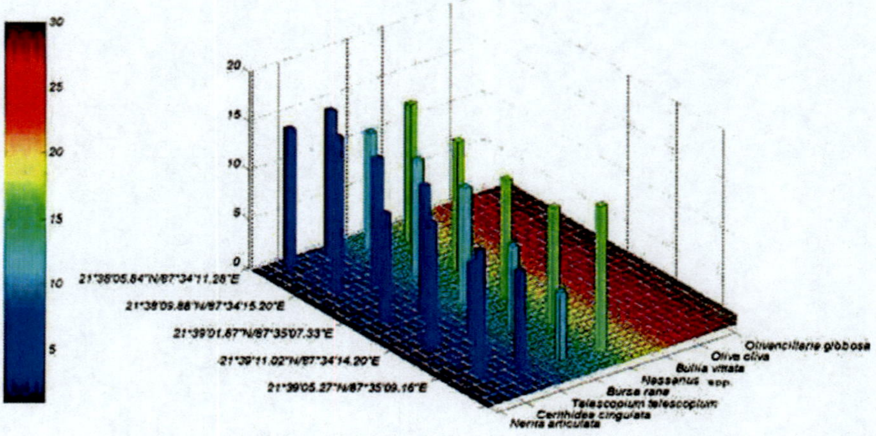


**Fig. 4.7** Population density of gastropod species in the Shankarpur coast of West Bengal; value for each coordinate represents the mean value of 10 quadrates

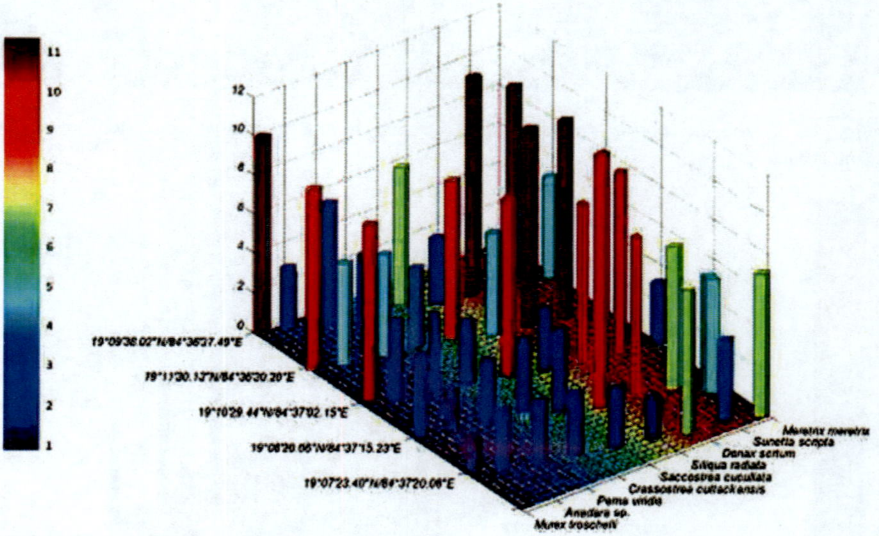


**Fig. 4.8** Population density of bivalve species in the Bahuda estuary of Odisha; value for each coordinate represents the mean value of 10 quadrates

as complexing ligands (organic and inorganic), particle surface for adsorption and colloidal matter etc. In the present study, such in-depth effort has not been attempted, but significant negative correlations between aquatic pH and dissolved heavy metals confirm the role of pH as one of the major factors influencing chemical speciation of the heavy metals in the present geographical locale. The lowering of pH might

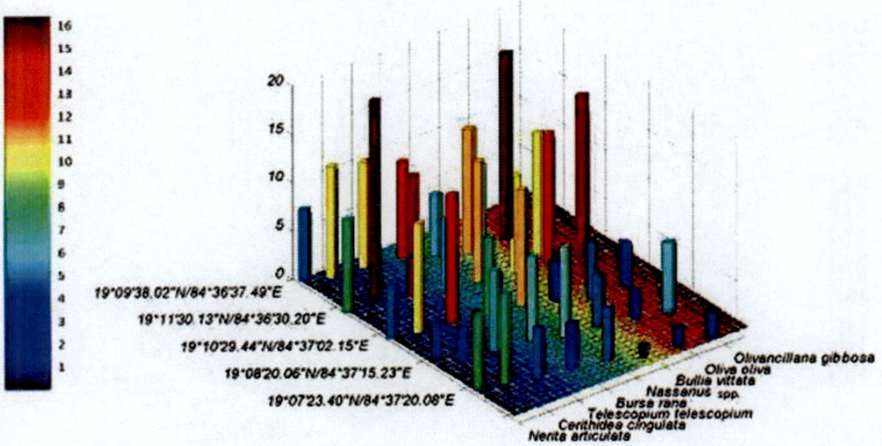


**Fig. 4.9** Population density of gastropod species in the Bahuda estuary of Odisha; value for each coordinate represents the mean value of 10 quadrates

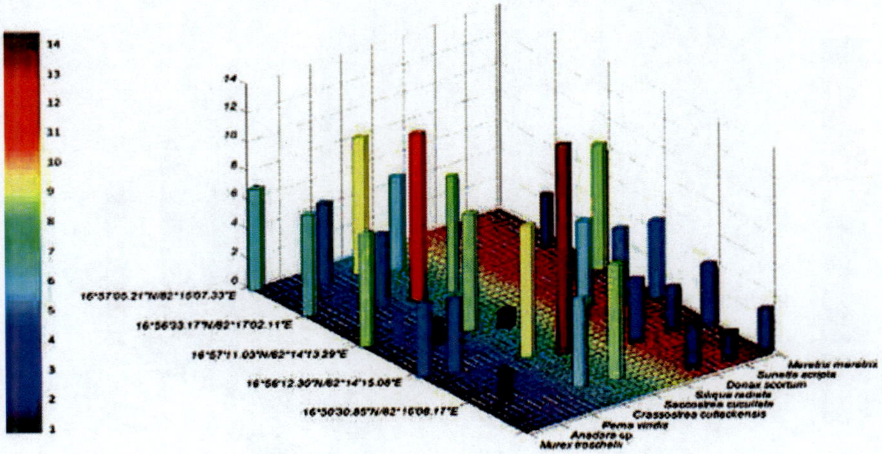


**Fig. 4.10** Population density of bivalve species in the East Godavari estuary of Andhra Pradesh; value for each coordinate represents the mean value of 10 quadrates

facilitate the dissolution of the precipitated form of metals and increase the amount of metallic ions in solutions, which has great role in the process of bioaccumulation.

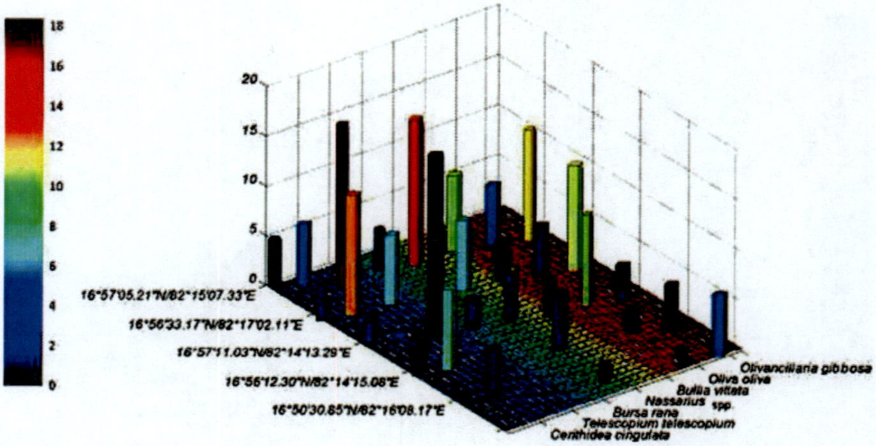


**Fig. 4.11** Population density of gastropod species in the East Godavari estuary of Andhra Pradesh; value for each coordinate represents the mean value of 10 quadrates

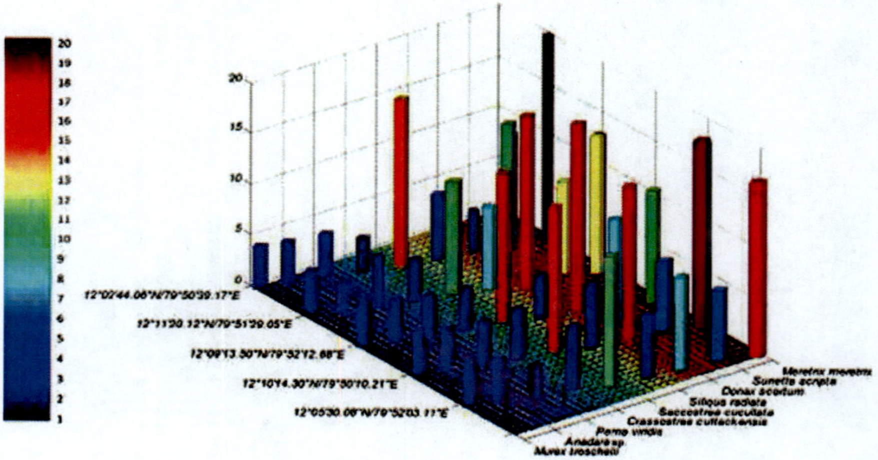


**Fig. 4.12** Population density of bivalve species in the Kalapet coast of Puducherry; value for each coordinate represents the mean value of 10 quadrates

## 4.3 Molluscs as Potential Sink of Carbon

The present work stands on the consideration of the molluscan community under the banner of blue carbon primarily because of their habitat in the marshy wetlands, mangroves, sea grass bed, saltmarsh grass bed, coastal zones, estuaries, river mouths and seas and also their ability to store carbon. Carbon is a major constituent of the calcareous hard shell over the body of molluscan organisms. However, acidification, a strong arm of climate change has posed a negative impact on this community by

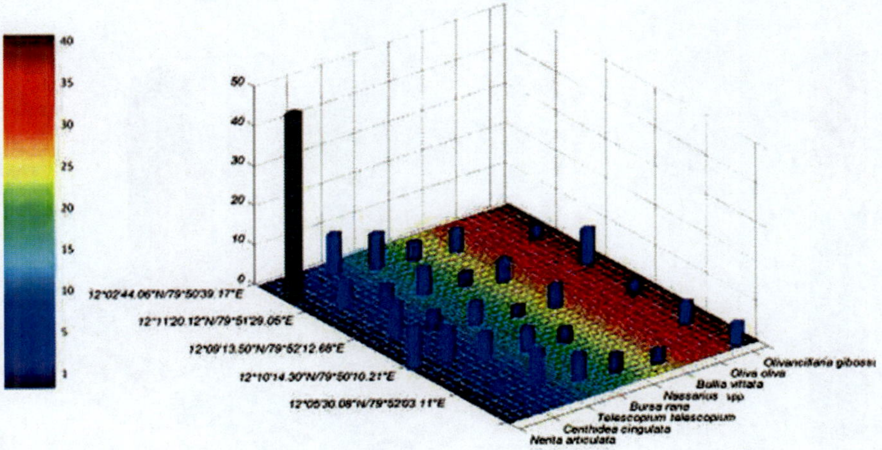


Fig. 4.13 Population density of gastropod species in the Kalapet coast of Puducherry; value for each coordinate represents the mean value of 10 quadrates

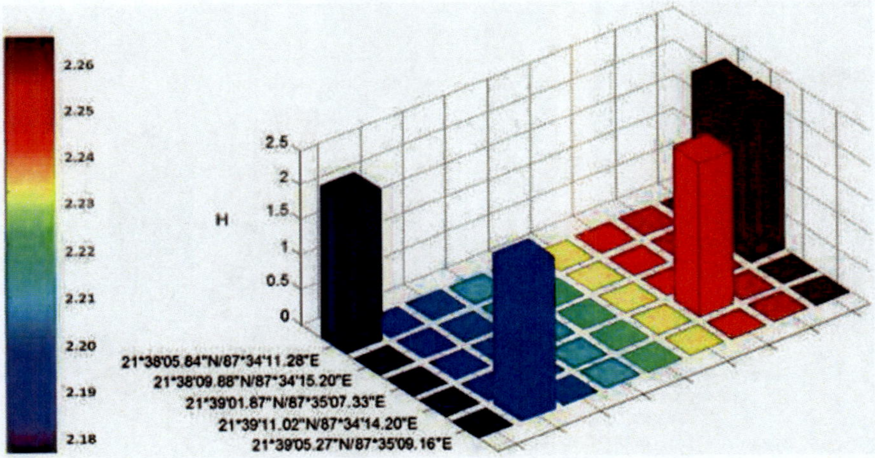


Fig. 4.14 Shannon—Weiner species diversity index of macrobenthic molluscan community in the Shankarpur coast of West Bengal

way of dissolving and thinning their calcareous shell. Filter-feeding oysters, clams or mussels sequester significant amounts of carbon by consuming phytoplankton and absorbing dissolved organic matter. Carbon is a primary component of oyster shell and it has been observed that for every kilogram of live clams or oysters grown in Florida waters approximately 114 g of carbon are removed from the water column and benthos. Stored carbon has also been documented in gastropod species (Jurkiewicz-Karnakowska 2005), which are mostly detritivores. On this background

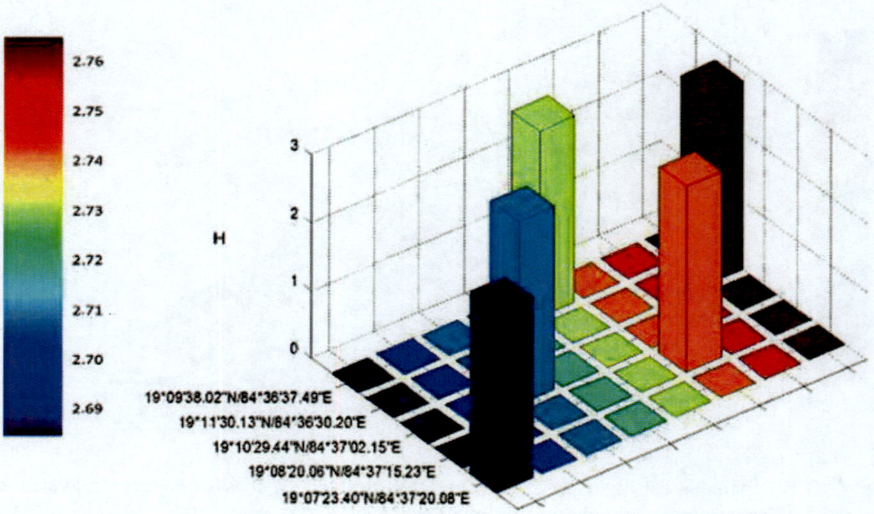


**Fig. 4.15** Shannon—Weiner species diversity index of macrobenthic molluscan community in the Bahuda estuary of Odisha

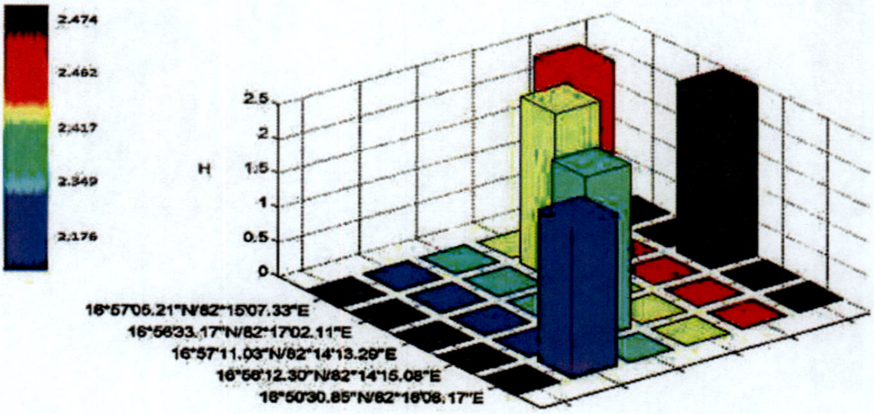


**Fig. 4.16** Shannon—Weiner species diversity index of macrobenthic molluscan community in the East Godavari estuary of Andhra Pradesh

the present section aims to quantify the stored carbon in major (dominant) macrobenthic species of molluscs preferably bivalves and gastropods sampled from the intertidal mudflats selected in two maritime states and one Union territory of India namely West Bengal (Hooghly estuary of Indian Sundarbans), Puducherry (Kalapet coast) and Goa (Mandovi estuary) during 2014 and 2015 (Tables 4.16, 4.17, 4.18, 4.19, 4.20, 4.21 and Figs. 4.20, 4.21, 4.22, 4.23, 4.24, 4.25).

The results for the present programme are summarized below in Table 4.22 shows

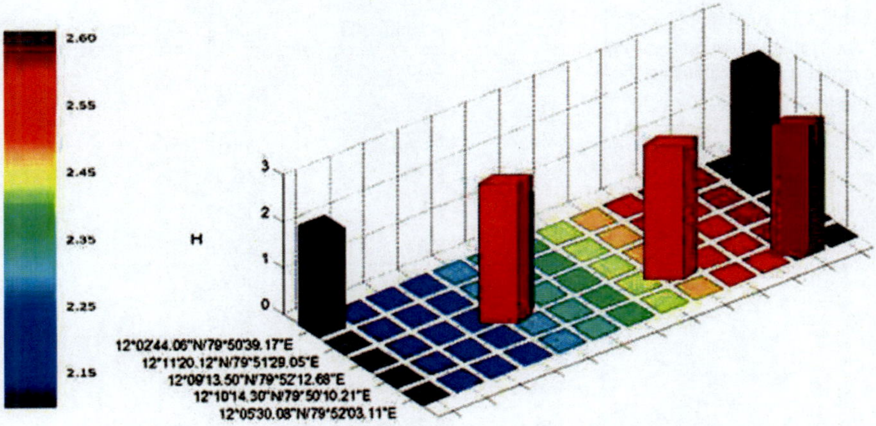


**Fig. 4.17** Shannon—Weiner species diversity index of macrobenthic molluscan community in the Kalapet coast of Puducherry

**Table 4.6** Probable adverse impact of acidification on shelled molluscs due to alteration of aquatic pH

| pH alteration | Impact |
|---|---|
| Lesser than 7.5 | • Increase of shell dissolution<br>• Decrease of growth rate<br>• Decrease of condition index |
| Less than 7 | • Reduction of filtration rate<br>• Increase of heart beat<br>• Reduction of feeding rate and flesh weight |
| Less than 6.5 | • Increased mortality<br>• Abnormal behavior<br>• Decrease of condition index |

the ANOVA results that depict the spatio-temporal variation of stored carbon in flesh (Fig. 4.21) and shell of the selected molluscan species.

The significant difference in stored carbon in the flesh of the selected macrobenthic molluscs is primarily due their food preference. *Saccostrea cucullata*, a dominant bivalve is common in almost all the coastal zones of Indian sub-continent. They are filter feeding bivalves and obtain their carbon from phytoplankton. The gastropods are also the store house of carbon as evidenced from the concentrations of total carbon (% dry weight) in their soft tissues and shells. One of the most important sources of carbon in marine and estuarine gastropods is their diets, which mainly are macroalgae or seaweeds on the mangrove vegetation (preferably on mangrove trunk/stem) or detritus on the intertidal mudflats. *Cerithedia obtusa* is fully dependent on algal matter for nutrition and *N. articulata* prefers both algal feed as well as detritus. *T. telescopium* and *C. cingulata* are detritivores.

**Table 4.7** Aquatic pH and condition index of *Saccostrea cucullata* sampled during the month of May (premonsoon season) from the Digha sea beach (21° 0.62′ 22″ N, 87° 50′ 66″ E) of coastal West Bengal

| Year | Aquatic pH | Condition index (CI) |
|---|---|---|
| 1984 | 8.33 | 16.35 |
| 1985 | 8.33 | 15.98 |
| 1986 | 8.33 | 15.88 |
| 1987 | 8.32 | 16.05 |
| 1988 | 8.31 | 15.42 |
| 1989 | 8.32 | 13.39 |
| 1990 | 8.31 | 14.65 |
| 1991 | 8.30 | 15.08 |
| 1992 | 8.30 | 13.55 |
| 1993 | 8.28 | 12.80 |
| 1994 | 8.26 | 12.95 |
| 1995 | 8.28 | 11.98 |
| 1996 | 8.27 | 12.60 |
| 1997 | 8.25 | 12.45 |
| 1998 | 8.26 | 11.90 |
| 1999 | 8.24 | 10.67 |
| 2000 | 8.24 | 10.50 |
| 2001 | 8.22 | 11.26 |
| 2002 | 8.21 | 11.45 |
| 2003 | 8.19 | 10.78 |
| 2004 | 8.20 | 9.88 |
| 2005 | 8.17 | 9.70 |
| 2006 | 8.20 | 9.66 |
| 2007 | 8.19 | 8.78 |
| 2008 | 8.18 | 8.59 |
| 2009 | 8.35 | 8.44 |
| 2010 | 8.16 | 8.38 |
| 2011 | 8.16 | 8.95 |
| 2012 | 8.15 | 8.20 |
| 2013 | 8.14 | 8.15 |
| 2014 | 8.17 | 7.69 |
| 2015 | 8.14 | 8.25 |
| 2016 | 8.14 | 6.90 |
| 2017 | 8.14 | 6.82 |
| 2018 | 8.13 | 6.35 |
| 2019 | 8.12 | 6.37 |
| 2020 | 8.32 | 6.20 |

**Table 4.8** Forecasting of aquatic pH in Digha using Autoregressive Neural Network Model

| Year | Predicted pH in May (premonsoon season) |
|---|---|
| 2021 | 8.17 |
| 2022 | 8.18 |
| 2023 | 8.19 |
| 2024 | 8.22 |
| 2025 | 8.21 |
| 2026 | 8.23 |
| 2027 | 8.21 |
| 2028 | 8.22 |
| 2029 | 8.21 |
| 2030 | 8.22 |
| 2031 | 8.21 |
| 2032 | 8.22 |
| 2033 | 8.21 |
| 2034 | 8.22 |
| 2035 | 8.21 |
| 2036 | 8.22 |
| 2037 | 8.21 |
| 2038 | 8.22 |
| 2039 | 8.21 |
| 2040 | 8.22 |
| 2041 | 8.21 |
| 2042 | 8.22 |
| 2043 | 8.21 |
| 2044 | 8.22 |
| 2045 | 8.21 |
| 2046 | 8.22 |
| 2047 | 8.21 |
| 2048 | 8.22 |
| 2049 | 8.21 |
| 2050 | 8.22 |

The shelled organisms preferably the molluscs are important store houses of carbon. A number of physical, chemical, and biological processes/factors carry out carbon sequestration in molluscs. The filter feeding bivalves like oysters source their carbon from their diet (phytoplankton) and ambient aquatic phase. Gastropods also play a significant role both in accumulation and circulation of carbon, especially in habitats where they are numerous and their biomass is high. However, studies concerning these problems are scarce.

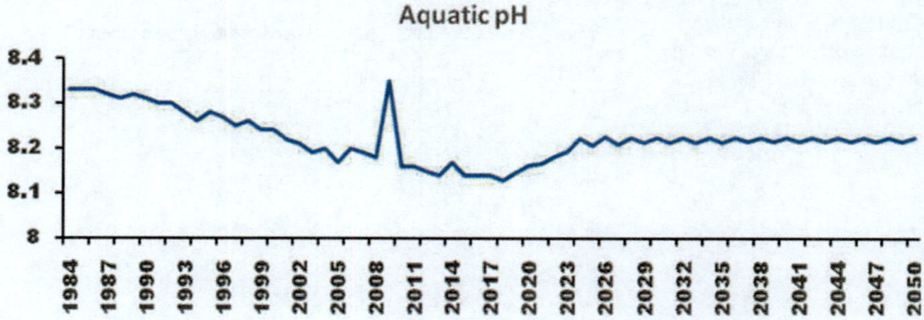


**Fig. 4.18** Predicted aquatic pH in Digha coast during May (premonsoon season) using Nonlinear Autoregressive Neural Network Model; real time data from 1984 to 2020 has been used as training in the model

**Table 4.9** Forecasting decrease of CI of *Saccostrea cucullata* using Non-linear Autoregressive Neural Network Model to forecast the CI

| Year | Predicted CI value of *Saccostrea cucullata* in May (premonsoon season) |
|---|---|
| 2021 | 4.06 |
| 2022 | 5.34 |
| 2023 | 6.03 |
| 2024 | 5.74 |
| 2025 | 4.98 |
| 2026 | 4.41 |
| 2027 | 4.65 |
| 2028 | 5.47 |
| 2029 | 5.82 |
| 2030 | 5.51 |
| 2031 | 4.92 |
| 2032 | 4.60 |
| 2033 | 4.91 |
| 2034 | 5.51 |
| 2035 | 5.69 |
| 2036 | 5.37 |
| 2037 | 4.90 |
| 2038 | 4.73 |
| 2039 | 5.07 |
| 2040 | 5.52 |
| 2041 | 5.58 |

(continued)

**Table 4.9** (continued)

| Year | Predicted CI value of *Saccostrea cucullata* in May (premonsoon season) |
|---|---|
| 2042 | 5.27 |
| 2043 | 4.90 |
| 2044 | 4.84 |
| 2045 | 5.17 |
| 2046 | 5.51 |
| 2047 | 5.50 |
| 2048 | 5.20 |
| 2049 | 4.91 |
| 2050 | 4.93 |

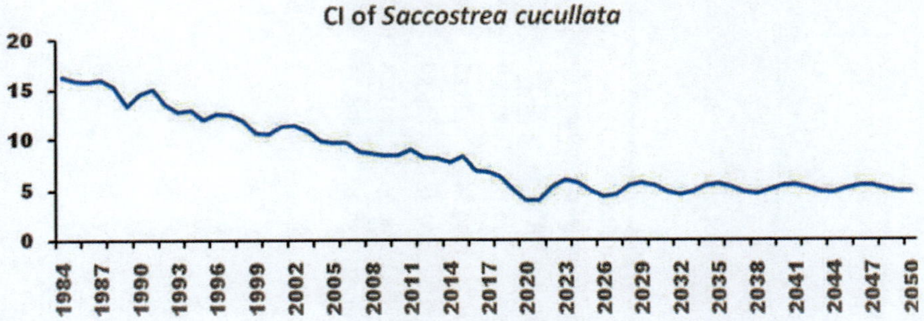


**Fig. 4.19** Predicted CI values of oyster sampled from Digha coast during May (premonsoon season) using Nonlinear Autoregressive Neural Network Model; real time data from 1984 to 2020 has been used as training in the model

Terrestrial gastropods ingest water from multiple sources, including dew, soil moisture, standing water, and precipitation, all of which contain some amount of dissolved inorganic carbon (DIC). Water is taken up through the foot of the gastropod by contact rehydration (Balakrishnan and Yapp 2004) and introduced to the hemolymph before being passed on to the extrapallial fluid. The soft fleshy meat and shells of gastropods and bivalves is store house of carbon as evidenced from the concentrations of total carbon (% dry weight) in the soft tissues and shells of these organisms.

There is a need to estimate the carbon storage potential of molluscs so that in the future policy programme on climate change this phyla constituting about 23% of identified marine organisms can be an important vertical.

**Table 4.10** Zn concentrations (in ppm dry wt.) in molluscan flesh collected from Hooghly estuary during June of every year

| Species | 2010 | 2011 | 2012 | 2013 | 2014 | 2015 | 2016 | 2017 | 2018 | 2019 |
|---|---|---|---|---|---|---|---|---|---|---|
| *Telescopium telescopium* | 411.05 | 413.84 | 417.37 | 411.87 | 419.18 | 423.09 | 427.12 | 417.43 | 418.95 | 427.56 |
| *Nerita articulata* | 213.38 | 216.65 | 219.90 | 215.76 | 230.76 | 226.85 | 232.76 | 231.89 | 233.05 | 235.71 |
| *Cerithedia cingulata* | 118.78 | 119.17 | 122.56 | 121.09 | 133.88 | 136.89 | 141.34 | 147.21 | 140.66 | 150.82 |
| *C. obtusa* | 124.55 | 130.99 | 134.11 | 137.05 | 140.66 | 141.88 | 143.90 | 149.06 | 152.67 | 159.25 |
| *Saccostrea cucullata* | 664.23 | 669.34 | 711.34 | 778.45 | 787.98 | 808.33 | 837.99 | 847.21 | 858.48 | 1110.34 |

**Table 4.11** Zn concentrations (in ppm dry wt.) in molluscan flesh collected from Matla estuary during June of every year

| Species | 2010 | 2011 | 2012 | 2013 | 2014 | 2015 | 2016 | 2017 | 2018 | 2019 |
|---|---|---|---|---|---|---|---|---|---|---|
| *Telescopium telescopium* | 219.66 | 213.50 | 189.54 | 176.32 | 168.88 | 163.95 | 155.87 | 139.43 | 141.12 | 127.94 |
| *Nerita articulata* | 176.44 | 170.05 | 166.58 | 162.96 | 160.18 | 153.22 | 141.78 | 126.90 | 133.85 | 114.20 |
| *Cerithedia cingulata* | 103.45 | 112.56 | 98.34 | 91.07 | 89.22 | 87.15 | 77.95 | 71.02 | 66.29 | 55.12 |
| *C. obtusa* | 107.35 | 111.20 | 100.66 | 114.67 | 103.88 | 99.67 | 89.22 | 84.28 | 73.17 | 62.08 |
| *Saccostrea cucullata* | 487.45 | 434.76 | 423.98 | 411.90 | 398.45 | 346.77 | 338.31 | 332.97 | 317.56 | 298.93 |

**Table 4.12** Cu concentrations (in ppm dry wt.) in molluscan flesh collected from Hooghly estuary during June of every year

| Species | 2010 | 2011 | 2012 | 2013 | 2014 | 2015 | 2016 | 2017 | 2018 | 2019 |
|---|---|---|---|---|---|---|---|---|---|---|
| *Telescopium telescopium* | 29.61 | 27.89 | 25.20 | 30.19 | 37.18 | 41.45 | 59.38 | 64.81 | 73.44 | 87.98 |
| *Nerita articulata* | 17.38 | 18.65 | 19.90 | 21.76 | 27.76 | 25.85 | 36.76 | 47.89 | 58.05 | 68.71 |
| *Cerithedia cingulata* | 14.56 | 15.09 | 14.99 | 16.56 | 20.98 | 18.02 | 28.34 | 39.42 | 49.67 | 53.14 |
| *C. obtusa* | 15.12 | 16.65 | 18.12 | 18.70 | 19.02 | 20.56 | 32.87 | 39.99 | 52.65 | 55.05 |
| *Saccostrea cucullata* | 47.88 | 55.34 | 65.15 | 70.05 | 77.34 | 50.10 | 61.97 | 74.08 | 80.53 | 91.22 |

**Table 4.13** Cu concentrations (in ppm dry wt.) in molluscan flesh collected from Matla estuary during June of every year

| Species | 2010 | 2011 | 2012 | 2013 | 2014 | 2015 | 2016 | 2017 | 2018 | 2019 |
|---|---|---|---|---|---|---|---|---|---|---|
| *Telescopium telescopium* | 21.09 | 25.85 | 22.28 | 28.32 | 27.61 | 24.66 | 23.78 | 21.90 | 25.78 | 20.27 |
| *Nerita articulata* | 14.33 | 12.50 | 11.76 | 11.84 | 14.39 | 12.62 | 11.35 | 10.55 | 9.21 | 8.56 |
| *Cerithedia cingulata* | 13.02 | 12.17 | 11.80 | 11.63 | 10.98 | 8.02 | 8.28 | 7.06 | 6.44 | 5.14 |
| *C. obtusa* | 14.75 | 13.70 | 15.55 | 15.70 | 13.33 | 9.71 | 11.94 | 11.60 | 10.65 | 7.05 |
| *Saccostrea cucullata* | 33.95 | 35.06 | 34.28 | 30.15 | 27.08 | 30.79 | 26.42 | 21.01 | 20.57 | 18.40 |

**Table 4.14** Pb concentrations (in ppm dry wt.) in molluscan flesh collected from Hooghly estuary during June of every year

| Species | 2010 | 2011 | 2012 | 2013 | 2014 | 2015 | 2016 | 2017 | 2018 | 2019 |
|---|---|---|---|---|---|---|---|---|---|---|
| *Telescopium telescopium* | 6.09 | 6.15 | 6.98 | 7.84 | 8.05 | 9.11 | 12.18 | 11.65 | 14.44 | 15.43 |
| *Nerita articulata* | 5.12 | 5.88 | 6.04 | 6.66 | 6.97 | 8.05 | 7.94 | 9.16 | 11.98 | 12.75 |
| *Cerithedia cingulata* | 3.76 | 3.11 | 4.99 | 5.30 | 5.51 | 6.08 | 6.49 | 6.35 | 7.09 | 7.45 |
| *C. obtusa* | 4.08 | 4.76 | 5.45 | 5.87 | 6.91 | 6.28 | 7.78 | 7.60 | 8.12 | 8.39 |
| *Saccostrea cucullata* | 10.34 | 16.72 | 16.41 | 17.23 | 20.90 | 18.55 | 22.30 | 15.65 | 23.92 | 25.63 |

**Table 4.15** Pb concentrations (in ppm dry wt.) in molluscan flesh collected from Matla estuary during June of every year

| Species | 2010 | 2011 | 2012 | 2013 | 2014 | 2015 | 2016 | 2017 | 2018 | 2019 |
|---|---|---|---|---|---|---|---|---|---|---|
| *Telescopium telescopium* | 5.12 | 4.85 | 3.70 | 3.84 | 2.05 | 2.11 | 2.02 | 1.96 | 1.87 | BDL |
| *Nerita articulata* | 3.56 | 2.27 | 2.88 | 1.75 | 1.32 | 1.05 | BDL | BDL | BDL | BDL |
| *Cerithedia cingulata* | 2.08 | BDL | BDL | BDL | BDL | BDL | BDL | BDL | BDL | BDL |
| *C. obtusa* | 2.97 | BDL | BDL | BDL | BDL | BDL | BDL | BDL | BDL | BDL |
| *Saccostrea cucullata* | 8.03 | 7.59 | 7.27 | 6.68 | 5.49 | 3.21 | 3.85 | 2.99 | 3.62 | 2.85 |

**Table 4.16** Concentrations of total carbon (% dry wt) in the soft tissues of the investigated gastropod and bivalve species (mean values and standard deviations) collected from three selected sampling sites during premonsoon 2014–2015

| Species | Station | | |
|---|---|---|---|
| | Hooghly estuary (Indian Sundarbans) | Mandovi estuary (Goa) | Kalapet coast (Puducherry) |
| | Carbon (%) | | |
| *Telescopium telescopium* | 36.02 ± 0.63 | 35.13 ± 0.68 | 34.98 ± 0.74 |
| *Nerita articulata* | 36.99 ± 0.64 | 36.38 ± 0.76 | 35.19 ± 0.74 |
| *Cerithedia cingulata* | 35.15 ± 0.66 | 33.28 ± 0.65 | 33.02 ± 0.72 |
| *C. obtusa* | 39.14 ± 0.91 | 38.46 ± 0.89 | 37.33 ± 0.86 |
| *Saccostrea cucullata* | 40.15 ± 0.92 | 39.83 ± 0.90 | 38.17 ± 0.87 |

**Table 4.17** Concentrations of total carbon (% dry wt) in the soft tissues of the investigated gastropod and bivalve species (mean values and standard deviations) collected from three selected sampling sites during monsoon 2014–2015

| Species | Station | | |
|---|---|---|---|
| | Hooghly estuary (Indian Sundarbans) | Mandovi estuary (Goa) | Kalapet coast (Puducherry) |
| | Carbon (%) | | |
| *Telescopium telescopium* | 35.23 ± 0.63 | 34.41 ± 0.68 | 33.20 ± 0.74 |
| *Nerita articulata* | 34.98 ± 0.64 | 33.05 ± 0.76 | 32.49 ± 0.74 |
| *Cerithedia cingulata* | 33.03 ± 0.66 | 32.14 ± 0.65 | 32.05 ± 0.72 |
| *C. obtusa* | 37.29 ± 0.91 | 35.30 ± 0.89 | 34.93 ± 0.86 |
| *Saccostrea cucullata* | 38.24 ± 0.92 | 36.21 ± 0.90 | 34.99 ± 0.87 |

**Table 4.18** Concentrations of total carbon (% dry wt) in the soft tissues of the investigated gastropod and bivalve species (mean values and standard deviations) collected from three selected sampling sites during postmonsoon 2014–2015

| Species | Station | | |
|---|---|---|---|
| | Hooghly estuary (Indian Sundarbans) | Mandovi estuary (Goa) | Kalapet coast (Puducherry) |
| | Carbon (%) | | |
| *Telescopium telescopium* | 35.00 ± 0.63 | 34.20 ± 0.68 | 34.20 ± 0.74 |
| *Nerita articulata* | 35.96 ± 0.62 | 34.06 ± 0.77 | 33.50 ± 0.75 |
| *Cerithedia cingulata* | 34.04 ± 0.67 | 33.15 ± 0.66 | 33.06 ± 0.73 |
| *C. obtusa* | 38.30 ± 0.92 | 36.31 ± 0.90 | 33.94 ± 0.87 |
| *Saccostrea cucullata* | 39.25 ± 0.93 | 37.22 ± 0.91 | 35.98 ± 0.88 |

**Table 4.19** Concentrations of total carbon (% dry wt) in the calcareous shell of the investigated gastropod and bivalve species (mean values and standard deviations) collected from three selected sampling sites during premonsoon 2014–2015

| Species | Station | | |
|---|---|---|---|
| | Hooghly estuary (Indian Sundarbans) | Mandovi estuary (Goa) | Kalapet coast (Puducherry) |
| | Carbon (%) | | |
| *Telescopium telescopium* | 12.98 ± 0.11 | 12.56 ± 0.12 | 12.33 ± 0.13 |
| *Nerita articulata* | 13.03 ± 0.12 | 13.00 ± 0.14 | 12.96 ± 0.15 |
| *Cerithedia cingulata* | 12.59 ± 0.09 | 12.43 ± 0.10 | 12.05 ± 0.12 |
| *C. obtusa* | 13.99 ± 0.18 | 13.83 ± 0.17 | 13.05 ± 0.16 |
| *Saccostrea cucullata* | 14.23 ± 0.19 | 13.97 ± 0.18 | 13.55 ± 0.17 |

**Table 4.20** Concentrations of total carbon (% dry wt) in the calcareous shell of the investigated gastropod and bivalve species (mean values and standard deviations) collected from three selected sampling sites during monsoon 2014–2015

| Species | Station | | |
|---|---|---|---|
| | Hooghly estuary (Indian Sundarbans) | Mandovi estuary (Goa) | Kalapet coast (Puducherry) |
| | Carbon (%) | | |
| *Telescopium telescopium* | 12.98 ± 0.11 | 12.56 ± 0.12 | 12.33 ± 0.13 |
| *Nerita articulata* | 13.03 ± 0.12 | 13.00 ± 0.14 | 12.96 ± 0.15 |
| *Cerithedia cingulata* | 12.59 ± 0.09 | 12.43 ± 0.10 | 12.05 ± 0.12 |
| *C. obtusa* | 13.99 ± 0.18 | 13.83 ± 0.17 | 13.05 ± 0.16 |
| *Saccostrea cucullata* | 13.09 ± 0.17 | 12.98 ± 0.16 | 12.57 ± 0.14 |

**Table 4.21** Concentrations of total carbon (% dry wt) in the soft tissues of the investigated gastropod and bivalve species (mean values and standard deviations) collected from three selected sampling sites during postmonsoon 2014–2015

| Species | Station | | |
|---|---|---|---|
| | Hooghly estuary (Indian Sundarbans) | Mandovi estuary (Goa) | Kalapet coast (Puducherry) |
| | Carbon (%) | | |
| *Telescopium telescopium* | 35.00 ± 0.63 | 34.20 ± 0.68 | 34.20 ± 0.74 |
| *Nerita articulata* | 35.96 ± 0.62 | 34.06 ± 0.77 | 33.50 ± 0.75 |
| *Cerithedia cingulata* | 34.04 ± 0.67 | 33.15 ± 0.66 | 33.06 ± 0.73 |
| *C. obtusa* | 38.30 ± 0.92 | 36.31 ± 0.90 | 33.94 ± 0.87 |
| *Saccostrea cucullata* | 39.25 ± 0.93 | 37.22 ± 0.91 | 35.98 ± 0.88 |

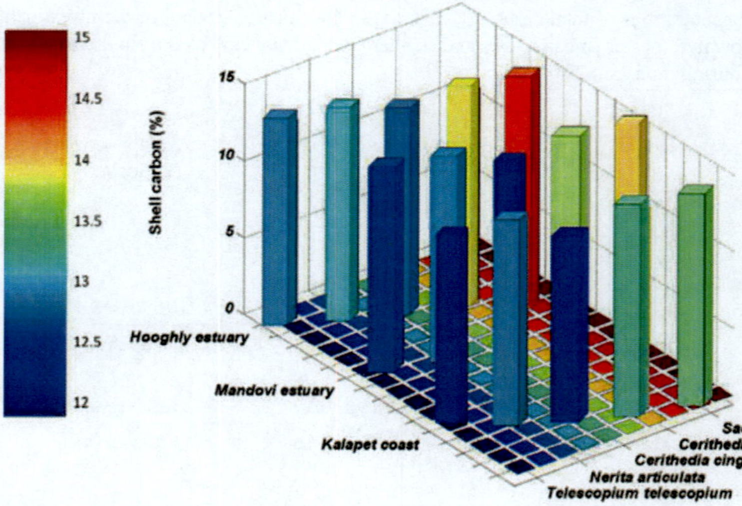


**Fig. 4.20** Stored carbon percentage in flesh of dominant molluscan species in premonsoon season during 2014–15

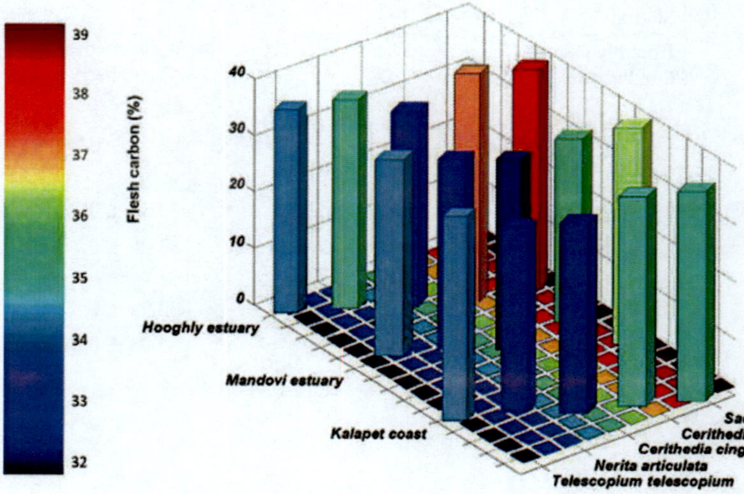


**Fig. 4.21** Stored carbon percentage in flesh of dominant molluscan species in monsoon season during 2014–15

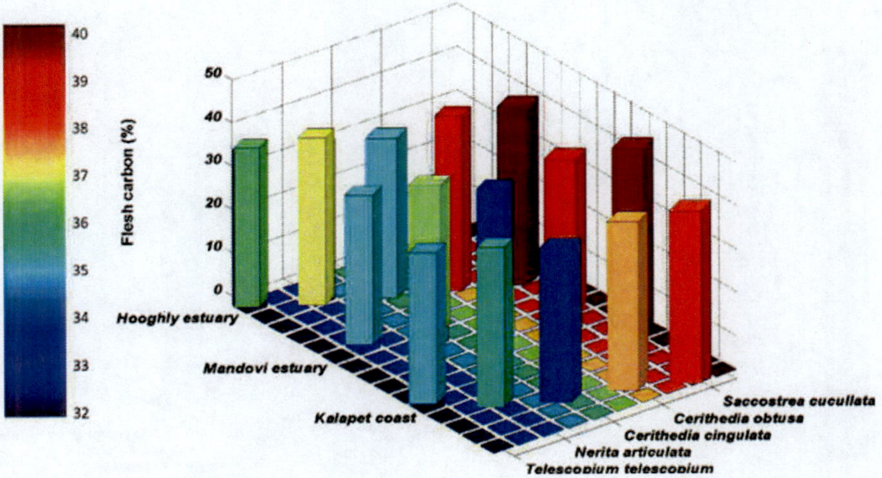


**Fig. 4.22** Stored Carbon Percentage in Flesh of Dominant Molluscan Species in postmonsoon Season during 2014–15

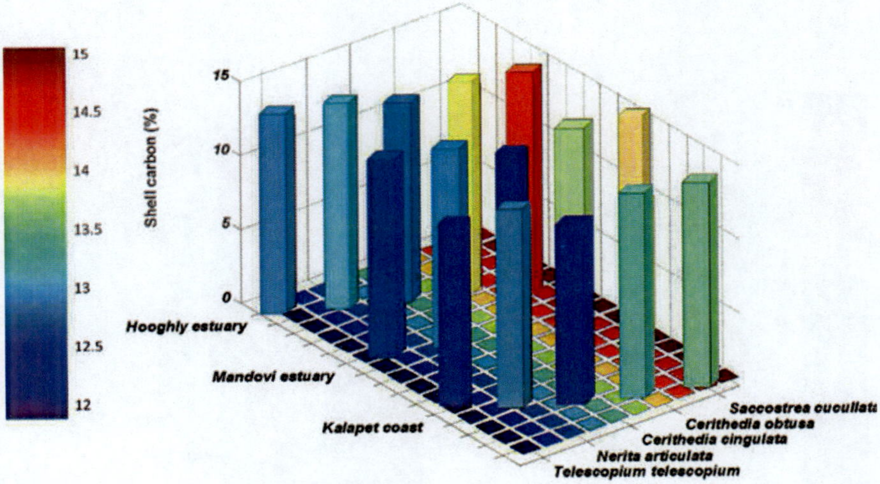


**Fig. 4.23** Stored carbon percentage in shell of dominant molluscan species in premonsoon season during 2014–15

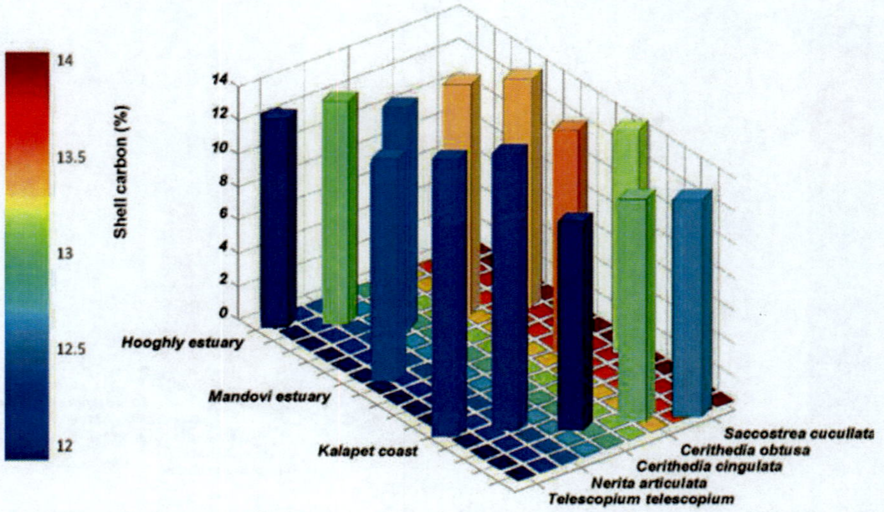


**Fig. 4.24** Stored carbon percentage in shell of dominant molluscan species in monsoon season during 2014–15

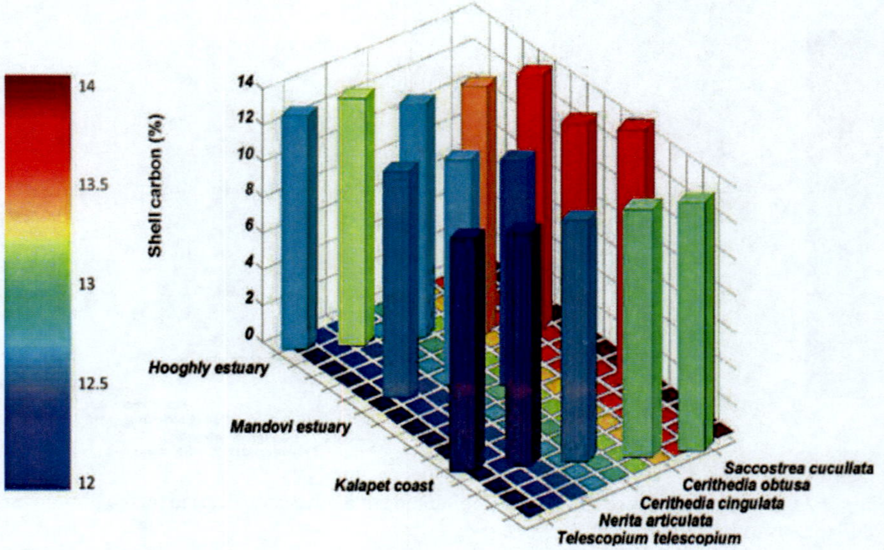


**Fig. 4.25** Stored carbon percentage in shell of dominant molluscan species in postmonsoon season during 2014–15

**Table 4.22** ANOVA results of flesh and shell carbon of selected molluscan species

| Parameter | $F_{obs}$ | $F_{crit}$ | *p*-value |
|---|---|---|---|
| Species | | | |
| *Telescopium telescopium* (flesh) | | | |
| Between stations | 8.939198 | 6.944272 | 0.033426 |
| Between seasons | 7.185884 | 6.944272 | 0.047404 |
| *Telescopium telescopium* (shell) | | | |
| Between stations | 59.69737 | 6.944272 | 0.001051 |
| Between seasons | 26.61842 | 6.944272 | 0.004884 |
| *Nerita articulata* (flesh) | | | |
| Between stations | 27.61633 | 6.944272 | 0.00456 |
| Between seasons | 38.73566 | 6.944272 | 0.002411 |
| *Nerita articulata* (shell) | | | |
| Between stations | 0.438143 | 6.944272 | 0.67286 |
| Between seasons | 17.72404 | 6.944272 | 0.010282 |
| *Cerithedia cingulata* (flesh) | | | |
| Between stations | 13.08253 | 6.944272 | 0.017584 |
| Between seasons | 12.33403 | 6.944272 | 0.019468 |
| *Cerithedia cingulata* (shell) | | | |
| Between stations | 0.528864 | 6.944272 | 0.625474 |
| Between seasons | 0.025349 | 6.944272 | 0.975125 |
| *Cerithedia obtusa* (flesh) | | | |
| Between stations | 10.75582 | 6.944272 | 0.024583 |
| Between seasons | 9.499951 | 6.944272 | 0.030246 |
| *Cerithedia obtusa* (shell) | | | |
| Between stations | 4.288894 | 6.944272 | 0.101137 |
| Between seasons | 14.99093 | 6.944272 | 0.013856 |
| *Saccostrea cucullata* (flesh) | | | |
| Between stations | 22.83252 | 6.944272 | 0.006487 |
| Between seasons | 24.72919 | 6.944272 | 0.005599 |
| *Saccostrea cucullata* (shell) | | | |
| Between stations | 4.581158 | 6.944272 | 0.092354 |
| Between seasons | 87.22928 | 6.944272 | 0.000502 |

## References

Balakrishnan M, Yapp CJ (2004) Flux balance models for the oxygen and carbon isotope compositions of land snail shells. Geochim Cosmochim Acta 68:2007–2024

Jurkiewicz-Karnakowska E (2005) Some aspects of nitrogen, carbon and calcium accumulation in Molluscs from the Zegrzyński reservoir ecosystem. Pol J Environ Stud 14(2):173–177

Loosanoff VL, Tommers FD (1947) Effect of low pH upon rate of water pumping of oysters, *Ostrea virginica.* Anat Rec 99:668–669

Mitra A, Zaman S (2015) Blue carbon reservoir of the blue planet. Publisher, Springer India, Book ID 322214_1_En Book, Book, ISBN- 978-81-322-2106-7; Copyright Holder Name Springer India Copyright Year 2015, pp 1–299. https://doi.org/10.1007/978-81-322-2107-4

# Chapter 5
# Impact of Acidification on Plankton

**Abstract** Plankton are unique sensors to climate change induced alteration in temperature, salinity and pH of coastal and estuarine waters. They are considered as one of the best proxies to understand the effect of climate change due to their very short life cycle, quick response to alteration of water quality and less exploitation by humans, unlike fishes. Hence the pulse of acidification can be clearly evaluated through change in plankton composition and productivity. The present chapter highlights how the community structure of plankton along with their productivity changes as an effect of acidification in estuarine water.

## 5.1 Plankton Community: Overview and Types

Phytoplankton community encompasses unicellular, free floating organisms whose sizes usually range between 1 μm to 1 mm. The cells of these organisms are characterized by the presence of chlorophyll pigment, which is the core pigment in the process of photosynthesis. Phytoplankton are mainly confined in the photic zone of the ocean as they require sunlight, nutrients and carbon dioxide to form glucose that provides these tiny organisms with energy to perform various life processes. Phytoplankton forms the base of the marine and estuarine food web or in other words they support half of the global primary production, which directly or indirectly forms the foundation of all marine life. The Potential Fishing Zone (PFZ), which is the region for maximum fish catch is detected on the basis of pigment characterization that is assessed through Remote Sensing Technique.

Phytoplankton, the tiny aquatic floral entities provide several ecosystem services as listed here:

(i) Phytoplanton form the foundation stone of World fishery. It has been estimated that the total annual primary production of the World seas is around $20 \times 10^9$ tonnes of carbon which has the capacity to yield 240 million tonnes of fish although our present harvest is only about 60 million tones.

(ii) The distribution and abundance of the commercially important fish and shellfish and their larvae are dependent on some species of the phytoplankton besides serving as their main food source. Among the species of diatoms,



A. Mitra and S. Zaman, *Estuarine Acidification*,
https://doi.org/10.1007/978-3-030-84792-0_5

*Fragilaria oceanica* and *Hemidiscus hardmannianus* have been recorded to indicate the abundance of the clupeid fish, Hilsa in the Hooghly estuary and oil sardine, *Sardinella longiceps* in the west coast of India. The 'white water phenomenon' due to abundance of the coccolithophores has been a good sign for the herring fisheries in the British waters. Similarly, the colonial diatom, *Fragilaria antartica* is known to indicate the abundance of the Antartic krill, *Euphausia superba.* Nowadays the distribution and abundance of fish is detected through remote sensing technology, where the abundance of phytoplankton and their pigments are used as proxy to forecast the density/abundance of fish population.

(iii) Phytoplankton are sources of fossil fuels. It has been reported that some species of phytoplankton store extra food as oil rather than starch to get an advantage in terms of buoyancy. When oil-storing forms phytoplankton die and descend to the bottom of the sea and undergo microbial degradation, they are often covered by huge sediment load and subjected to enormous pressure. Such conditions are supposed to change the deposited phytoplankton into fossil fuel after hundreds and millions of ten years. Thus human World is benefitted from these tiny organisms even after their death.

(iv) The diatoms are commercially important as 'diatomaceous Earth' (diatomite or kielselghur), a congregation of dead silicon-rich diatom frustules in the sea beds like protozoan oozes. This material is employed in the filtration of fruit juice, syrup and varnish; in the boilers, electric ovens and refrigerators as insulators; as polishing agent for precious metals; in the separation of paraffin wax from petroleum and in the manufacture of concrete. Because of the presence of the hydrocarbon compound 'diatomin' the diatoms serve as the indicators of rich petroleum grounds. Present day scientists are of the opinion that these tiny free-floating producers are the chief precursors of the petroleum rich fields of Venezuela at Los Angeles.

(v) The dinoflagellates also at times play an important role in the fishery sector as by virtue of their power of luminescence they can emit 'cold living light' which though lasts for about one-tenth of a second, helps in identifying certain fish shoals during night.

(vi) Phytoplankton can also be used to reflect the water quality in terms of certain pollutants other than sewage (that contribute considerable amount of nitrate and phosphate in the ambient water). It has been reported that an increase in cell volume occurs in certain species (like *Dunaliella tertiolecta* and *Phaeodactylum tricornutum*) due to copper pollution (Stauber and Florence 1987).

(vii) Phytoplankton play an important role in maintaining the global carbon cycle. The elemental composition of phytoplankton is C:N:P = 106:16:1, which is commonly referred as the Red Field Ratio' (Redfield 1934, 1958). This ratio clearly explains that about 100 units of carbon is delivered to the deep sea for every sixteen units of nitrogen and one unit of phosphorous. As such, the biological pump delivers carbon from the atmosphere to the deep sea, where it is concentrated and sequestered for centuries.

(viii) Diatoms are important components contributing to the stability of intertidal mudflats. They excrete polysaccharides, which trap sediment sediment grains and stabilize sediment (http://www.nioo.know.hl/cemo/ecoflat/work.htm).

(ix) Mass culture of phytoplankton may be undertaken in closed systems of the space capsules. The nitrogenous wastes of the travelers could be diluted and circulated as algal nutrients and the respired air as a source of carbon dioxide. In exchange, the travelers could utilize the oxygen released by the photosynthesis of algae and benefit from the food value cultured phytoplankton.

(x) Phytoplankton has a major role to play in the pharmaceutical sector of blue economy. They can serve as a good source of omega-3 and omega-6 polyunsaturated fatty acids (PUFAs) and can be alternative to fish oil, which has been the main human source of eicosapentaenoic acid (EPA) and docosahexaenoic acid (DHA).

The recent inclination of the scientists towards the phytoplankton community is mainly due to their paramount importance in various spheres of life, as pointed below (http://www.bigelow.or/sci_overview).

1. Oceans cover more than 70% of Earth's surface, containing all but 3% of the World's water and this saline water mass supports huge quantum of phytoplankton.
2. Dissolved gases in seawater regulate the composition of our atmosphere through phytoplankton compartment.
3. The World's phytoplankton generate half of the oxygen we breathe and absorb half of the carbon dioxide that may be contributing to global warming.
4. Phytoplankton constitute the base of the marine and estuarine food web and are the dominant producers of the aquatic ecosystem.
5. Phytoplankton act as the sink of anthropogenic wastes (specially nitrate and phosphate) that are contributed by sewage and agricultural operations.

Phytoplankton community may be divided into five major categories as listed in Table 5.1.

Depending on the nature of valves and pattern of ornamentation in the valve surface, the diatoms are grouped into **centric** and **pennate** diatoms. The major differences between these two groups are given in Table 5.2.

On the basis of size, the phytoplankton may be grouped under five categories (Table 5.3).

(a) Phytoplankton may also be classified on the basis of the cell characteristics (Table 5.4).

The zooplankton occupy the tier next to phytoplankton in the marine and estuarine food web. They either graze on phytoplankton or feed on other members of zooplankton. Foraminifera and radiolarians are members of zooplankton community who possess only one cell. Copepods and euphausiids are widely distributed zooplankton in the marine and estuarine environments. Zooplankton exhibit the phenomenon of **vertical migration** in which some species migrate towards the

**Table 5.1** Different categories of phytoplankton

| Category | Salient features | Example |
|---|---|---|
| Diatoms | (1) Presence of shell/frustules, which, is composed of translucent silica<br>(2) The shell resembles a 'pillbox' bottom and lid; the lid is called the epitheca and the bottom is known as hypotheca<br>(3) Many species of diatoms exhibit sculptured shells that look like beautiful ornaments<br>(4) On the basis of nature of values and pattern of ornamentation on the surface of the value the diatoms are grouped into centric and pinnate diatom, the differences between which are highlighted in Table 5.2 | *Thallassiosira pacifica, Stephanopyxis nipponica* |
| Dinoflagellates | (1) Unicellular organisms<br>(2) Same are without any cover while others are armoured with plates of cellulose<br>(3) Presence of two flagella for locomotion<br>(4) Many species of dinoflagellates exhibit bioluminescence | *Skeletonema costatum* and *Coscinodiscus eccentricus* |
| Coccolithophores | (1) Extremely tiny having a range between 5 and 20 microns<br>(2) Flagella are present in several forms<br>(3) Body covered with tiny calcified circular plates having wide variety of designs<br>(4) Coccolithophores are the preferred diet for filter feeding animals like bivalves | *Isochrysis galbana* and *Coccolithus* sp. |
| Blue green algae | (1) Prokaryotic in nature<br>(2) Unicellular, or multicellular organisms<br>(3) Presence of phycocyanin which imparts blue colour to the organism<br>(4) Contributes greatly in 'bloom' formation | *Trichodesmium erythraeum* |
| Green algae | (1) Eukaryotic in nature<br>(2) Dominate in the coastal water, rich in nutrients<br>(3) Presence of chloroplasts which imparts green colour<br>(4) Noted for high protein, carotenoid, vitamin and mineral content | *Chlorella marina* and *Chlorella salina* |

**Table 5.2** Differences between centric and pennate diatoms

| Point | Centric diatom | Pennate diatom |
|---|---|---|
| Cell shape | Discoid, solenoid or cylindrical | Elongated and fusiform, oval, sigmoid or roughly circular |
| Ornamentation | Radial in nature i.e., the arrangement of markings is radiating from the centre | Bilateral in nature i.e., the arrangement of the markings is on either side of the apical (main) axis |

**Table 5.3** Classification of phytoplankton on the basis of size

| Plankton category | Maximum dimension (μm) |
|---|---|
| Ultraplankton | <2 |
| Nanoplankton | 2–20 |
| Microplankton | 20–200 |
| Macroplankton | 200–2000 |
| Megaplankton | >2000 |

**Table 5.4** Classification of phytoplankton on the basis of cell characteristics

| Class | Common name | Area(s) of predominance | Common genera |
|---|---|---|---|
| Cyanophyceae (cyanobacteria) | Blue-green algae | Tropical | *Oscillatoria, Synechococcus* |
| Rhodophyceae | Red algae | Cold temperate | *Rhodella* |
| Cryptophyceae | Cyptomonads | Coastal | *Cryptomonas* |
| Chrysophyseae | Chrysomonads<br>Silicoflagellates | Coastal<br>Cold waters | *Aureococcus*<br>*Dictyocha* |
| Bacillariophyceae (Diatomophyceae) | Diatoms | All waters, especially coastal waters | *Coscinodiscus, Chaetoceros, Rhizosolenia* |
| Raphidophyceae | Chloromonads | Brackish | *Heterosigma* |
| Xanthophyceae | Yellow-green algae | Brackish | Very rare |
| Eustigmatophyceae | Yellow-green algae | Estuarine | Very rare |
| Prymnesiophyceae | Coccolithophorids<br>Prymnesiomonads | Oceanic<br>Coastal | *Emiliania*<br>*Isochrysis*<br>*Prymnesium* |
| Euglenophyceae | Euglenoids | Coastal | *Eutreptiella* |
| Prasinophyceae | Prasinomonads | All waters | *Tetraselmis*<br>*Micromonas* |
| Chlorophyceae | Green algae | Coastal | Rare |
| Pyrrophyceae (Dinophyceae) | Dinoflagellates | All waters, especially warm | *Ceratium*<br>*Gonyaulax*<br>*Protoperidinium* |

sea surface during the night time and return back to their original depth during day. The span of daily migration varies between 10 and 500 m and this change of zone is keenly related to their nutritional requirement or maintenance of their light level. The zooplankton are mainly filter feeding in nature and are the prey of many shore-dwelling species. The zooplankton also exhibits various adaptations to keep them in floating condition. In many protozoa, the cilia act as blade of the rotor of a helicopter which enables the ciliates to move up and down the water. The scyphomedusae floats and moves in the direction of current by the rhythmic contraction and relaxation of their umbrella. Although these are large and heavy, but 99% of their mass is due to the presence of a jelly-like substance, due to which their density is reduced. *Vellela* sp. has chambers filled with air that keeps it afloat. The zoea of many arthropods have well developed uropods and these help them to remain in floating condition or buoy up by swimming. The molluscan larvae are important component of zooplankton community. The *Lanthina* sp. remains afloat by creating froath of bubbles, which contract the surface film of water and the animal hangs on froth. The zooplankton usually produce 3 to 5 generations in warm water where there is plenty of food supply and optimum temperature. At higher latitudes, where the phytoplankton growth is restricted within a brief period of time, the zooplankton produce only a single generation in a year.

The most abundant members of marine zooplankton are crustaceans called copepods. In coastal waters, the population density of copepods can be as high as 100,000 individuals per cubic metre of water. The main reasons for such large numbers are the tremendous reproductive capacity of these animals and the rich food supply. Copepods are the primary consumers of diatoms. A single copepod can consume as many as 120,000 diatoms per day.

There are major contributions of zooplankton to the marine and estuarine environments as highlighted here in points.

1. Among the various levels of production in the sea, the secondary production contributed by zooplankton is an important linkage between the primary and tertiary productions. The zooplankton mainly consumes the primary producers and forms the major food source for tertiary producers. An important food for the Antarctic baleen whales is the krill, which are important zooplankton in the Antarctic region.
2. In certain parts of the world, the zooplankton are directly consumed as food, e.g., certain species of *Mysids* are consumed as food in West Indies. A deep-water copepod (*Euchaeta norvegica*) is delicious food item in West Indies.
3. In many places of the world, zooplankton are treated as source of bioactive substances.
4. Certain species of zooplankton are used as bio-indicators of water quality.
5. The zooplankton contribute substantial biogenic material to ooze formation, which has wide application in instrument related industry where the ooze is used as thermal insulators, chromatographic column filters etc.

Zooplankton are found in almost all the layers of the photic zone of the ocean. They are potentially limited by two factors in the coastal and estuarine zones. Firstly by turbidity which can limit phytoplankton production and thus restrict the ration supply for the zooplankton community, and secondly by currents which, particularly in small estuaries are dominated by high river flow that usually carries the zooplankton out to the sea. The zooplankton biomass can increase the fishery productivity because they chiefly consume the primary producers (phytoplankton) and form the major food source for members of higher trophic levels in which several species of osteichthyes and chondrichthyes exist. Zooplankton are classified according to their habitat, depth distribution, size and duration of planktonic life (Tables 5.5, 5.6, 5.7 and 5.8).

**Table 5.5** Classification of zooplankton on the basis of habitat

| Type | Description |
|---|---|
| Oceanic plankton | These are marine zooplankton that inhabit beyond the continental shelf |
| Neritic plankton | These zooplankton inhabit waters overlying continental shelves. These waters are often very productive as they receive the runoff from the adjacent landmasses that triggers the phytoplankton growth in these regions |
| Brackish water plankton | These zooplankton inhabit estuarine regions, where there is a continuous mixing of fresh water and sea water. The zooplankton species of this category have wide range of tolerance to different dilution factors. Such zooplankton are very common in the shrimp culture farms and form important diet of the prawns |

**Table 5.6** Classification of zooplankton on the basis of depth distribution

| Type | Description |
|---|---|
| Neuston | The zooplankton of this category are restricted at the top few millimeters (usually 10 mm) of the surface micro layer |
| Pleuston | These are widely distributed at the surface of the sea (with parts of the body sometime projecting above the water) |
| Epipelagic | These are distributed between 0 and 300 m water column e.g. siphonophores, arrow worms etc. |
| Mesopelagic | The zooplankton of this category are restricted within the depth 300–1000 m e.g. euphausiids, chaetognath etc. |
| Bathypelagic | These are restricted within the depth 1000 and 3000 m e.g., foraminifera, euphausiids etc. |
| Abyssopelagic | The waters overlying the vast abyssal plains of the ocean are inhabited by a variety of zooplankton species, which are often referred to as abyssopelagic zooplankton. These zooplankton are thus restricted between 3000 and 4000 m |

*Source* Srinivasan and Santhanan (1998)

**Table 5.7** Classification of zooplankton on the basis of size

| Type | Size range |
|---|---|
| Nannozooplankton | <20 μm |
| Microzooplankton | 20–200 μm |
| Mesozooplankton | 200 μm–2 mm |
| Macrozooplankton | 2–20 mm |
| Megazooplankton | >20 mm |

*Source* Srinivasan and Santhanam (1998)

**Table 5.8** Classification of zooplankton on the basis of duration of planktonic life

| Type | Description |
|---|---|
| Holoplankton | This group includes organisms which are planktonic throughout their life cycle e.g., tintinnids, cladocerans, copepods, chaetognaths etc. |
| Meroplankton | This group encompasses those organisms which remain planktonic only for a portion of their life cycle e.g., larvae of benthic invertebrates and fish larvae (ichthyoplankton) |

*Source* Santhanam and Srinivasan (1998)

## 5.2 Common Plankton of Indian Sundarban Estuaries

The estuaries of Indian Sundarbans are noted for their unique spatio-temporal variations of salinity (Mitra 2000, 2013; Mitra et al. 2009, 2011; Banerjee et al. 2013; Sengupta et al. 2013; Chakraborty et al. 2013; Mitra and Zaman 2014, 2015,2016. This variation has resulted to a wide range of adaptability to plankton community.

The list of phytoplankton species has increased more than 150 in recent times in Indian Sundarban estuaries (Tables 5.9, 5.10 and 5.11).

In aquatic sub-system of deltaic Sundarbans, the zooplankton community comprises a heterogenous assemblage of animals covering many taxonomic groups including copepods, mycids, lucifers, gammarid, cladocerae, ostracods, cumacea, hydromedusae, ctenophores and chaetognaths among holoplankters and polychaete larvae, molluscan and echinoderm larvae, crustacean larvae and fish eggs and larvae among meroplankters Among the zooplankton, calaniod copepods constitute the major bulk, followed by cyclopoids, copepods and harpecticoids (Chaudhuri and Choudhury 1994). In general, higher abundance of zooplankton is encountered during the premonsoon periods. The main cause of the high species diversity index of zooplankton during the premonsoon months may be attributed to the coincidence of the breeding period of the coastal and estuarine fishes in the months of late February and March in the tropical mangrove ecosystem (Mitra 2000). A wide range of species under zooplankton community is available in the Indian Sundarban waters (Table 5.12).

We conducted a field survey during June 2017, 2018 and 2019 on ichthyoplankton abundance and aquatic pH at ten different stations in Indian Sundarbans (Table

**Table 5.9** Checklist of diatoms in Indian Sundarban estuarine water

| S. No. | Species |
|---|---|
| 1 | *Coscinodiscus eccentricus* |
| 2 | *Coscinodiscus jonesianus* |
| 3 | *Coscinodiscus lineatus* |
| 4 | *Coscinodiscus radiates* |
| 5 | *Coscinodiscus gigas* |
| 6 | *Coscinodiscus oculus-iridis* |
| 7 | *Coscinodiscus concinnus* |
| 8 | *Coscinodiscus perforates* |
| 9 | *Coscinodiscus asteromphalus* |
| 10 | *Coscinodiscus thorii* |
| 11 | *Coscinodiscus granii* |
| 12 | *Cyclotella* sp. |
| 13 | *Cyclotella striata* |
| 14 | *Cyclotella stylorum* |
| 15 | *Thalassiosira subtilis* |
| 16 | *Thalassiosira* sp. |
| 17 | *Thalassiosira decipiens* |
| 18 | *Thalassiosira punctigera* |
| 19 | *Thalassiosira hyaline* |
| 20 | *Thalassiosira eccentric* |
| 21 | *Skeletonema costatum* |
| 22 | *Paralia sulcata* |
| 23 | *Planktoniella sol* |
| 24 | *Planktoniella blanda* |
| 25 | *Rhizosolenia setigera* |
| 26 | *Rhizosolenia alata* |
| 27 | *Rhizosolenia hebetate* |
| 28 | *Rhizosolenia styliformis* |
| 29 | *Rhizosolenia robusta* |
| 30 | *Rhizosolenia stolterfothii* |
| 31 | *Rhizosolenia cylindrus* |
| 32 | *Rhizosolenia shrubsolei* |
| 33 | *Rhizosolenia imbricata:* |
| 34 | *Lauderia annulata* |
| 35 | *Bacteriastrum* sp. |
| 36 | *Bacteriastrum delicatulum* |
| 37 | *Bacteriastrum varians* |

(continued)

Table 5.9 (continued)

| S. No. | Species |
|---|---|
| 38 | *Bacteriastrum comosum* |
| 39 | *Bacteriastrum hyalinum* |
| 40 | *Chaetoceros dydymus* |
| 41 | *Chaetoceros curvisetus* |
| 42 | *Chaetoceros diversus* |
| 43 | *Chaetoceros messanensis* |
| 44 | *Chaetoceros peruvianus* |
| 45 | *Chaetoceros eibenii* |
| 46 | *Chaetoceros lorenzianus* |
| 47 | *Chaetoceros compressus* |
| 48 | *Chaetoceros decipiens* |
| 49 | *Chaetoceros atlanticus* |
| 50 | *Chaetoceros subtilis* |
| 51 | *Chaetoceros convolutes* |
| 52 | *Chaetoceros holsaticum* |
| 53 | *Chaetoceros gracile* |
| 54 | *Chaetoceros cinctum* |
| 55 | *Chaetoceros affinis* |
| 56 | *Chaetoceros danicus* |
| 57 | *Chaetoceros constrictus* |
| 58 | *Ditylum sol* |
| 59 | *Ditylum brightwelli* |
| 60 | *Lithodesmium undulatum* |
| 61 | *Triceratium favus* |
| 62 | *Triceratium reticulatum* |
| 63 | *Triceratium* sp. |
| 64 | *Biddulphia sinensis* |
| 65 | *Biddulphia mobiliensis* |
| 66 | *Biddulphia regia* |
| 67 | *Eucampia zodiacus* |
| 68 | *Hemidiscus cuneiformis* |
| 69 | *Climacosphenia elongate* |
| 70 | *Fragilaria oceanic* |
| 71 | *Rhaphoneis amphiceros* |
| 72 | *Thalassionema nitzchioides* |
| 73 | *Thalassionema* sp. |
| 74 | *Thalassiothrix longissima* |

(continued)

**Table 5.9** (continued)

| S. No. | Species |
|---|---|
| 75 | *Thalassiothrix fraunfeldii* |
| 76 | *Thalassiothrix* sp. |
| 77 | *Thalassiothrix nitzchioides* |
| 78 | *Asterionella japonica* |
| 79 | *Asterionellopsis glacialis* |
| 80 | *Diatoma vulgare* |
| 81 | *Diatoma* sp. |
| 82 | *Cocconeis* sp. |
| 83 | *Cocconeis soutellrem* |
| 84 | *Gyrosigma balticum* |
| 85 | *Gyrosigma* sp. |
| 86 | *Pleurosigma normanii* |
| 87 | *Pleurosigma elongatum* |
| 88 | *Pleurosigma directum* |
| 89 | *Diploneis smithii* |
| 90 | *Navicula longa* |
| 91 | *Navicula rhombic* |
| 92 | *Navicula pennata* |
| 93 | *Navicula* sp. |
| 94 | *Nevicula cancellata* |
| 95 | *Cymbella marina* |
| 96 | *Nitzschia sigma* |
| 97 | *Nitzschia longissima* |
| 98 | *Nitzschia closterium* |
| 99 | *Nitzschia striata* |
| 100 | *Nitzschia seriata* |
| 101 | *Pseudo-nitzschia pungens* |
| 102 | *Pseudo-nitzschia australis* |
| 103 | *Pinnularia alpine* |
| 104 | *Pinnularia* sp. |
| 105 | *Pinnularia Trevelyan* |
| 106 | *Bacillaria paradoxa* |
| 107 | *Bacillaria paxilifer* |
| 108 | *Triceratium spinosum* |
| 109 | *Phaeocystis* sp. |
| 110 | *Amphora hyaline* |
| 111 | *Amphipleura* sp. |

(continued)

Table 5.9 (continued)

| S. No. | Species |
|---|---|
| 112 | *Amphiprora constrica* |
| 113 | *Halosphera viridis* |
| 114 | *Halosphera* sp. |
| 115 | *Corethron crinophyllum* |
| 116 | *Guinerdia flacida* |
| 117 | *Hemiaulus sinensis* |
| 118 | *Melosira numuloides* |
| 119 | *Melosira granulate* |
| 120 | *Melosira variance* |
| 121 | *Melosira* sp. |
| 122 | *Bellerochea malleus* |
| 123 | *Ceratualina pelagic* |
| 124 | *Leptocylindrius danicum* |
| 125 | *Leptocylindrius minimus* |
| 126 | *Asteromphalus* sp. |
| 127 | *Hyalodiscus* sp. |
| 128 | *Actinocyclus octanarius* |
| 129 | *Actinoptychus* sp. |
| 130 | *Eunotia* sp. |
| 131 | *Tabellaria* sp. |
| 132 | *Grammatophora marina* |
| 133 | *Closterium* sp. |
| 134 | *Striatella* sp. |
| 135 | *Striatella unipunctata* |
| 136 | *Diplosalis* sp. |
| 137 | *Licmophora chrenbergii* |

*Source* Mitra and Zaman (2015)

5.13 and Fig. 5.1) to interpret the impact of pH on ichthyoplankton community comprising of fish juveniles, which cannot move against the current and are basically meroplankton (Fig. 5.2).

The standing stock of ichthyoplankton (N) per haul was determined during 2017, 2018 and 2019 using standard procedure and expressed in gm. Simultaneously the aquatic pH was also determined with a portable pH meter (sensitivity = ±0.01). The data for each of the 10 selected stations are presented in Table 5.14.

The data of Table 5.14 reflects three important facts (i) both ichthyoplankton standing stock (N) and pH decreased with time (ii) ichthyoplankton standing stock (N) was relatively high in stations like Lothian Island, Prentice Island and Sajnekhali, and (iii) pH values were almost uniform in Lothian Island, Prentice Island and

**Table 5.10** Checklist of dinoflagellates in Indian Sundarban estuarine water

| S. No. | Species |
|---|---|
| 1 | *Prorocentrum gracile* |
| 2 | *Prorocentrum micans* |
| 3 | *Prorocentrum concavum* |
| 4 | *Ceratium furca* |
| 5 | *Ceratium fusus* |
| 6 | *Ceratium teres* |
| 7 | *Ceratium minutum* |
| 8 | *Ceratium tripose* |
| 9 | *Ceratium trichoceros* |
| 10 | *Ceratium horridum* |
| 11 | *Ceratium inflatum* |
| 12 | *Ceratocorys horrida* |
| 13 | *Protoperidinium* sp. |
| 14 | *Protoperidinium crassipes* |
| 15 | *Protoperidinium depressum* |
| 16 | *Protoperidinium ovatum* |
| 17 | *Protoperidinium pellucidum* |
| 18 | *Protoperidinium conicum* |
| 19 | *Pyrocystis fusiformis* |
| 20 | *Pyrocystis* sp. |
| 21 | *Dinophysis caudata* |
| 22 | *Dinophysis acuta* |
| 23 | *Dinophysis norvegica* |
| 24 | *Dinophysis* sp. |
| 25 | *Alexandrium* sp. |
| 26 | *Preperidinium meunieri* |

*Source* Mitra and Zaman (2015)

Sajnekhali. The reasons behind these outputs may be many, but our first order analysis points towards the presence of rich mangrove vegetation at Lothian Island, Prentice Island and Sajnekhali compared to other stations, which imparted a buffering capacity to adjacent estuarine water in terms of pH and nutrient rich water, mostly derived from mangrove litter and detritus. The abundance of mangroves in these stations is presented in Fig. 5.3.

Ten stations selected in the present work can be categorized into two major sectors in the study area namely (i) western sector encompassing stations 1 to 5 and (ii) central sector with stations 6–10 (see Fig. 5.1). These two sectors are significantly different in terms of salinity, pH, and mangrove vegetation, which resulted in the variation of ichthyoplankton in these two sectors (Table 5.15). The western sector

**Table 5.11** Checklist of other algae in Indian Sundarban estuarine water

| S. No. | Species |
|---|---|
| 1 | *Trichodesmium* sp. |
| 2 | *Trichodesmium thiebautii* |
| 3 | *Oscillatoria* sp. |
| 4 | *Oscillatoria limosa* |
| 5 | *Gleocapsa* sp. |
| 6 | *Stigonema* sp. |
| 7 | *Cylindrospermopsis raciborskii* |
| 8 | *Anabaena* sp. |
| 9 | *Dictyocha* sp. |
| 10 | *Ditylum sol* |
| 11 | *Chlorella salina* |
| 12 | *Amphora hyaline* |
| 13 | *Netrium* sp. |
| 14 | *Drapamaldia* sp. |
| 15 | *Gleocapsa* sp. |

*Source* Mitra and Zaman (2015)

of Indian Sundarbans exhibited greater abundance of ichthyoplankton of commercially important fishes like *Tenualosa* sp., *Liza* sp., *Scatophagus* sp. etc., whereas ichthyoplankton of trash fish variety like *Coilia* sp., *Thryssa* sp. etc., dominated the estuarine water of central Indian Sundarbans.

To sum up it can be stated that mangrove vegetation acts as a unique buffer against pH fall and serves as a nursery ground for ichthyoplankton by offering safe shelter through wave attenuation by root systems and nutrition derived from mangrove detritus (Fig. 5.4).

Hence, it is the need of the hour to conserve and propagate this unique forest of halophytes to retard the pace of acidification and subsequent impact on plankton community.

## 5.3 Plankton: Effect of Acidification

The growing population coupled with industrial development at large scale has scratched the stability of the environment. The level of carbon dioxide has increased even in the atmosphere of Indian Sundarbans through seasons (Fig. 5.5), probably due to massive destruction of mangroves (Fig. 5.6), the major sink of carbon dioxide in the region.

**Table 5.12** Zooplankton in Indian Sundarban estuarine water (ichthyoplankton community has not been considered in the list)

| Class | Species |
|---|---|
| Acantharia | *Acanthometron* sp. |
| Spirotrichea | *Leprotintinnus nordqvisti* |
| | *Tintinnopsis beroidea* |
| | *T. butschlii* |
| | *T. cylindrica* |
| | *T. mortensenii* |
| | *T. tocantinensis* |
| | *T. tubulosa* |
| | *T. uruguayensis* |
| | *Codonellopsis ostenfeldi* |
| | *Dictyocysta seshaiyai* |
| | *Metacylis jorgenseni* |
| | *Rhabdonella* sp. |
| | *Amphorellopsis* sp. |
| | *Eutintinnus tenue* |
| | *Favella philippinensis* |
| Polythalamea | *Globigerina bulloides* |
| | *G. rubescense* |
| | *Globigerina* sp. |
| | *Asterorotalia trispinosa* |
| | *Quinqueloculina* sp. |
| Hydrozoa | *Cladonema* sp. |
| | *Sarsia* sp. |
| | *Podocoryne* sp. |
| | *Obelia* sp. |
| | *Phialella quadrata* |
| | *Liriope tetraphylla* |
| | *Aglaura hemistoma* |
| | *Aequorea vitrina* |
| | *Agalma elegans* |
| | *Sulculeolaria* sp. |
| | *Diphyes chamissonis* |
| | *D. dispar* |
| | *Diphyes* sp. |
| | *Lensia* sp. |
| | *Eudoxoides mitra* |
| | *Muggiaea* sp. |

(continued)

**Table 5.12** (continued)

| Class | Species |
|---|---|
| | *Abylopsis* sp. |
| | *Bassia bassensis* |
| Tentaculata | *Pleurobrachia pileus* |
| Nuda | *Beroe* sp. |
| Gastropoda | *Atlanta* sp. |
| | *Janthina* sp. |
| | *Limacina bulimoides* |
| | *Limacina inflata* |
| | *Cresis acicula* |
| | *Hyalocylix striata* |
| Branchiopoda | *Evadne tergestina* |
| | *Penilia avirostris* |
| Maxillipoda | *Pleuromamma* sp. |
| | *Acartia. centrura* |
| | *A. danae* |
| | *A. erythraea* |
| | *A. spinicauda* |
| | *A. negligens* |
| | *A. southwelli* |
| | *Acartia* sp. |
| | *Candacia catula* |
| | *Candacia* sp. |
| | *Paracandacia truncata* |
| | *Paracandacia* sp. |
| | *Centropages alcocki* |
| | *C. furcatus* |
| | *C. calaninus* |
| | *C. dorsispinatus* |
| | *C. orsinii* |
| | *C. tenuiremis* |
| | *Centropages* sp. |
| | *Calanopia minor* |
| | *C. elliptica* |
| | *Labidocera acuta* |
| | *L. detruncata* |
| | *L. minuta* |
| | *L. pectinata* |

(continued)

**Table 5.12** (continued)

| Class | Species |
|---|---|
| | *Labidocera* sp. |
| | *Pontella fera* |
| | *P. securifer* |
| | *Pontellina platychela* |
| | *Temora discaudata* |
| | *T. turbinata* |
| | *T. stylifera* |
| | *Tortanus barbatus* |
| | *T. forcipatus* |
| | *T. gracilis* |
| | *Mesocalanus tenuicornis* |
| | *Canthocalanus pauper* |
| | *Nannocalanus minor* |
| | *Undinula vulgaris* |
| | *Acrocalanus gracilis* |
| | *A. longicornis* |
| | *A. gibber* |
| | *Acrocalanus* sp. |
| | *Calocalanus pavo* |
| | *Paracalanus aculeatus* |
| | *P. parvus* |
| | *Paracalanus* sp. |
| | *Eucalanus attenuates* |
| | *E. monachus* |
| | *Eucalanus* sp. |
| | *Subeucalanus crassus* |
| | *S. mucronatus* |
| | *S. pileatus* |
| | *S. subcrassus* |
| | *S. subtenuis* |
| | *Undeuchaeta* sp. |
| | *Metacalanus aurivilli* |
| | *Clausocalanus arcuicornis* |
| | *Euchaeta concinna* |
| | *E. marina* |
| | *Euchaeta* sp. |
| | *Scolecithrix danae* |

(continued)

**Table 5.12** (continued)

| Class | Species |
|---|---|
| | *Scolecithricella minor* |
| | *Pseudodiaptomus aurivilli* |
| | *P. serricaudatus* |
| | *P. annandalei* |
| | *Pseudodiaptomus* sp. |
| | *Haloptilus* sp. |
| | *Oithona nana* |
| | *O. oculata* |
| | *O. setigera* |
| | *O. similis* |
| | *O. simplex* |
| | *O. tenuis* |
| | *O. rigida* |
| | *O. brevicornis* |
| | *O. spinirostris* |
| | *Oithona* sp. |
| | *Miracia efferata* |
| | *Macrosetella gracilis* |
| | *M. oculata* |
| | *Microsetella norvegica* |
| | *M. rosea* |
| | *Clytemnestra scutellata* |
| | *Euterpina acutifrons* |
| | *Longipedia weberi* |
| | *Oncaea conifera* |
| | *O. media* |
| | *O. mediterranea* |
| | *O. venusta* |
| | *Oncaea* sp. |
| | *Copilia quadrata* |
| | *Sapphirina auronitens* |
| | *S. maculosa* |
| | *S. ovatolanceolata* |
| | *Sapphirina* sp. |
| | *Corycaeus agilis* |
| | *C. andrewsi* |
| | *C. catus* |

(continued)

**Table 5.12** (continued)

| Class | Species |
|---|---|
| | *C. danae* |
| | *C. erythraeus* |
| | *C. lautus* |
| | *C. longistylis* |
| | *C. ovalis* |
| | *C. robustus* |
| | *C. speciosus* |
| | *Corycaeus* sp. |
| | *Farranula carinata* |
| | *F. concinna* |
| | *F. curta* |
| | *F. gibbula* |
| | *F. gracilis* |
| Ostracoda | *Macrocypridina castanea* |
| | *Conchoecia elegans* |
| | *Euconchoecia chierchiae* |
| | *Mesopodopsis orientalis* |
| | *Leucothoe spinicarpa* |
| | *Hyperia* sp. |
| | *Caprella* sp. |
| | *Metacaprella* sp. |
| | *Orchestoidea* sp. |
| | *Campylaspis costata* |
| | *Euphausia tenera* |
| | *Euphausia* sp. |
| | *Lucifer hanseni* |
| | *Sergestes* sp. |
| Sagittoidea | *Pseudosagitta maxima* |
| | *Sagitta bedoti* |
| | *S. bipunctata* |
| | *S. enflata* |
| | *Sagitta* sp. |
| | *Adinosagitta bedfordii* |
| Appendicularia | *Oikopleura dioica* |
| | *O. parva* |
| | *Fritillaria* sp. |
| Thaliacea | *Doliolum* sp. |
| | *Salpa fusiformis* |

**Table 5.13** Selected sampling stations (for the determination of average ichthyoplankton standing stock and mangrove density) in Indian Sundarbans with salient features

| Study Site | Latitude and Longitude |
|---|---|
| Harinbari (Stn. 1) | 21° 46′ 53.07″ N and 88° 04′ 22.88″ E |
| Chemaguri (Stn. 2) | 21° 39′ 42.88″ N and 88° 08′ 49.01″ E |
| Sagar South (Stn. 3) | 21° 37′ 49.90″ N and 88° 04′ 0.51″ E |
| Lothian Island (Stn. 4) | 21° 39′ 08.04″ N and 88° 19′ 8.47″ E |
| Prentice Island (Stn. 5) | 21° 42′ 43.31″ N and 88° 17′ 3.62″ E |
| Canning (Stn. 6) | 22° 19′ 03.20″ N and 88° 41′ 04.43″ E |
| Sajnekhali (Stn. 7) | 22° 06′ 34.19″ N and 88° 48′ 15.78″ E |
| Chotomollakhali (Stn. 8) | 22° 10′ 40.00″ N and 88° 54′ 26.71″ E |
| Satjelia (Stn. 9) | 22° 05′ 17.86″ N and 88° 52′ 49.51″ E |
| Pakhiralaya (Stn. 10) | 22° 07′ 07.23″ N and 88° 48′ 29.00″ E |

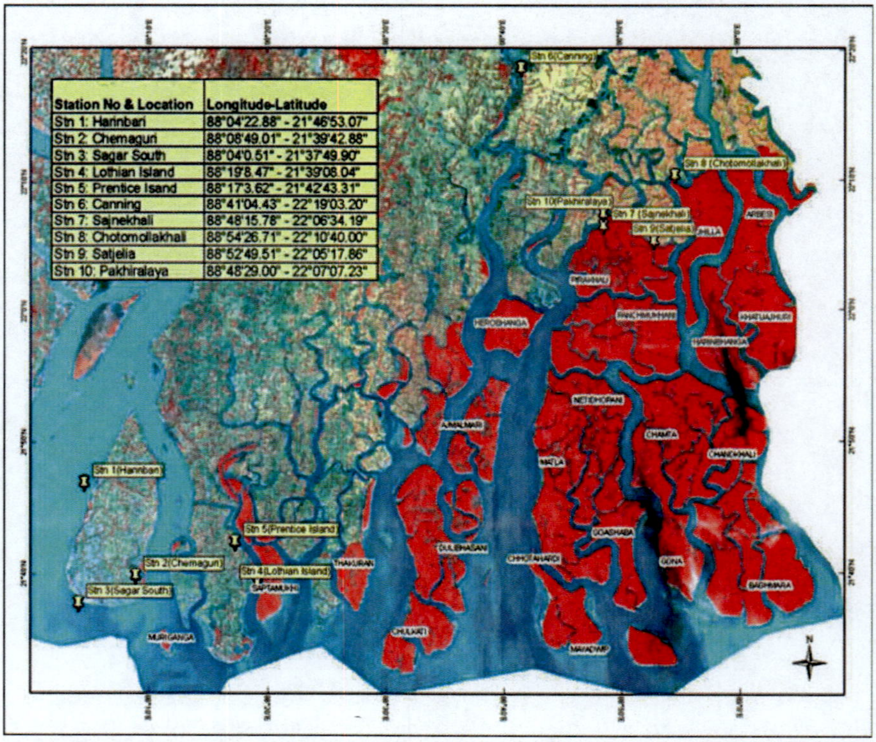


**Fig. 5.1** Map showing the ten stations in Indian Sundarbans selected for the determination of inter-relationship between ichthyoplankton standing stock and aquatic pH

**Fig. 5.2** Ichthyoplankton in the estuarine water of Indian Sundarbans

**Table 5.14** Spatio-temporal variation of average standing stock of ichthyoplankton (N) per haul (in gm) and surface water pH recorded during high tide

| S. No. | Station name | 2017 | | 2018 | | 2019 | |
|---|---|---|---|---|---|---|---|
| | | N | pH | N | pH | N | pH |
| 1 | Harinbari (Stn. 1) | 467 | 8.17 | 401 | 8.17 | 380 | 8.15 |
| 2 | Chemaguri (Stn. 2) | 454 | 8.26 | 358 | 8.24 | 307 | 8.24 |
| 3 | Sagar South (Stn. 3) | 491 | 8.28 | 397 | 8.27 | 295 | 8.24 |
| 4 | Lothian Island (Stn. 4) | 566 | 8.27 | 481 | 8.27 | 428 | 8.27 |
| 5 | Prentice Island (Stn. 5) | 593 | 8.27 | 467 | 8.27 | 439 | 8.27 |
| 6 | Canning (Stn. 6) | 402 | 8.05 | 361 | 8.02 | 134 | 8.00 |
| 7 | Sajnekhali (Stn. 7) | 565 | 8.24 | 405 | 8.23 | 365 | 8.23 |
| 8 | Chotomollakhali (Stn. 8) | 305 | 8.13 | 302 | 8.09 | 217 | 8.06 |
| 9 | Satjelia (Stn. 9) | 366 | 8.11 | 313 | 8.09 | 207 | 8.07 |
| 10 | Pakhiralaya (Stn. 10) | 417 | 8.11 | 395 | 8.08 | 287 | 8.07 |

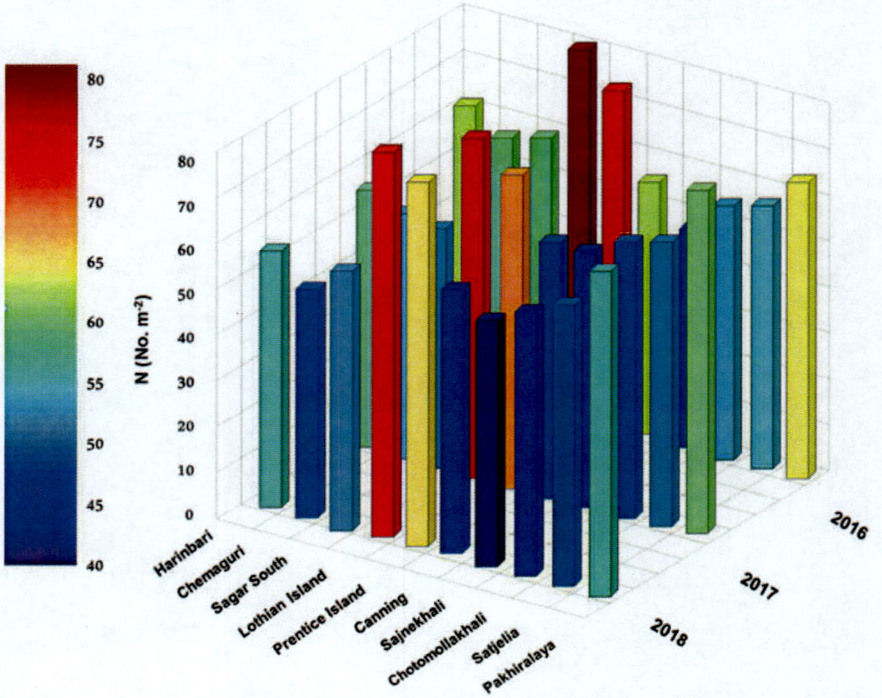


**Fig. 5.3** Spatio-temporal variation of population density of mangroves (in No./100 $m^2$) in the 10 selected stations during 2016, 2017 and 2018

The estuaries of Indian Sundarbans synchronized with the pace of rising atmospheric carbon dioxide (5.13) and exhibited a simultaneous decrease of aquatic pH (Tables 5.16 and 5.17).

A significant negative relationship witnessed between the two variables (near surface atmospheric carbon dioxide level and aquatic pH) clearly confirms the pulse of acidification in the estuaries of Indian Sundarbans.

It is interesting to note in this context that the pace of acidification differs considerably between the western and eastern Indian Sundarbans, the former having a higher pace than the latter. This may be attributed to contrasting anthropogenic-cum-environmental set-ups. The western Indian Sundarbans is exposed to high degree of aquatic pollution from the chain of factories and industries situated at the bank of Hooghly estuary (Table 5.18), which mostly release untreated/partially treated wastes.

The eastern Indian Sundarbans, on contrary, is basically a Reserve Forest (RF) area, where there are minimum anthropogenic footprints. The presence of rich mangrove vegetation in this sector is also the factor to reduce the rate of acidification (Fig. 5.7).

**Table 5.15** Abundance of ichthyoplankton species in 10 gm composite sample collected from two sectors of Indian Sundarbans

| S. No. | Species | Western Sector (composite samples of Stations 1–5) | Central Sector (composite samples of Stations 6–10) |
|---|---|---|---|
| 1 | *Coilia* sp. | ++ | + |
| 2 | *Thryssa hamiltonii (Gray)* | + | +++ |
| 3 | *Thryssa baelama* | + | +++ |
| 4 | *Torquigener oblongus* | + | + |
| 5 | *Rhinomugil corsula* | + | + |
| 6 | *Mugil cephalus* | +++ | + |
| 7 | *Sillaginopsis panijus* | +++ | + |
| 8 | *Sillago sihama* | +++ | + |
| 9 | *Sillago soringa* | +++ | + |
| 10 | *Zenarchopterus dispar* | + | + |
| 11 | *Glossogobius quiris* | + | + |
| 12 | *Pseudapocryptes lanceolatus* | + | +++ |
| 13 | *Eupleurogrammus glossodon* | + | + |
| 14 | *Pseudorhombus* sp. | + | + |
| 15 | *Pisodonophis boro* | + | + |
| 16 | *Tenualosa ilisha* | ++ | +++ |
| 17 | *Cynoglossus arel* | + | +++ |
| 18 | *Cynoglossus* sp. | + | +++ |
| 19 | *Leiognathus blochii* | + | ++ |
| 20 | *Leiognathus equalus* | + | + |
| 21 | *Scatophagus argus* | + | + |
| 22 | *Liza parsia* | +++ | + |
| 23 | *Liza tade* | +++ | + |
| 24 | *Stolephorus commersonii* | + | + |
| 25 | *Stolephorus baganensis* | + | + |
| 26 | *Stolephorus kammalensis* | + | + |
| 27 | *Lutjanus johni* | + | + |
| 28 | *Setipinna taty* | + | +++ |
| 29 | *Lagocephalus lunaris* | + | + |

(continued)

**Table 5.15** (continued)

| S. No. | Species | Western Sector (composite samples of Stations 1–5) | Central Sector (composite samples of Stations 6–10) |
|---|---|---|---|
| 30 | *Escualosa thoracata* | + | + |
| 31 | *Epinephelus tauvina* | + | + |
| 32 | *Epinephelus coioides* | + | + |
| 33 | *Sphyraena* sp. | + | + |
| 34 | *Pomadasys* sp. | + | + |
| 35 | *Sardinella longiceps* | + | + |
| 36 | *Periophthalmus* sp. | + | + |
| 37 | *Macrognathus* sp. | + | + |
| 38 | *Tetradon cutcutia* | + | + |
| 39 | *Bregmaceros meclellandii* | + | + |
| 40 | *Ichthyocampus carce* | + | + |
| 41 | *Stigmatogobius* sp. | + | + |
| 42 | *Channa* sp. | + | + |
| 43 | *Kurtus indicus* | + | + |
| 44 | *Harpodon nehereus* | + | +++ |
| 45 | *Moringua raitaborua* | + | + |
| 46 | *Suggrundus rodricensis* | + | + |

'+' abundance < 10%; '++' abundance = 10–20%; '+++' abundance = 21–80%

The phytoplankton diversity (measured in terms of Shannon-Weiner Species Diversity Index) in these two sectors also exhibited different spatio-temporal scores (Table 5.19), which may be due variations in Total Suspended Solid (TSS), nitrate and phosphate (Tables 5.20, 5.21 and 5.22 respectively). For interested readers, with a little bit of knowledge on programming, we have highlighted the computation of Shannon-Weiner Species Diversity Index of phytoplankton by using C programming language (*Vide* Annexure 5). This is an imperative procedural language which is designed to be compiled using a relatively straight forward compiler, to provide low level access to memory, and to provide language, constructs the map efficiently to machine instructions. This language requires minimum run-time support.

Similar trends were also observed in Haldia and Bali Island (Tables 5.23, 5.24, 5.25 and 5.26) where significant negative correlation values were observed between phytoplankton diversity and suspended solids. The presence of more suspended solid on the water mass prohibits the primary producer, phytoplankton in getting sunlight at optimum level due to which their survival and growth are adversely impacted. The gradual decrease of Shannon- Weiner Index values in all the selected stations (Tables 5.19 and 5.23) with the passage of time is a clear indication of the adverse role played by increased suspended solid and decreased pH, both of which are

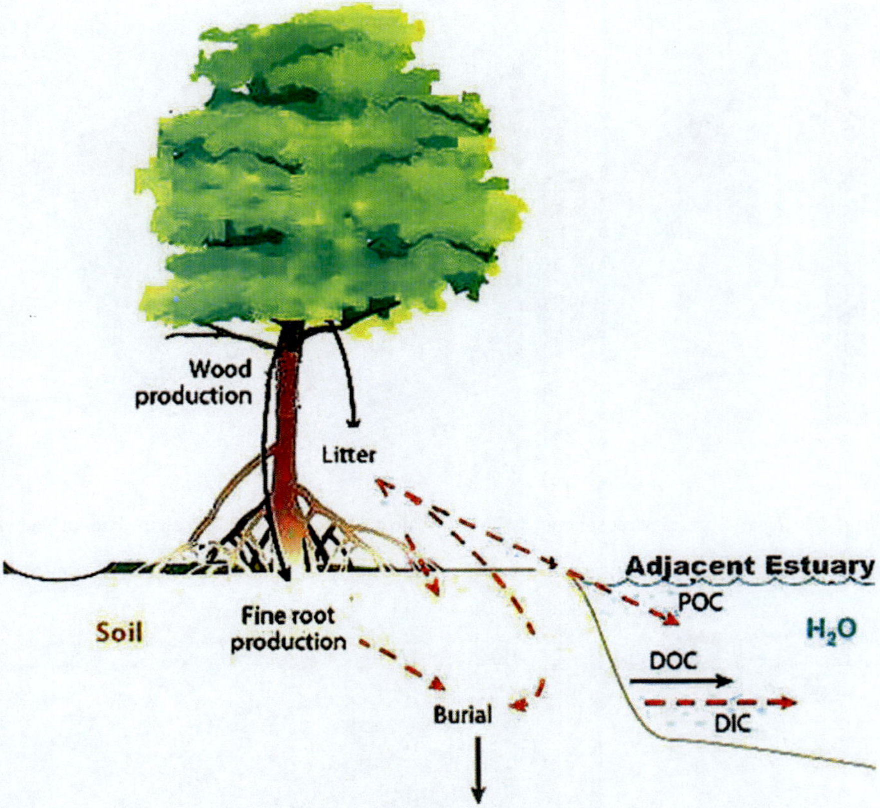


**Fig. 5.4** Mangrove vegetation provides nutrients to adjacent water body that supports the ichthyoplankton in the system

anthropogenically sourced in the region. The decrease in phytoplankton diversity and standing stock results in the un-utilization of nutrients (nitrate and phosphate) in the aquatic phase due to which the levels increase with time.

Acidification is the net result of climate change induced carbon dioxide rise in the atmosphere. With the rise of carbon dioxide, surface waters have become more susceptible to acidification. Anthropogenic carbon dioxide from the burning of fossil fuels has increased the uptake of the green house gas, causing the pH to drop from 8.2 to 8.1 (Chu 2015).

Ocean acidification has high probability to pose negative impact on the marine life. It is well known that coral bleaching, reduced calcification and thinning of oyster shells are the direct effects of acidification. However, the species abundance and diversity of phytoplankton are also impacted by the pH level of the aquatic ecosystem, (Fabry et al. 2008). The uptake of Fe (iron) by phytoplankton, such as diatoms and coccolithophores, decreases with increasing acidification (Shi et al. 2010). On this background, we carried out an outdoor experiment in the Kadwip region of western

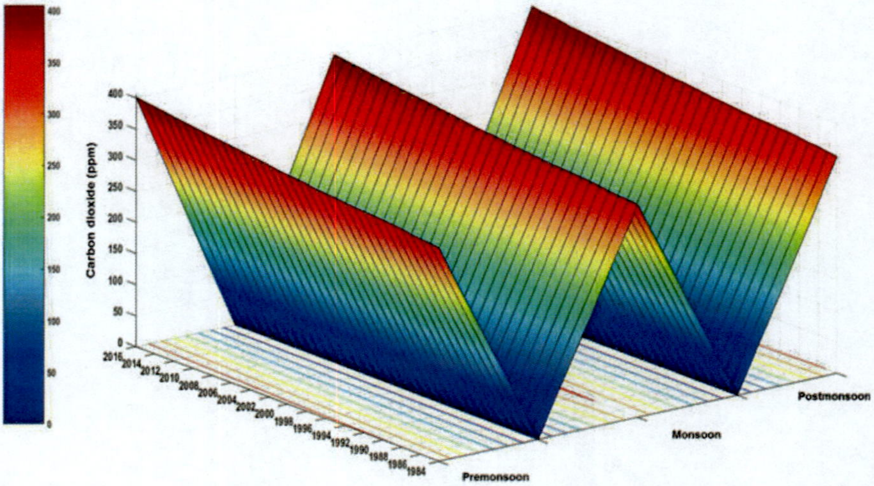


**Fig. 5.5** Near Surface atmospheric carbon dioxide (in ppm) in the Kakdwip area of Indian Sundarbans

**Fig. 5.6** Cutting of mangroves for fuel wood and deforestation for settlement in Indian Sundarbans

**Table 5.16** Temporal variation of near surface atmospheric carbon dioxide through seasons at Kakdwip in Western Indian Sundarbans

| Year | Premonsoon | Monsoon | Postmonsoon |
|---|---|---|---|
| 1984 | 345.07 | 342.87 | 348.38 |
| 1985 | 346.55 | 344.35 | 350.86 |
| 1986 | 347.97 | 345.77 | 353.24 |
| 1987 | 349.71 | 347.51 | 354.12 |
| 1988 | 352.23 | 350.03 | 356.54 |
| 1989 | 353.88 | 351.68 | 358.79 |
| 1990 | 355.06 | 352.86 | 359.37 |
| 1991 | 356.38 | 354.18 | 360.69 |
| 1992 | 357.08 | 354.88 | 361.39 |
| 1993 | 357.8 | 355.6 | 362.2 |
| 1994 | 359.29 | 357.09 | 363.6 |
| 1995 | 361.11 | 358.91 | 365.42 |
| 1996 | 362.88 | 360.68 | 367.19 |
| 1997 | 363.99 | 361.79 | 368.3 |
| 1998 | 366.64 | 364.44 | 370.95 |
| 1999 | 368.72 | 366.52 | 373.03 |
| 2000 | 369.9 | 367.7 | 375.22 |
| 2001 | 371.5 | 369.3 | 375.81 |
| 2002 | 373.5 | 371.3 | 377.81 |
| 2003 | 376.05 | 373.85 | 380.36 |
| 2004 | 377.87 | 375.67 | 382.18 |
| 2005 | 379.9 | 377.7 | 384.21 |
| 2006 | 382.02 | 379.82 | 386.33 |
| 2007 | 383.76 | 381.56 | 388.07 |
| 2008 | 385.87 | 383.67 | 390.18 |
| 2009 | 387.37 | 385.17 | 391.68 |
| 2010 | 389.65 | 387.45 | 393.96 |
| 2011 | 391.53 | 389.33 | 395.84 |
| 2012 | 393.54 | 391.34 | 397.19 |
| 2013 | 396.28 | 394.08 | 400.59 |
| 2014 | 398.21 | 396.01 | 402.52 |
| 2015 | 400.5 | 398.3 | 403.86 |
| 2016 | 402.31 | 400.09 | 406.4 |

A = Premonsoon, B = Monsoon and C = Postmonsoon

**Table 5.17** Temporal variation of surface water pH through seasons at different tidal phase at Kakdwip in Western Indian Sundarbans

| | A1 | A2 | B1 | B2 | C1 | C2 |
|---|---|---|---|---|---|---|
| 1984 | 8.32 ± 0.03 | 8.31 ± 0.03 | 8.29 ± 0.03 | 8.28 ± 0.03 | 8.31 ± 0.03 | 8.29 ± 0.03 |
| 1985 | 8.31 ± 0.03 | 8.3 ± 0.03 | 8.29 ± 0.03 | 8.27 ± 0.03 | 8.31 ± 0.03 | 8.29 ± 0.03 |
| 1986 | 8.32 ± 0.03 | 8.31 ± 0.03 | 8.29 ± 0.03 | 8.28 ± 0.03 | 8.31 ± 0.03 | 8.3 ± 0.03 |
| 1987 | 8.31 ± 0.03 | 8.3 ± 0.03 | 8.28 ± 0.03 | 8.26 ± 0.03 | 8.29 ± 0.03 | 8.28 ± 0.03 |
| 1988 | 8.31 ± 0.03 | 8.3 ± 0.03 | 8.28 ± 0.03 | 8.26 ± 0.03 | 8.3 ± 0.03 | 8.29 ± 0.03 |
| 1989 | 8.31 ± 0.03 | 8.3 ± 0.03 | 8.27 ± 0.03 | 8.26 ± 0.03 | 8.31 ± 0.03 | 8.3 ± 0.03 |
| 1990 | 8.31 ± 0.03 | 8.3 ± 0.03 | 8.27 ± 0.03 | 8.26 ± 0.03 | 8.3 ± 0.03 | 8.28 ± 0.03 |
| 1991 | 8.29 ± 0.03 | 8.28 ± 0.03 | 8.27 ± 0.03 | 8.25 ± 0.03 | 8.28 ± 0.03 | 8.27 ± 0.03 |
| 1992 | 8.3 ± 0.03 | 8.29 ± 0.03 | 8.26 ± 0.03 | 8.25 ± 0.03 | 8.29 ± 0.03 | 8.26 ± 0.03 |
| 1993 | 8.3 ± 0.03 | 8.29 ± 0.03 | 8.27 ± 0.03 | 8.24 ± 0.02 | 8.29 ± 0.03 | 8.26 ± 0.03 |
| 1994 | 8.29 ± 0.03 | 8.28 ± 0.03 | 8.27 ± 0.03 | 8.24 ± 0.02 | 8.29 ± 0.03 | 8.27 ± 0.03 |
| 1995 | 8.29 ± 0.03 | 8.28 ± 0.03 | 8.26 ± 0.03 | 8.23 ± 0.02 | 8.28 ± 0.03 | 8.26 ± 0.03 |
| 1996 | 8.28 ± 0.03 | 8.27 ± 0.03 | 8.25 ± 0.03 | 8.24 ± 0.02 | 8.27 ± 0.03 | 8.26 ± 0.03 |
| 1997 | 8.28 ± 0.03 | 8.27 ± 0.03 | 8.25 ± 0.03 | 8.23 ± 0.02 | 8.27 ± 0.03 | 8.25 ± 0.03 |
| 1998 | 8.28 ± 0.02 | 8.27 ± 0.03 | 8.25 ± 0.03 | 8.23 ± 0.02 | 8.27 ± 0.03 | 8.26 ± 0.03 |
| 1999 | 8.26 ± 0.03 | 8.25 ± 0.03 | 8.23 ± 0.02 | 8.23 ± 0.02 | 8.25 ± 0.03 | 8.23 ± 0.02 |
| 2000 | 8.27 ± 0.03 | 8.25 ± 0.03 | 8.23 ± 0.02 | 8.22 ± 0.02 | 8.26 ± 0.03 | 8.24 ± 0.02 |
| 2001 | 8.27 ± 0.03 | 8.26 ± 0.03 | 8.24 ± 0.02 | 8.22 ± 0.02 | 8.27 ± 0.03 | 8.24 ± 0.02 |
| 2002 | 8.25 ± 0.03 | 8.24 ± 0.02 | 8.24 ± 0.02 | 8.22 ± 0.02 | 8.25 ± 0.03 | 8.23 ± 0.02 |
| 2003 | 8.27 ± 0.03 | 8.24 ± 0.02 | 8.24 ± 0.02 | 8.22 ± 0.02 | 8.26 ± 0.03 | 8.24 ± 0.02 |
| 2004 | 8.26 ± 0.03 | 8.25 ± 0.02 | 8.23 ± 0.02 | 8.22 ± 0.02 | 8.25 ± 0.03 | 8.22 ± 0.02 |
| 2005 | 8.24 ± 0.02 | 8.23 ± 0.02 | 8.23 ± 0.02 | 8.21 ± 0.02 | 8.23 ± 0.02 | 8.22 ± 0.02 |
| 2006 | 8.24 ± 0.02 | 8.23 ± 0.02 | 8.22 ± 0.02 | 8.2 ± 0.02 | 8.23 ± 0.02 | 8.21 ± 0.02 |
| 2007 | 8.23 ± 0.02 | 8.22 ± 0.02 | 8.22 ± 0.02 | 8.21 ± 0.02 | 8.22 ± 0.02 | 8.21 ± 0.02 |
| 2008 | 8.25 ± 0.03 | 8.22 ± 0.02 | 8.21 ± 0.02 | 8.21 ± 0.02 | 8.22 ± 0.02 | 8.2 ± 0.02 |
| 2009 | 8.24 ± 0.05 | 8.24 ± 0.04 | 8.21 ± 0.02 | 8.2 ± 0.02 | 8.23 ± 0.02 | 8.22 ± 0.02 |
| 2010 | 8.21 ± 0.02 | 8.19 ± 0.02 | 8.2 ± 0.02 | 8.19 ± 0.02 | 8.22 ± 0.02 | 8.21 ± 0.02 |
| 2011 | 8.2 ± 0.02 | 8.19 ± 0.02 | 8.19 ± 0.02 | 8.18 ± 0.02 | 8.19 ± 0.02 | 8.18 ± 0.02 |
| 2012 | 8.17 ± 0.02 | 8.15 ± 0.02 | 8.16 ± 0.02 | 8.15 ± 0.02 | 8.16 ± 0.02 | 8.15 ± 0.02 |
| 2013 | 8.13 ± 0.02 | 8.13 ± 0.02 | 8.12 ± 0.02 | 8.1 ± 0.02 | 8.13 ± 0.02 | 8.11 ± 0.02 |
| 2014 | 8.11 ± 0.02 | 8.1 ± 0.02 | 8.1 ± 0.02 | 8.1 ± 0.02 | 8.11 ± 0.02 | 8.08 ± 0.01 |
| 2015 | 8.13 ± 0.02 | 8.1 ± 0.02 | 8.09 ± 0.01 | 8.07 ± 0.01 | 8.1 ± 0.02 | 8.09 ± 0.01 |
| 2016 | 8.12 ± 0.02 | 8.08 ± 0.01 | 8.09 ± 0.01 | 8.04 ± 0.01 | 8.1 ± 0.02 | 8.07 ± 0.01 |

A1 = Premonsoon high tide condition, B1 = Monsoon high tide condition and C1 = Postmonsoon high tide condition, A2 = Premonsoon low tide condition, B2 = Monsoon low tide condition and C2 = Postmonsoon low tide condition

**Table 5.18** List of major industries along the lower stretch of the Hooghly estuary

| S. No. | Name of Industry | Product |
|---|---|---|
| 1 | Indian Oil Corporation Ltd., Haldia | L.P.G., Motor Gasoline, Naptha, ATF, MTO, HSD, JBO, Kerosene, Furnace Oil, Lubes, Bitumen |
| 2 | KoPT/Haldia Dock Complex | Port Services |
| 3 | Tata Chemicals Ltd., Haldia | Industrial Phosphate and Acids |
| 4 | Exide Industries Ltd., Haldia | Automotive Batteries, Heavy Duty Batteries, Containers, Special Types of Separators, etc. |
| 5 | Shaw Wallace, Haldia | Pesticides |
| 6 | MCC PTA India Corpn. Pvt. Ltd., Haldia | P.T.A |
| 7 | Haldia Petrochemicals Ltd., Haldia | LLDPE, HDPE, Naptha Cracker etc. |
| 8 | IOCL, Paradip-Haldia Oil Pipeline | Petroleum Storage and Transportation |
| 9 | IOC Petronas Ltd., Haldia | L.P.G |
| 10 | Shamon Ispat Ltd. | Steel Rolling |
| 11 | Dhunseri Petrochem and Tea Ltd. | PET Resin |
| 12 | Greenways Shipping Agencies Pvt. Ltd. | Containers Freight Station (CFS) |
| 13 | IOC Ltd., Haldia | Petroleum Storage |
| 14 | Hindustan Petroleum Corporation Ltd | Petroleum and allied products |
| 15 | Bharat Petroleum Corporation Ltd., Haldia | Petroleum and allied products |
| 16 | Hindustan Unilever Limited | Detergents |
| 17 | Marcus Oils and Chemical Pvt. Ltd. | Polyehylene Waxes |
| 13 | IOC Ltd., Haldia | Petroleum Storage |
| 18 | Ruchi Soya Industries Ltd. | Edible Oil |
| 19 | Manaksia Ltd. | Aluminum and Steel |
| 20 | Sanjana Cryogenic Storages Ltd. | Ammonia Storage and handling terminal |
| 21 | R. D. B. Rasayans Ltd. | PP Jumbo Bag and Small bag |
| 22 | Reliance Industries Limited | Storage and handling Petroleum Product |
| 23 | Adani Wilmar Ltd. | PEdible Oil Refinery |
| 24 | Electrosteel Castings Ltd. | Coke Oven Plant, sponge iron plant, power plant |
| 25 | URAL India Ltd. | Automobile |

(continued)

**Table 5.18** (continued)

| S. No. | Name of Industry | Product |
|---|---|---|
| 26 | K.S. Oils Ltd. | Edible Oil Refinery |
| 27 | DPM Net Pvt. Ltd. | Fishing net |
| 28 | Hooghly Met Coke and Power Co. Ltd. | Coke Oven Plant |
| 29 | Ruchi Infrastructure Pvt. Ltd. | 3rd Party liquid storage tank terminal |
| 30 | Shree Renuka Sugars Ltd. | Sugar Refinery and Food Complex |
| 31 | Gokul Refoils and Solvent Ltd. | Edible Oil Refinery |
| 32 | Emami Biotech Ltd. | Bio-diesel Plant |
| 33 | Ennore Coke Private Ltd. | Coke Oven Plant |
| 34 | West Bengal Waste Management Ltd. | Industrial waste/municipal waste management complex |
| 35 | Lalbaba Seamless Tubes Pvt. Ltd. | Seamless Tube |
| 36 | Modern India Con-cast Ltd. | Ferro Alloy Plant |
| 37 | Rohit Ferro Tech Ltd. | Ferro Alloy |

**Fig. 5.7** Dense mangrove vegetation in eastern Indian Sundarbans

**Table 5.19** Temporal variation of phytoplankton species diversity (Shannon Weiner Species Diversity Index) at Chemaguri (in Western Indian Sundarbans) and Jhilla (in Eastern Indian Sundarbans)

| Year | Chemaguri (Western Indian Sundarbans) | Jhilla forest (Eastern Indian Sundarbans) |
|---|---|---|
| | 21° 38′ 25.86″ N 88° 08′ 53.55″ E | 22° 09′ 51.53″ N 88° 57′ 57.07″ E |
| 1984 | 2.614 | 3.088 |
| 1985 | 2.587 | 3.087 |
| 1986 | 2.543 | 3.085 |
| 1987 | 2.511 | 3.083 |
| 1988 | 2.497 | 3.080 |
| 1989 | 2.478 | 3.079 |
| 1990 | 2.452 | 3.077 |
| 1991 | 2.311 | 3.075 |
| 1992 | 2.297 | 3.072 |
| 1993 | 2.846 | 3.069 |
| 1994 | 2.777 | 3.064 |
| 1995 | 2.718 | 3.062 |
| 1996 | 2.679 | 3.060 |
| 1997 | 2.637 | 3.059 |
| 1998 | 2.599 | 3.055 |
| 1999 | 2.543 | 3.054 |
| 2000 | 2.507 | 3.051 |
| 2001 | 2.476 | 3.050 |
| 2002 | 2.401 | 3.048 |
| 2003 | 2.347 | 3.047 |
| 2004 | 2.296 | 3.043 |
| 2005 | 2.243 | 3.040 |
| 2006 | 2.176 | 3.039 |
| 2007 | 2.119 | 3.037 |
| 2008 | 2.017 | 3.033 |
| 2009 | 2.915 | 3.135 |
| 2010 | 2.142 | 3.061 |
| 2011 | 1.907 | 3.058 |
| 2012 | 1.843 | 3.051 |
| 2013 | 1.612 | 3.039 |
| 2014 | 1.407 | 3.012 |
| 2015 | 1.245 | 2.986 |
| 2016 | 1.016 | 2.925 |
| 2017 | 0.963 | 2.919 |
| 2018 | 0.875 | 2.917 |

**Table 5.20** Temporal variation of Total Suspended Solid (in mg/l) at Chemaguri (in Western Indian Sundarbans) and Jhilla (in Eastern Indian Sundarbans)

| Year | Chemaguri (Western Indian Sundarbans) | Jhilla forest (Eastern Indian Sundarbans) |
|---|---|---|
| 1984 | 127 | 98 |
| 1985 | 131 | 102 |
| 1986 | 133 | 104 |
| 1987 | 129 | 100 |
| 1988 | 130 | 101 |
| 1989 | 138 | 109 |
| 1990 | 143 | 114 |
| 1991 | 147 | 118 |
| 1992 | 140 | 111 |
| 1993 | 146 | 117 |
| 1994 | 150 | 121 |
| 1995 | 153 | 124 |
| 1996 | 159 | 130 |
| 1997 | 162 | 133 |
| 1998 | 165 | 136 |
| 1999 | 168 | 139 |
| 2000 | 158 | 129 |
| 2001 | 167 | 138 |
| 2002 | 161 | 132 |
| 2003 | 166 | 137 |
| 2004 | 170 | 141 |
| 2005 | 172 | 143 |
| 2006 | 175 | 146 |
| 2007 | 177 | 148 |
| 2008 | 171 | 142 |
| 2009 | 230 | 201 |
| 2010 | 184 | 155 |
| 2011 | 186 | 157 |
| 2012 | 180 | 151 |
| 2013 | 181 | 152 |
| 2014 | 189 | 160 |
| 2015 | 193 | 164 |
| 2016 | 195 | 166 |
| 2017 | 197 | 168 |
| 2018 | 192 | 163 |

**Table 5.21** Temporal variation of nitrate (in μg at/l) at Chemaguri (in Western Indian Sundarbans) and Jhilla (in Eastern Indian Sundarbans)

| Years | Chemaguri (Western Indian Sundarbans) | Jhilla forest (Eastern Indian Sundarbans) |
|---|---|---|
| 1984 | 19.76 | 11.9 |
| 1985 | 21.54 | 13.68 |
| 1986 | 18.85 | 10.99 |
| 1987 | 20.66 | 12.8 |
| 1988 | 22.08 | 14.22 |
| 1989 | 24.71 | 16.85 |
| 1990 | 25.63 | 17.77 |
| 1991 | 26.22 | 18.36 |
| 1992 | 26.89 | 19.03 |
| 1993 | 27.13 | 19.27 |
| 1994 | 24.69 | 16.83 |
| 1995 | 25.44 | 17.58 |
| 1996 | 27.89 | 20.03 |
| 1997 | 29.02 | 21.16 |
| 1998 | 29.76 | 21.9 |
| 1999 | 30.35 | 22.49 |
| 2000 | 30.95 | 23.09 |
| 2001 | 31.08 | 23.22 |
| 2002 | 31.23 | 23.37 |
| 2003 | 31.75 | 23.89 |
| 2004 | 32.68 | 24.82 |
| 2005 | 33.01 | 25.15 |
| 2006 | 33.9 | 26.04 |
| 2007 | 34.15 | 26.29 |
| 2008 | 34.66 | 26.8 |
| 2009 | 40.21 | 32.35 |
| 2010 | 33.98 | 26.12 |
| 2011 | 34.85 | 26.99 |
| 2012 | 35.02 | 27.16 |
| 2013 | 35.11 | 27.25 |
| 2014 | 36.76 | 28.9 |
| 2015 | 37.39 | 29.53 |
| 2016 | 37.56 | 29.7 |
| 2017 | 38.07 | 30.21 |
| 2018 | 38.19 | 30.33 |

**Table 5.22** Temporal variation of phosphate (in μg at/l) at Chemaguri (in Western Indian Sundarbans) and Jhilla (in Eastern Indian Sundarbans)

| Years | Chemaguri (Western Indian Sundarbans) | Jhilla forest (Eastern Indian Sundarbans) |
|---|---|---|
| 1984 | 0.98 | 0.87 |
| 1985 | 1.05 | 0.94 |
| 1986 | 1.16 | 1.05 |
| 1987 | 1.25 | 1.14 |
| 1988 | 1.39 | 1.28 |
| 1989 | 1.88 | 1.77 |
| 1990 | 1.93 | 1.82 |
| 1991 | 2.04 | 1.93 |
| 1992 | 2.11 | 2 |
| 1993 | 2.19 | 2.08 |
| 1994 | 2.37 | 2.26 |
| 1995 | 2.65 | 2.54 |
| 1996 | 2.73 | 2.62 |
| 1997 | 2.81 | 2.7 |
| 1998 | 2.56 | 2.45 |
| 1999 | 2.69 | 2.58 |
| 2000 | 2.93 | 2.82 |
| 2001 | 3.02 | 2.91 |
| 2002 | 3.15 | 3.04 |
| 2003 | 3.44 | 3.33 |
| 2004 | 3.58 | 3.47 |
| 2005 | 3.72 | 3.61 |
| 2006 | 3.86 | 3.75 |
| 2007 | 3.9 | 3.79 |
| 2008 | 3.97 | 3.86 |
| 2009 | 6.05 | 5.94 |
| 2010 | 4.18 | 4.07 |
| 2011 | 3.95 | 3.84 |
| 2012 | 4.05 | 3.94 |
| 2013 | 4.23 | 4.12 |
| 2014 | 4.36 | 4.25 |
| 2015 | 4.49 | 4.38 |
| 2016 | 4.58 | 4.47 |
| 2017 | 4.61 | 4.5 |
| 2018 | 4.75 | 4.64 |

**Table 5.23** Temporal variation of phytoplankton species diversity at Haldia (adjacent to Western Indian Sundarbans) and Bali Island (in Central Indian Sundarbans)

| Year | Haldia (adjacent to Western Indian Sundarbans) | Bali Island (Central Indian Sundarbans) |
|---|---|---|
| | 22° 01′ 18.30″ N<br>88° 03′ 11.40″ E | 22° 04′ 35.17″ N<br>88° 44′ 55.70″ E |
| 1984 | 1.717 | 2.832 |
| 1985 | 1.69 | 2.831 |
| 1986 | 1.646 | 2.829 |
| 1987 | 1.614 | 2.827 |
| 1988 | 1.6 | 2.824 |
| 1989 | 1.581 | 2.823 |
| 1990 | 1.555 | 2.821 |
| 1991 | 1.414 | 2.819 |
| 1992 | 1.400 | 2.816 |
| 1993 | 1.949 | 2.813 |
| 1994 | 1.88 | 2.808 |
| 1995 | 1.821 | 2.806 |
| 1996 | 1.782 | 2.804 |
| 1997 | 1.74 | 2.803 |
| 1998 | 1.702 | 2.799 |
| 1999 | 1.646 | 2.798 |
| 2000 | 1.61 | 2.795 |
| 2001 | 1.579 | 2.794 |
| 2002 | 1.504 | 2.792 |
| 2003 | 1.45 | 2.791 |
| 2004 | 1.399 | 2.787 |
| 2005 | 1.346 | 2.784 |
| 2006 | 1.279 | 2.783 |
| 2007 | 1.222 | 2.781 |
| 2008 | 1.12 | 2.777 |
| 2009 | 2.018 | 2.879 |
| 2010 | 1.245 | 2.805 |
| 2011 | 1.01 | 2.802 |
| 2012 | 0.946 | 2.795 |
| 2013 | 0.715 | 2.783 |
| 2014 | 0.51 | 2.756 |
| 2015 | 0.348 | 2.73 |

(continued)

**Table 5.23** (continued)

| Year | Haldia (adjacent to Western Indian Sundarbans) | Bali Island (Central Indian Sundarbans) |
|---|---|---|
| | 22° 01′ 18.30″ N 88° 03′ 11.40″ E | 22° 04′ 35.17″ N 88° 44′ 55.70″ E |
| 2016 | 0.119 | 2.669 |
| 2017 | 0.066 | 2.663 |
| 2018 | 0.022 | 2.661 |

Indian Sundarbans (21° 52′ 35.7″ N and 88° 11′ 55.0″ E) considering water bodies with varying pH levels. Our experimental design consisted of raising the pH of the three selected ponds by applying lime. Three ponds with aquatic pH 7.5, 7.8 and 8.1 were prepared for the experimental purpose by stocking the same source water having salinity 10.58 psu, nitrate 28.32 μg at $L^{-1}$, $PO_4$ 2.95 μg at $L^{-1}$ and $SiO_3$ 145.75 μg at $L^{-1}$. The entire experiment was carried out during March 2020 to May 2020, the period of complete lockdown due to COVID-19 pandemic. We selected this period intentionally to minimize the external sources of pollution in our experimental ponds. During this three months period there was almost minimum anthropogenic influence due to complete closures of industries, fishing activities, fish landing stations and tourism units, which are common in the present study area and could provide external inputs to our experimental systems. At the end of the experiment, the water samples were collected to enumerate the phytoplankton density and diversity. The water samples were collected within bucket of a known volume, filtered through the bolting silk cloth and the plankton was concentrated. Centrifugation was carried out to concentrate the sample. The final volume of plankton concentrate was recorded to achieve the result of plankton density in terms of cells/litre or cells/$m^3$. In the next step, counting of plankton through Sedgwick Rafter was carried out. 1 ml of plankton sample obtained from the stock through the pipette was transferred to the Sedgwick Rafter counting cell. The sample for counting in this chamber was spread evenly in the form of a thin layer and this was done by placing a cover slip diagonally across the counting cell and the sample was then introduced at one of its corner. The total number of phytoplankton (standing stock) present in a litre of water sample was calculated using the formula:

$$\mathbf{N = nv/V}$$

where,

N= total number plankton cells per litre of water filtered.

n = average number of plankton cells in 1 ml of plankton sample.

v = volume of plankton concentrate (ml)

V= volume of total water filtered (l).

**Table 5.24** Temporal variation of Total Suspended Solid (in mg/l) at Haldia (adjacent to Western Indian Sundarbans) and Bali (in Central Indian Sundarbans)

| Year | Haldia (adjacent to Western Indian Sundarbans) | Bali Island (Central Indian Sundarbans) |
|---|---|---|
| 1984 | 176 | 115 |
| 1985 | 180 | 119 |
| 1986 | 182 | 121 |
| 1987 | 178 | 117 |
| 1988 | 179 | 118 |
| 1989 | 187 | 126 |
| 1990 | 192 | 131 |
| 1991 | 196 | 135 |
| 1992 | 189 | 128 |
| 1993 | 195 | 134 |
| 1994 | 199 | 138 |
| 1995 | 202 | 141 |
| 1996 | 208 | 147 |
| 1997 | 211 | 150 |
| 1998 | 214 | 153 |
| 1999 | 217 | 156 |
| 2000 | 207 | 146 |
| 2001 | 216 | 155 |
| 2002 | 210 | 149 |
| 2003 | 215 | 154 |
| 2004 | 219 | 158 |
| 2005 | 221 | 160 |
| 2006 | 224 | 163 |
| 2007 | 226 | 165 |
| 2008 | 220 | 159 |
| 2009 | 279 | 218 |
| 2010 | 233 | 172 |
| 2011 | 235 | 174 |
| 2012 | 229 | 168 |
| 2013 | 230 | 169 |
| 2014 | 238 | 177 |
| 2015 | 242 | 181 |
| 2016 | 244 | 183 |
| 2017 | 246 | 185 |
| 2018 | 241 | 180 |

**Table 5.25** Temporal variation of nitrate (in μg at/l) at Haldia (adjacent to Western Indian Sundarbans) and Bali (in Central Indian Sundarbans)

| Years | Haldia (adjacent to Western Indian Sundarbans) | Bali Island (Central Indian Sundarbans) |
|---|---|---|
| 1984 | 23.21 | 14.76 |
| 1985 | 24.99 | 16.54 |
| 1986 | 22.3 | 13.85 |
| 1987 | 24.11 | 15.66 |
| 1988 | 25.53 | 17.08 |
| 1989 | 28.16 | 19.71 |
| 1990 | 29.08 | 20.63 |
| 1991 | 29.67 | 21.22 |
| 1992 | 30.34 | 21.89 |
| 1993 | 30.58 | 22.13 |
| 1994 | 28.14 | 19.69 |
| 1995 | 28.89 | 20.44 |
| 1996 | 31.34 | 22.89 |
| 1997 | 32.47 | 24.02 |
| 1998 | 33.21 | 24.76 |
| 1999 | 33.8 | 25.35 |
| 2000 | 34.4 | 25.95 |
| 2001 | 34.53 | 26.08 |
| 2002 | 34.68 | 26.23 |
| 2003 | 35.2 | 26.75 |
| 2004 | 36.13 | 27.68 |
| 2005 | 36.46 | 28.01 |
| 2006 | 37.35 | 28.9 |
| 2007 | 37.6 | 29.15 |
| 2008 | 38.11 | 29.66 |
| 2009 | 43.66 | 35.21 |
| 2010 | 37.43 | 28.98 |
| 2011 | 38.3 | 29.85 |
| 2012 | 38.47 | 30.02 |
| 2013 | 38.56 | 30.11 |
| 2014 | 40.21 | 31.76 |
| 2015 | 40.84 | 32.39 |
| 2016 | 41.01 | 32.56 |
| 2017 | 41.52 | 33.07 |
| 2018 | 41.64 | 33.19 |

**Table 5.26** Temporal variation of phosphate (in μg at/l) at Haldia (adjacent to Western Indian Sundarbans) and Bali Island (in Central Indian Sundarbans)

| Years | Haldia (adjacent to Western Indian Sundarbans) | Bali (Central Indian Sundarbans) |
|---|---|---|
| 1984 | 1.8 | 1.08 |
| 1985 | 1.87 | 1.15 |
| 1986 | 1.98 | 1.26 |
| 1987 | 2.07 | 1.35 |
| 1988 | 2.21 | 1.49 |
| 1989 | 2.7 | 1.98 |
| 1990 | 2.75 | 2.03 |
| 1991 | 2.86 | 2.14 |
| 1992 | 2.93 | 2.21 |
| 1993 | 3.01 | 2.29 |
| 1994 | 3.19 | 2.47 |
| 1995 | 3.47 | 2.75 |
| 1996 | 3.55 | 2.83 |
| 1997 | 3.63 | 2.91 |
| 1998 | 3.38 | 2.66 |
| 1999 | 3.51 | 2.79 |
| 2000 | 3.75 | 3.03 |
| 2001 | 3.84 | 3.12 |
| 2002 | 3.97 | 3.25 |
| 2003 | 4.26 | 3.54 |
| 2004 | 4.4 | 3.68 |
| 2005 | 4.54 | 3.82 |
| 2006 | 4.68 | 3.96 |
| 2007 | 4.72 | 4 |
| 2008 | 4.79 | 4.07 |
| 2009 | 6.87 | 6.15 |
| 2010 | 5 | 4.28 |
| 2011 | 4.77 | 4.05 |
| 2012 | 4.87 | 4.15 |
| 2013 | 5.05 | 4.33 |
| 2014 | 5.18 | 4.46 |
| 2015 | 5.31 | 4.59 |
| 2016 | 5.4 | 4.68 |
| 2017 | 5.43 | 4.71 |
| 2018 | 5.57 | 4.85 |

The units of standing crop are N/l or N $\times$ $10^3/m^3$.

**Shannon Wiener Species Diversity Index (H)** of 1949 was computed by using the formula

$$(\overline{\mathbf{H}}) = -\sum_{i=1}^{S} \frac{n_i}{N} \log_e \frac{n_i}{N}$$

where,

$n_i$ = importance probability for each species.

N = total of importance values.

The results of the experiment are presented in Table 5.27.

The Shannon Weiner Species Diversity Index is a reflection of environmental health. It is observed that the index follows the trend pH 8.1 (3.6346) > pH 7.8 (3.5855) > pH 7.5 (3.4827), which implies that with the decrease of pH, the stress increases more on the phytoplankton community, if other factors like salinity and nutrient levels are kept more or less uniform.

The subsequent level of Chlorophyll *a* in the order 10.55 mg $m^{-3}$ in pond with aquatic pH = 8.1 > 7.25 mg $m^{-3}$ in pond with aquatic pH = 7.8 > 4.60 mg $m^{-3}$ in pond with aquatic pH = 7.5 confirms the adverse impact of acidification on the primary productivity of the estuarine ecosystem. This is also confirmed through fluorescence microscopic images. Maximum fluorescence was observed in pond with aquatic pH = 8.1, followed by pond with aquatic pH = 7.8 and pond with aquatic pH = 7.5 (Fig. 5.8a–c).

Zooplankton constitute an important tier in the marine and estuarine food web. The effect of climate change on the zooplankton community of estuarine system has not been studied by many researchers. A study conducted by Taucher et al. (2017) documented the dominancy of calanoid copepod *Pseudocalanaus acuspes* in simulated condition of ocean acidification. However, majority of studies have confirmed the resilient property of copepod to increasing carbon dioxide level (Weydmann et al. 2012; McConville et al. 2013; Hildebrandt et al. 2014). Several researchers have pointed negative physiological effects of ocean acidification on *Pseudocalanaus acuspes* by increasing respiration and decreasing ingestion and fecundity (Thor and Oliva 2015; Thor and Dupont 2015). Our work on the impact of lowering of aquatic pH on the ichthyoplankton community (a major component of zooplankton) as discussed in Table 5.14 confirms with confidence the adverse role of estuarine acidification on the floating juvenile community of bony fishes. It can also be inferred from the entire discussion that in the present geographical locale the role of anthropogenic activities in triggering the process of acidification cannot be ruled out due to presence of an industrial hub in the lower stretch of the Hooghly estuary preferably at Haldia (refer Table 5.18), which mostly discharge un-treated or partially treated wastes in the Hooghly estuary (Mitra and Zaman 2020).

**Table 5.27** Standing stock of the phytoplankton (N × $10^3/m^3$) in the selected brackish water ponds with variable pH

| S. No. | Species | pH = 8.1 | pH = 7.8 | pH = 7.5 |
|---|---|---|---|---|
| 1 | *Coscinodiscus eccentricus* | 29.12 | 26.92 | 17.77 |
| 2 | *Coscinodiscus jonesianus* | 23.32 | 22.27 | 17.27 |
| 3 | *Coscinodiscus lineatus* | 21.55 | 19.47 | 20.15 |
| 4 | *Coscinodiscus radiates* | 26.09 | 24.47 | 20.92 |
| 5 | *Coscinodiscus gigas* | 17.65 | 12.87 | 15.47 |
| 6 | *Coscinodiscus oculusiridis* | 14.12 | 12.47 | 12.07 |
| 7 | *Planktoniella sol* | 15.72 | 14.92 | 13.02 |
| 8 | *Cyclotella striata* | 18.02 | 15.22 | 13.32 |
| 9 | *Thalassiosira subtilis* | 20.22 | 15.97 | 18.17 |
| 10 | *Ceratium tripos* | 10.19 | 9.47 | 7.62 |
| 11 | *Skeletonema costatum* | 10.82 | 9.07 | 8.12 |
| 12 | *Paralia sulcata* | 12.52 | 11.32 | 11.12 |
| 13 | *Rhizosolenia crassipina* | 10.29 | 9.02 | 8.02 |
| 14 | *Rhizosolenia setigeara* | 8.15 | 8.27 | 9.12 |
| 15 | *Rhizosolenia alata* | 10.89 | 9.77 | 7.22 |
| 16 | *Ceratium teres* | 8.75 | 6.72 | 5.82 |
| 17 | *Ceratium trichoceros* | 10.02 | 9.37 | 6.32 |
| 18 | *Bacteriastrum delicatulum* | 11.59 | 10.22 | 7.82 |
| 19 | *Bacteriastrum varians* | 13.35 | 11.12 | 9.47 |
| 20 | *Bacteriastrum comosum* | 9.99 | 8.67 | 8.07 |
| 21 | *Chaetoceros didymus* | 10.65 | 8.87 | 7.02 |
| 22 | *Chaetoceros peruvianus* | 8.89 | 7.37 | 7.37 |
| 23 | *Chaetoceros compressus* | 8.45 | 7.37 | 7.87 |
| 24 | *Ditylum sol* | 12.39 | 10.42 | 10.02 |
| 25 | *Triceratium favus* | 16.79 | 15.07 | 13.77 |
| 26 | *Triceratium reticulatum* | 14.72 | 11.67 | 11.07 |
| 27 | *Biddulphia sinensis* | 11.12 | 10.02 | 8.92 |
| 28 | *Biddulphia mobiliensis* | 12.89 | 11.32 | 9.82 |
| 29 | *Hemidiscus hardmannianus* | 11.42 | 9.27 | 7.82 |
| 30 | *Climacosphenia elongate* | 6.37 | 6.08 | 6.13 |
| 31 | *Fragillaria oceanic* | 12.15 | 9.97 | 9.77 |
| 32 | *Rhaphoneis amphiceros* | 6.2 | 5.97 | 5.94 |
| 33 | *Thalassionema nitzschioides* | 9.85 | 8.37 | 8.32 |
| 34 | *Thalassiothrix longissima* | 13.79 | 11.47 | 11.47 |
| 35 | *Thalassiothrix fraunfeldii* | 12.75 | 11.47 | 9.37 |

(continued)

**Table 5.27** (continued)

| S. No. | Species | pH = 8.1 | pH = 7.8 | pH = 7.5 |
|---|---|---|---|---|
| 36 | *Asterionella japonica* | 7.89 | 6.92 | 6.62 |
| 37 | *Ceratium extensum* | 9.55 | 7.52 | 6.57 |
| 38 | *Gyrosigma balticum* | 10.92 | 9.37 | 8.52 |
| 39 | *Pleurosigma normanii* | 16.82 | 13.32 | 13.42 |
| 40 | *Pleurosigma elongatum* | 10.55 | 7.87 | 6.77 |
| 41 | *Diploneis smithii* | 13.62 | 11.42 | 10.27 |
| 42 | *Cymbella marina* | 11.49 | 9.87 | 7.97 |
| 43 | *Nitzschia sigma* | 8.72 | 7.97 | 7.72 |
| 44 | *Nitzschia closterium* | 10.29 | 8.52 | 7.62 |
| 45 | *Ceratium furca* | 8.89 | 7.72 | 6.72 |
| 46 | *Trichodesmium erythraea* | 26.45 | 22.87 | 25.47 |
| 47 | *Chlorella marina* | 13.15 | 9.91 | 11.82 |
| | Total (N) | 619.14 | 535.59 | 491.05 |
| | Shannon Weiner Index (H) | 3.6346 | 3.5656 | 3.4827 |

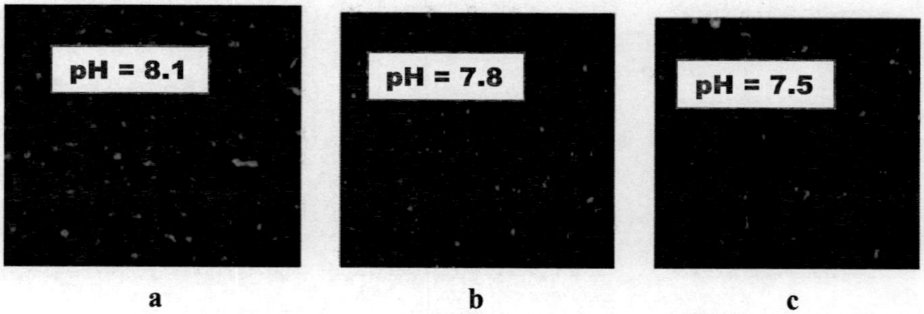


**Fig. 5.8** Fluorescence microscopic view of phytoplankton standing stock in three ponds with varying pH

## 5.4 Take Home Messages

A. Phytoplankton community encompasses unicellular, free floating organisms whose sizes usually range between 1 μm and 1 mm. The cells of these organisms are characterized by the presence of chlorophyll pigment, which is the core pigment in the process of photosynthesis. Phytoplankton are mainly confined in the photic zone of the ocean as they require sunlight, nutrients and carbon dioxide to form glucose that provides these tiny organisms with energy to perform various life processes. Phytoplankton forms the base of the marine and estuarine food web or in other words they support half of the global primary production, which directly or indirectly forms the foundation of all marine life. The Potential Fishing Zone (PFZ), which is the region for maximum fish catch is detected on the basis of pigment characterization that is assessed through Remote Sensing Technique. The estuaries of Indian Sundarbans are noted for their unique spatio-temporal variations of salinity. This variation has resulted to a wide range of adaptability to plankton community. The list of phytoplankton species has increased more than 150 in recent times in Indian Sundarban estuaries.

B. The zooplankton occupy the tier next to phytoplankton in the marine and estuarine food web. They either graze on phytoplankton or feed on other members of zooplankton. Foraminfera and radiolarians are members of zooplankton community who possess only one cell. Copepods and euphausiids are widely distributed zooplankton in the marine and estuarine environments. Zooplankton exhibit the phenomenon of vertical migration in which some species migrate towards the sea surface during the night time and return back to their original depth during day. The amount of daily migration varies between 10 and 500 m and this change of zone is keenly related to their nutritional requirement or maintenance of their light level. The zooplankton are mainly filter feeding in nature and are the prey of many shore-dwelling species. In aquatic sub-system of deltaic Sundarbans, the zooplankton community comprises a heterogenous assemblage of animals covering many taxonomic groups including copepods, mycids, lucifers, gammarid, cladocerae, ostracods, cumacea, hydromedusae, ctenophores and chaetognaths among holoplankters and polychaete larvae, molluscan and echinoderm larvae, crustacean larvae and fish eggs and larvae among meroplankters. Among the zooplankton, calaniod copepods constitute the major bulk, followed by cyclopoids, copepods and harpecticoids.

C. Acidification is the net result of climate change induced carbon dioxide rise in the atmosphere. With the rise of carbon dioxide, surface waters have become more susceptible to acidification. Anthropogenic carbon dioxide from the burning of fossil fuels has increased the uptake of the green house gas, causing the pH to drop from 8.2 to 8.1. Ocean acidification has high probability to pose negative impact on the marine life. It is well known that coral bleaching, reduced calcification and thinning of oyster shells are the direct effects of acidification. However, the species abundance and diversity of plankton are also adversely impacted by the pH level of the aquatic ecosystem as evidenced by our indoor experiments and field data.

## Annexure 5: Program to Evaluate Shannon-Weiner Species Diversity Index

deci Shannon.c

```c

#include <stdio.h>
#define MAX_SIZE 1000 // Maximum array size
#include<math.h>
int main()
{
    float arr[MAX_SIZE]; // Declare an array of MAX_SIZE
    int  i,N,sum=0,j,temp=-1;
float l, k,s=0,s1=0;
    /* Input array size */
    printf("Enter total number of species: ");
    scanf("%d", &N);

    /* Input elements in array */

    for(i=0; i<N; i++)
    {
       printf("Enter No. of individuals of species: ", N,i+1);
        scanf("%f", &arr[i]);
    }


    /*
     * Print all elements of array
     */

   for(i=0;i<N;i++)
   {
    sum=sum+arr[i];
   }
```

deci Shannon.c

```c
    l=sum;
     printf("\n Percentage of species are: ");
     for(i=0; i<N; i++)
     {
         printf("\n  No. %d  percentage is %f ", i+1,arr[i]*100/l);
     }



   for(i=0;i<N;i++)
   {

     j=arr[i];
    k=j/l;

   s=s+k*log(k)*temp;
  s1=s1+pow(k,2);

           }









      if(s>0 && s<=0.5)
{
    printf("\n SHANNON INDEX IS %lf WHICH IS WORST",s);
}

```

```
else if(s>0.5 && s<=1.5)
{


    printf("\nSHANNON INDEX IS %lf WHICH IS MODERATE",s);
}
else if(s>1.5 && s<=2.5)
{


    printf("\nSHANNON INDEX IS %lf WHICH IS GOOD",s);
}

else if(s>2.5 && s<=3.5 )
{


printf("\nSHANNON INDEX IS %lf WHICH IS BETTER",s);

}
else
{
    printf("\nSHANNON WEINER INDEX IS %lf WHICH IS BEST",s);
}
printf("............................................. \n");
printf("\n EVENNESS INDEX is %f",s/log(N));
printf("\n INDEX OF DOMINANCE=%f",s1);
```

## References


Banerjee K, Sengupta K, Raha A, Mitra A (2013) Salinity based allometric equations for biomass estimation of Sundarban mangroves. Biomass Bioenerg 56:382–391

Chakraborty S, Zaman S, Pramanick P, Raha AK, Mukhopadhyay N, Chakravartty D, Mitra A (2013) Acidification of Sundarbans mangrove estuarine system. Discovery Nature 6(14):14–20. ISSN: 2319-5703

Chaudhuri AB, Choudhury A (1994) In: Mangroves of the Sundarbans, vol I: India. IUCN- The World Conservation Union

Chu Jennifer, MIT News Office (2015) Ocean acidification may cause dramatic changes to phytoplankton. Retrieved 07 Nov 2016, from http://news.mit.edu/2015/ocean-acidification-phytoplankton-0720

Fabry VJ, Seibel BA, Feely RA, Orr JC (2008) Impacts of ocean acidification on marine fauna and ecosystem processes. ICES J of Mar Sci 65:414–432

Hildebrandt N, Niehoff B, Sartoris FJ (2014) Long-term effects of elevated $CO_2$ and temperature on the Arctic calanoid copepods *Calanus glacialis* and *C. hyperboreus.* Mar Pollu Bull 80(1±2):59±70. https://doi.org/10.1016/j.marpolbul.2014.01.050 PMID: 24529340

McConville K, Halsband C, Fileman ES, Somerfield PJ, Findlay HS, Spicer JI (2013) Effects of elevated $CO_2$ on the reproduction of two calanoid copepods. Mar Pollut Bull 73(2):428±34. https://doi.org/10.1016/j.marpolbul.2013.02.010 PMID: 23490345

Mitra A (2000) The Northeast coast of the Bay of Bengal and deltaic Sundarbans. In: Sheppard C (ed) Seas at the millennium – an environmental evaluation (Chap 62). Elsevier Science, London, pp 143–157

Mitra A (2013) In: Sensitivity of mangrove ecosystem to changing climate by Dr. Abhijit Mitra. Springer, New Delhi, Heidelberg, New York, Dordrecht, London, edition (August 31, 2013); ISBN-10: 8132215087; ISBN-13: 978-8132215080, copyright Springer, India 2013; ISBN 978-81-322-1509-7 (eBook)

Mitra A, Gangopadhyay A, Dube A, Andre CKS, Banerjee K (2009) Observed changes in water mass properties in the Indian Sundarbans (Northwestern Bay of Bengal) during 1980–2007. Curr Sci 97(10):1445–1452

Mitra A, Sengupta K, Banerjee K (2011) Standing biomass and carbon storage of above ground structures in dominant mangrove trees in the Sundarbans. Ecol Manage 261:1325–1335

Mitra A, Zaman S (2014) Carbon sequestration by coastal floral community. The Energy and Resources Institute (TERI) TERI Press, India, ISBN 978-81-7993-551-4

Mitra A, Zaman S (2015) Blue carbon reservoir of the blue planet. Springer, Berlin. ISBN 978-81-322-2106-7

Mitra A, Zaman S (2016) Basics of marine and estuarine ecology. Springer, Berlin, pp 1–481. ISBN 978-81-322-2705-2; https://doi.org/10.1007/978-81-322-2707-6

Mitra A, Zaman S (2020) Environmental science—a ground zero observation on the Indian Sub-continent. Springer Nature Switzerland AG, ISBN 978-3-030-49130-7, ISBN 978-3-030-49131-4 (eBook https://doi.org/10.1007/978-3-030-49131-4)

Redfield AC (1934). In: Danial RJ (ed) James Johnstone Memorial Volume. University of Liverpool Press, Liverpool, 176 p

Redfield AC (1958) The biological control of chemical factors in the environment. Am Sci 46:205–207

Sengupta K, Roy Chowdhury M, Bhattacharya SB, Raha A, Zaman S, Mitra A (2013) Spatial variation of stored carbon in *Avicennia alba* of Indian Sundarbans. Discovery Nat 3(8):19–24

Shannon CE, Wiener W (1949) The mathematical theory of communication. University of Illinois Press, Urbana, p 177

Shi D, Xu Y, Hopkinson BM, Morel FMM (2010) Effect of ocean acidification on iron availability to marine phytoplankton. Sci 327:676–679

Srinivasan A, Santhanam R (1998) Seasonal distribution and density of copepods in the sewage polluted coastal waters of Tuticorin, India. The biodiversity crisis and crustacea: Proceeding of the fourth international crustacean congress, Amsterdam, The Netherlands, pp 463–469

Stauber JL, Florence TM (1987) Mechanism of toxicity of ionic copper and copper complexes to algae. Mar Biol 94:511–519

Taucher J, Haunost M, Boxhammer T, Bach LT, AlgueroÂ-Muñiz M, Riebesell U (2017) Influence of ocean acidification on plankton community structure during a winter-to-summer succession: An imaging approach indicates that copepods can benefit from elevated $CO_2$ *via* indirect food web effects. PLoS ONE 12(2):e0169737. https://doi.org/10.1371/journal.pone.0169737

Thor P, Oliva EO (2015) Ocean acidification elicits different energetic responses in an Arctic and a boreal population of the copepod *Pseudocalanus acuspes*. Mar Biol 162(4):799–807

Thor P, Dupont S (2015) Transgenerational effects alleviate severe fecundity loss during ocean acidification in a ubiquitous planktonic copepod. Glob Change Biol 21(6):2261–2271. https://doi.org/10.1111/gcb.12815 PMID: 25430823

Weydmann A, Soreide JE, Kwasniewski S, Widdicombe S (2012) Influence of $CO_2$-induced acidification on the reproduction of a key Arctic copepod *Calanus glacialis*. J Exp Mar Biol Ecol 428:39–42

## *Internet References*

http://www.bigelow.org/sci_overview
http://www.nioo.know.hl/cemo/ecoflat/work.htm

# Chapter 6
# Impact of Acidification on Ecosystem Services of Sundarban Estuaries

**Abstract** Estuaries, being the zone of interaction between the sea and the river offer a wide spectrum of ecosystem services under the categories of provisioning services (fishes, minerals, oil and natural gas, products from mangroves, marsh grass etc.), regulating services (water quality maintenance, flood control, bioremediation, carbon sink etc.), supporting services (biodiversity, life cycle stages of fish and several aquatic organisms) and cultural services (religious convergence and rituals). Acidification poses an adverse impact on these ecosystem services primarily by disturbing the estuarine food webs where several categories of finfish and shellfish species are intricately linked. Lowering of pH also triggers the transference of the precipitated heavy metals from the underlying sediment bed to the overlying water column and thus the regulating services are affected. This chapter throws light on various disturbances caused by acidification on several ecosystem services provided by estuaries to mankind considering the mangrove dominated Indian Sundarbans as the case study zone.

## 6.1 Concept of Ecosystem Services: Focus Sundarban Mangroves

Ecosystem comprising of biotic and abiotic components provide a wide range of services to mankind. In addition to meet the daily needs of human beings, the floral community of ecosystem is the source of several life saving drugs. Disasters like super cyclones, landslides, tsunamis etc. are often checked by ecosystem and plays great role in reducing their intensities. However, the services are mostly unmarked and therefore the concept is not known to many. For this reason, we have carried out a respondent analysis in Sect. 6.2 of this chapter to get the people's feedback involving individuals across different tiers of the society. Today the concept of ecosystem services has become an important model for linking the functioning of ecosystems to societal benefits. With the progress of civilization, human beings have attempted to depend on nature for the supply of items like food, timber, fuel wood, fiber etc. (Foley et al. 2005; Kareiva et al. 2007; Monfreda et al. 2008; Ramankutty et al. 2008). In coastal areas and islands, people are depending on mangroves for procuring

 
A. Mitra and S. Zaman, *Estuarine Acidification*,
https://doi.org/10.1007/978-3-030-84792-0_6

fishes, honey, fodder, timber, wax etc. However these attempts have been masked as nature simultaneously generate multiple ecosystem services that interrelate in complex dynamic ways (Peterson et al. 2003; Chan et al. 2006; Rodriguez et al. 2006; Brauman et al. 2007). It has been found that the scale of different ecosystem services is highly variable in nature, e.g., few ecosystem services like food and timber have great values compared to services like pollination or bioremediation.

Ecosystem services have been defined from various points of views. According to Millennium Ecosystem Assessment (2003, 2005) the benefits that people obtain from ecosystem including provisioning services such as food, timber, fresh water fiber; cultural services that provide non-material benefits, such as places for recreation, meditation and inspiration; and regulating services such as flood control and climate regulation together constitute the ecosystem services. This definition of ecosystem services encompasses gifts of nature like aesthetic beauty, carbon storage or oxygen generation etc.

Several research papers have explored the spatial patterns of provision of multiple services across landscapes by mangroves. This unique floral community acts as agent of shoreline protection in some areas, where as the same species plays significant role in bioremediation in some other zone (preferably in the coastal industrial belts).

In the domain of ecosystem services 'driver' is a very important term, which is basically a factor often directly modified by human management and affects one or more ecosystem services (Table 6.1). It has been observed that certain service of ecosystem affects the other through synergistic or antagonistic mechanism, e.g., afforestation enhance the process of carbon sequestration but at the same time the evapotranspiration increases the growth of the trees, which ultimately decreases the water availability. Thus it is an example of antagonistic mechanism. Again by maintaining forest patches close to agricultural land generates two services simultaneously namely pollination and crop production. This is an example of synergistic mechanism. It is therefore important to analyze the drivers and interactions among the multiple services and find a road map to create synergies and avoid antagonistic process by manipulating ecosystem service drivers.

Lot of works have been done on the valuation methods, data and classification systems for ecosystem services for terrestrial ecosystems compared to the mangrove ecosystems in the coastal and estuarine regions of the world. Peer review literature on global economic valuations of coastal forest like mangrove ecosystems, sea grass ecosystems is rather limited. Mangrove ecosystems are widely distributed at the land-sea interface and it has been projected that the number of people living within 10 km of significant mangrove areas might rise to 120 million (Van Bochove et al. 2014).

There are certain confusion to pinpoint the exact ecosystem services offered by mangroves because the ecosystem is often overlapped or combined with tidal marshes, sea grasses community, coral reefs etc. (Figs. 6.1 and 6.2).

Table 6.2 lists the ecosystem services categories identified by the mangrove Delphi, grouped according to the Common International Classification of Ecosystem Services (CICES).

**Table 6.1** Examples of ecosystem service relationships

| Sector | Driver | Service A | Service B | Shared driver | Response type | Interaction type | Synergy or trade-off | References |
|---|---|---|---|---|---|---|---|---|
| 1. | Trail building | Cultural tourism | Agricultural production | No | – | None | None | Brscic (2006) |
| 2. | Fertilizer use | Crop production | Water quality | Yes | Opposite | None | Trade-off | Carpenter et al. (1998) |
| 3. | Wolf re-introduction (For Indian Sundarban mangroves, the present author considers it as deer re-introduction) | Nature tourism | Flood plain maintenance | Yes | Similar | None | Synergy | Wolf et al. (2007) |
| 4. | Restoring riparian vegetation | Flood control | Crop production | No | – | Uni-directional, positive | Synergy | Kramer et al. (1997) |
| 5. | Maintaining forest patches close to coffee plantations (For Indian Sundarban mangroves, the present author considers salt tolerant paddy plantations) | Pollination | Crop production | No | – | Uni-directional, positive | Synergy | Ricketts et al. (2008) |

(continued)

**Table 6.1** (continued)

| Sector | Driver | Service A | Service B | Shared driver | Response type | Interaction type | Synergy or trade-off | References |
|---|---|---|---|---|---|---|---|---|
| 6. | Wetland restoration | Flood control | Water quality | Yes | Similar | Uni-directional, positive | Synergy | Zedler (2003) |
| 7. | Afforestation | Carbon sequestration | Water quantity | Yes | Opposite | Uni-directional, negative | Trade-off | Engel et al. (2005) |
| 8. | Marine protected area development | Regulation of algae growth | Tourism | Yes | Similar | Uni-directional, positive | Synergy | Hughes et al. (2005) |
| 9. | Dry spells | C sequestration/soil organic matter | Crop yield | No | – | Bi-directional, positive | Synergy | Enfors et al. (2008) |
| 10. | Pesticide spraying | Wood production | Pest control | No | – | Bi-directional, negative | Trade-off | Clark et al. (1979) |
| 11. | Cloud forest land clearing | Moisture retention | Carbon sequestration and tree growth | Yes | Similar | Bi-directional, positive | Synergy | del-Val et al. (2006) |

**Fig. 6.1** Overlapping of mangrove associate species (*Suaeda* sp.) with true mangrove flora

The total spectrum of ecosystem services offered by mangroves may be broadly divided into four major categories as presented in Fig. 6.3.

The valuation method for mangrove ecosystem services has not been standardized fully. Economist generally divide the total economic value of mangrove into three major groups namely (i) Direct use value (ii) Indirect use value and (iii) Non-use value. Direct use values refer to consumptive and non-consumptive uses that encompass physical interaction with the mangroves and their services as for example revenues generated from fishery, timber, fuel wood, fodder, honey, wax, recreation and transport. Indirect use values include regulatory ecological functions like flood control, storm protection, nutrient retention and erosion control. Non-use values include existence and bequest values of mangroves. Table 6.3 summarizes these three types of services as well as the methods most commonly used in their valuation.

Over the past few decades, research articles on mangrove ecosystem primarily focused on four major services namely (1) nursery services of mangrove forest, (2) fuel, fodder and timber production as sources of livelihood for adjacent coastal population and island dwellers, (3) Shoreline stabilization and land-building capacity and (4) carbon storage and subsequent sequestration. In recent time considerable progress has been made in the field of bioprospecting of mangrove derived microbes. Mangrove actinomycetes are the promising source for secondary metabolites.

**Fig. 6.2** Overlapping of mangrove trees with saltmarsh grass (*Porteresia coarctata*)

The various forms of ecosystem services provided by mangroves are highlighted here in points.

1. Mangrove vegetations and their associates are economically very important for their products like timber, fire-wood, honey, wax, alcohol, tannins (Table 6.4) and even extracts having surprising medicinal properties.
2. Mangroves are thought to possess the ability to control coastal water quality. The complexity of the mangrove forest habitat increases the residence time, which assists in the assimilation of inorganic nutrients and traps suspended particulate matter. The mangroves also function as flood control barrier and binder of sediment particles (http://www.fao.org/gpa/sediments/habitat.htm).
3. The ecosystem forms an ideal ecological asset because the production of leaf litter and detrital matter from mangrove plants fulfill the nutritional requirements of prawn juveniles, adult shrimps, molluscs and fishes of high economic value. Because of this, mangrove ecosystem is recognized as the world's most potential nursery. There are three main factors to justify the nursery role of the mangrove habitat. They are:

   - High levels of water turbidity, which increases the survival rate of larvae through reduced perception distance of predators.

**Table 6.2** Ecosystem service categories identified by the mangrove Delphi, grouped according to the Common International Classification of Ecosystem Services (CICES) v4.3 (http://cices.eu/)

| Section | Division | Group | Class | Class type | Delphi technique categories |
|---|---|---|---|---|---|
| Provisioning | Nutrition | Biomass | Wild animals and their outputs | Animals by amount, type | Fisheries (food) |
| | | | | | Honey |
| | | | Animals from in-situ aquaculture | Animals by amount, type | Fisheries (aquaculture) |
| | Materials | Biomass | Fibres and other materials from plants, algae and animals for direct use or processing | Material by amount, type, use, media (land, soil, freshwater, marine) | Wood and timber |
| | | | Materials from plants, algae and animals for agricultural use | | Fodder |
| | Energy | Biomass-based energy sources | Plant-based resources | By amount, type, source | Energy resources |
| Regulation and maintenance | Mediation of waste, toxics and other nuisances | Mediation by biota | Bio-remediation by micro-organisms, algae, plants, and animals | By amount, type, use, media (land, soil, freshwater, marine) | Water bioremediation |
| | | Mediation by ecosystems | Filtration/sequestration/ storage/ accumulation by ecosystems | By amount, type, use, media (land, soil, freshwater, marine) | Pollution abatement, Environmental risk indicator |
| | Mediation of flows | Mass flows | Mass stabilization and control of erosion rates | By reduction in risk, area protected | Protection from sedimentation |
| | | | Buffering and attenuation of mass flows | | Protection from salt intrusion |

(continued)

**Table 6.2** (continued)

| Section | Division | Group | Class | Class type | Delphi technique categories |
|---|---|---|---|---|---|
| | | Gaseous/air flows | Storm protection | By reduction in risk, area protected | Coastal protection |
| | Maintenance of physical, chemical, biological conditions | Lifecycle maintenance, habitat and gene pool protection | Maintaining nursery populations and habitats | By amount and source | Fisheries (nursery) |
| | | Soil formation and composition | Decomposition and fixing processes | | Carbon sequestration |
| | | Atmospheric composition and climate regulation | Global climate regulation by reduction of greenhouse gas concentrations | By amount, concentration or climatic parameter | Carbon sequestration |
| Cultural | Physical and intellectual interactions with biota, ecosystems, and land-/seascapes (environmental settings) | Intellectual and representative interactions | Entertainment | | Ecotourism and recreation |
| | | | Aesthetic | | Aesthetic value |

The first five columns belong to the CICES framework and the results of the Delphi technique are included in the last column. https://doi.org/10.1371/journal.pone.0107706.t001

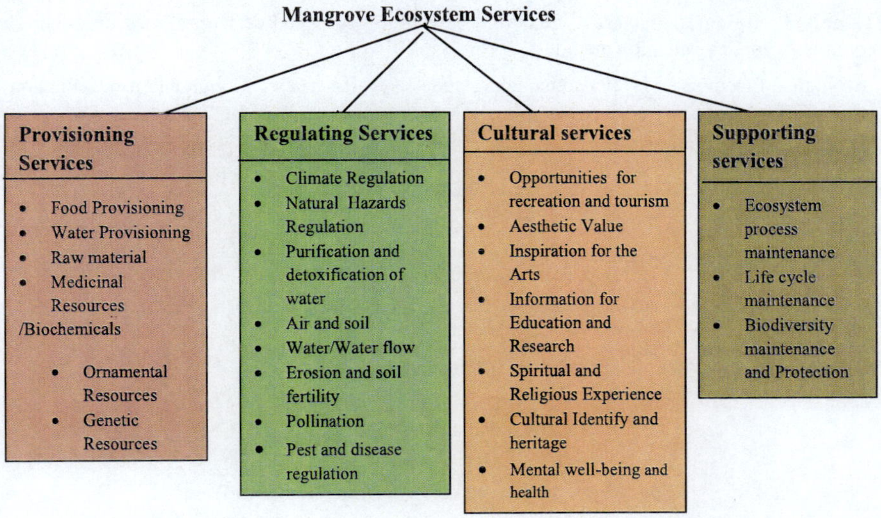


**Fig. 6.3** Ecosystem services of mangroves

- Tidal mixing, nutrient trapping and fresh water inflow that result in considerable primary productivity and provide base of a food web from zooplankton to post larval fishes and juveniles.
- The physical and structural complexity of the habitat itself provides a variety of niches favourable to survival of juveniles. Niche segregation often leads to minimization of inter-and intra-specific competition resulting in a high survival rate of finfish and shellfish juveniles.

4. The vibrating mangrove ecosystem provides nutritional inputs to adjacent shallow channels and bay system that constitute the primary habitat of a large number of aquatic species, algae of commercial importance, seaweeds, phytoplankton etc. Mangroves and mangrove habitats contribute significantly to the global carbon cycle. Twilley et al. (1992) estimated the total global mangrove biomass to be approximately 8.70 gigatons dry weight (which is equivalent to 4.00 gigatons of carbon). Accurate biomass estimation however needs the measurement of the volumes of individual trees.
5. Mangroves trap debris and silt, leading to the stabilization of the near shore environment. (http://www.reefrelief.org/Documents/mangrove.html).
6. The highly specialized mangrove ecosystem also acts as the protector of the coastal landmass from storm surges, tropical cyclone, high winds, tidal bores, seawater seepage and intrusion.
7. Bioaccumulation of heavy metals by certain mangrove species reveals a most surprising feature about these plants as they can act as bio-purifier or bio-filter. Certain mangrove species are highly efficient in detecting or assessing the change of ambient environment. The concentration of heavy metal pollutants

**Table 6.3** Ecological mangrove functions, economic goods and services, types of value, and commonly applied valuation methods

| Ecological function | Economic goods and services | Value type | Commonly used valuation method(s)* |
|---|---|---|---|
| Flood and flow control | Flood protection | Indirect use | RCM<br>MP |
| Storm buffering/ sediment | Storm protection | Indirect use | RCM<br>PF |
| Water quality maintenance/nutrient retention | Improved water quality | Indirect use | CVM |
| | Waste disposal | Direct use | RCM |
| Habitat and nursery for plant and animal species | Commercial fishing and hunting | Direct use | MP<br>NFI |
| | Recreational fishing and hunting | Direct use | TCM<br>CVM |
| | Harvesting of natural materials | Direct use | MP<br>NFI |
| | Energy resources | Direct use | MP<br>NFI |
| Biodiversity | Appreciation of species existence | Non-use | CVM |
| Carbon sequestration | Reduced global warming | Indirect use | RCM |
| Natural environment | Recreation, tourism | Direct use | CVM<br>TCM |
| | Existence, bequest, option values | Non-use | CVM |

*Source* Brander et al. (2006) who adapted with modifications from Barbier (1991), Barbier et al. (1997), Brouwer (2000) and Woodward and Wui (2001)

*Abbreviations represent: market prices (MP), production function method (PFM), travel cost method (TCM), contingent valuation method (CVM), replacement cost method (RCM), and net factor income (NFI)

in different parts of mangrove plants may act as a pathfinder for water quality monitoring programme (Mitra et al. 2004).

8. Mangroves filter ground water and storm water run-off that often contain harmful pesticides, herbicides and hydrocarbons. Mangroves also recharge underground water table by collecting rain water and slowly releasing it into the underground reservoir.
9. Mangrove prop roots protect and offer habitat for mammals, amphibians, reptiles, countless unique plants, juvenile fish and invertebrates that filter water such as sponges, barnacles, oysters, mussels, crabs, shrimps etc.
10. Mangroves are ideal nesting grounds for many water birds such as the great white heron, reddish egrets, roseate spoonbills, white-crowned pigeons, cuckoos and frigate birds. The excretory materials of these birds serve as the

**Table 6.4** Some traditional uses of true mangrove species

| Mangrove species | Uses |
|---|---|
| *Aegiceras corniculatum* | Bark used as fish poison. Also contains tannin |
| *Avicennia alba* | Used as fodder and fuel |
| *Bruguiera caryophylloides* | Wood used for firewood and timber. Bark used for tannin production |
| *Bruguiera gymnorrhiza* | Wood used for house posts. Bark used as tannins. Also provides fuel |
| *Bruguiera parviflora* | Source of fuel. Leaves and barks contain tannin |
| *Bruguiera sexangula* | Timber used in house building |
| *Ceriops tagal* | Used for keel of boats and house posts. Provides good fuel charcoal. Bark is rich in tannin, used for dyeing fishing nets |
| *Rhizophora mucronata* | Bark is used for tannin and cattle fodder |
| *Sonneratia apetala* | Wood used in house building, packing cases and yield excellent fuel |
| *Xylocarpus granatum* | Yields gum, resin which are used in local medicines. Bark contains tannin |

*Source* Mangroves of the Sundarbans, Vol. 1: India, by A. B. Chaudhuri and A. Choudhury, IUCN publication, 1994

fertilizers of the adjacent water bodies on which the primary production of the aquatic phase depends.

11. Mangrove forests are the housing complexes for bees, birds, mammals and reptiles from which honey, wax, food etc. are obtained.
12. The molluscan species in the mangrove ecosystem (under gastropoda and bivalvia) are the rich sources of lime (Fig. 6.4).
13. Mangrove leaves are used as fodder and green manure. The cyanobacterial strains present on the forest floor of mangrove ecosystem are important sources of biofertilizer.
14. Extracts from mangrove and mangrove dependent species have proven activity against some animal and plant pathogens. Moreover, mangrove extracts kill larvae of the mosquitoes *e.g.*, a pyrethrin like compound in stilt roots of *Rhizophora apiculata* shows strong mosquito larvicidal activity (Thangam 1990).
15. Bioactive compound (ecteinascidin) extracted from the mangrove ascidian *Ecteinascidia turbinata* have shown strong in vivo activity against a variety of cancer cells.
16. Phenols and flavonoids in mangroves leaves serve as UV-screening compounds. Hence, mangroves can tolerate solar UV radiation and create a UV-free, under-canopy environment (Moorthy 1995).
17. Bark of *Ceriops* sp. is an excellent source of tannin and a decoction of it is used to stop haemorrhage and as an application to malignant ulcers. Flowers of this plant are a rich source of honey and bee wax.

**Fig. 6.4** Oysters in the mangrove ecosystem are rich sources of high quality lime

18. Mangrove ecosystem affords recreation to hunters, fishermen, bird-watchers, photographers and others who treasure natural areas. However, the intrusive actions of noisy jet-skis, happy campers and others, which disturb nesting and breeding areas, chop down mangroves and otherwise damage this fragile environment, threaten its existence.
19. Mangroves are potential vegetation from the point of carbon storage and sequestration. Despite the large number of case studies dealing with various aspects of organic matter cycling in mangrove systems (Kristensen 2008), there is very limited consensus on the carbon sequestering potential of mangroves. Not only the vegetative and reproductive parts of the mangrove tree species, but the soil below on which the mangroves thrive is also a unique carbon reservoir. The term blue carbon encompasses this carbon pool in the biotic and abiotic components in the mangrove forests (Fig. 6.5).
20. Oxygen production is one of the most common but under-researched benefits of mangroves. Most of the mangrove forests are usually oxygen-neutral or oxygen-negative system as they produce equal or less oxygen compared to what they consume. The dead mangrove trees and the mangrove litter along with detritus consume oxygen to carry out the process of decomposition,

**Fig. 6.5** Mangrove forests with adjacent mudflats—a potential sink of carbon

which pushes the mangrove system toward oxygen-negative. In order to make the system oxygen-positive, it is essential to retard the process of decomposition. An in-depth study conducted on the oxygen production potential of mangroves shows that the dominant mangrove species in Indian Sundarbans can release on an averge about 7.7 tonnes of oxygen per hecatre in 1 year, which amounts to INR 841,533 for filling 1295 cylinders of 6 L capacity. The net oxygen release (tonnes/ha/yr) of mangroves is a function of carbon sequestration (10 years considered in this study) and Table 6.5 presents the net oxygen release (tonnes/ha/yr) of five dominant species namely *Sonneratia*

**Table 6.5** Sector-wise net $O_2$ release (in tonnes $ha^{-1}$ $y^{-1}$) by dominant mangrove floral species in Indian Sundarbans

| Sector | Carbon sequestration (tonnes $ha^{-1}$ $y^{-1}$) | Net $O_2$ release (tonnes $ha^{-1}$ $y^{-1}$) | No. of $O_2$ cylinders needed to fill up by the $O_2$ gas (capacity 6 lt.) | Total price of cylinder (INR) |
|---|---|---|---|---|
| Central | 3.014 | 8.047 | 1341 | 871,650 |
| Western | 2.800 | 7.476 | 1246 | 809,900 |
| Eastern | 2.916 | 7.785 | 1297 | 843,050 |

*apetala*, *Avicennia marina*, *Avicennia alba*, *Avicennia officinalis*, and *Exocecaria agallocha*. The values are average of five species (for each of the three sectors in Indian Sundarbans.

21. Mangrove based fungi degrade the vegetation and play major roles in ecological and biogeochemical processes. They contribute significantly to the degradation process of mangrove–derived organic matter and thereby generate food/nutrients for benthic fauna. The major phyla under fungi are Basidiomycota, Ascomycota, Glomeromycota, Microsporidia, Blastocladiomycota, Neocallimastigomycota and Chytridiomycota.
22. Flora and fauna of the estuaries are today used as sources of bioactive substances, which have immense importance in the pharmaceutical industries. Examples of CAL (Carcinoscorpious Amoebocyte Lysate) and TAL (Tachypleus Amoebocyte Lysate) are very relevant in this context. They are derived from the blue blood of horseshoe crabs (*Carcinoscorpious rotundifolia* and *Tachypleus gigas*) found abundantly in the estuaries of West Bengal and Odisha, which are two maritime states in the North-East coast of the Indian sub-continent. CAL and TAL are used for bacterial endotoxin test.
23. Several mangrove medicinal plants widely distributed in the coastal and estuarine regions were tested for antiviral, antifungal and mosquito larvicidal activities. In the Parangipettai coast of India (South east coast), 113 medicinal plants have been recorded. Out of 36 plants, 3 plants showed high anti-HIV activity. The anti-HIV activity of coastal population was also tested in comparison with their terrestrial counterparts. In general, coastal populations exhibited better anti-HIV activity than terrestrial ones. 23 plants were screened for antibacterial activity against 13 human pathogenic bacterial strains, of which 23 extracts showed anti-bacterial activity against one or more bacteria. 25 plants were screened for antifungal activity against 6 fungi. All the plant extracts examined showed activity against one or more fungi. 2 plant species exhibited high antifungal activity. In general, the coastal samples showed higher activity than terrestrial ones. 11 plants were tested for mosquito larvicidal activity against two mosquito species namely, *Culex quinquefasciatus* and *Aedes aegypti*, of which, 5 samples were effective. In addition to these services, the mangrove flora thriving in the coastal and estuarine regions are rich sources of antioxidants.
24. In recent time considerable progress has been made in the field of bioprospecting of mangrove derived microbes. Mangrove actinomycetes are the promising source for secondary metabolites. Novel indolocarbazoles identified from a mangrove streptomycete present unprecedented cyclic N-glycosidic linkages between 1,3-carbon atoms of the glycosyl moiety and two indole nitrogen atoms of the indolocarbazole core with anticancer activity.
25. The vegetative structures of mangroves offer considerable resistance to the magnitude of natural disasters. Features of mangrove vegetation like presence of pneumatophores or complex prop root systems, the density of the forest and the diameter of the tree stems regulate the magnitude of damage caused by storms and cyclones. The spatial dimensions of the forest, in particularly

**Fig. 6.6** Pneumatophores of mangroves act as structures against wave attenuation

how far it reaches inland from the shore, are considered some of the most important determinants of protective capabilities with regards to tsunamis and cyclones. Mangroves are also capable to reduce wind related damage from large cyclone, storms by directly attenuating the energy of wind that passes through there dense canopies. The dense pneumatophores of mangroves serve as natural agents of wave attenuation (Fig. 6.6).

India with coast line of approximately 7517 km including the Island territories has a mangrove cover of 6749 km$^2$.

The Indian mangrove habitats may be classified into three distinct categories (Table 6.6).

Considering the geomorphologic characters the mangrove habitat in India can be sub-divided into three categories (Table 6.7).

In Indian subcontinent a total of 82 very common true mangrove and associate species, distributed in 52 genera and 36 families have been recorded (Table 6.8).

In Indian sub-continent, mangrove biodiversity is extremely rich in regions like Indian Sundarbans and Bhitarkanika. In these two maritime states in the northeast part of the country, some 36 species of true mangroves have been documented.

The Indian Sundarbans appears as a continuous green carpet in aerial surveys, consisting of dense low trees ranging from 3 to 5 m in height comprising of *Excoecaria agallocha* and *Avicennia* spp. *Rhizophora apiculata* and *R. mucronata* are found in the creeks, often in association with *Nypa fruticans*. Three common

**Table 6.6** Classification of Indian mangroves habitats

| Sl No | Category | Description | Example |
|---|---|---|---|
| (a) | Deltaic mangrove habitat | High tidal actions along with tidal amplitude are the characteristic features associated with deltaic mangrove habitat. In this habitat strong tidal currents are seen which are responsible for the dispersion of sediments brought to the coasts by the rivers and the main river channels | Sundarban mangrove forest in West Bengal, mangrove forest in Mahanadi delta in Odisha, mangrove forest in the Bhavnagar estuary in Gujarat |
| (b) | Coastal mangrove habitat | The tidal range is not very high in the coastal mangrove habitat. This habitat is formed by the deposition of terrigenous sands, silts and clays brought by the river discharges | It includes the inter- tidal coastlines, minor river mouths, sheltered bays and back water areas of the west coast |
| (c) | Island mangrove habitat | This type of mangrove habitat is seen at the mouths where rivers are seen to border the open sea | It includes shallow but protected inter tidal zones of bay island and Lakshadweep Atoll |

**Table 6.7** Geo-morphological character- based categorization of Indian mangroves

| Sl No | Category | Example |
|---|---|---|
| (a) | East coast mangrove habitats | Sundarban mangrove forest (West Bengal), Bhitarkanika mangrove forest (Odisha), Mahanadi delta mangrove forest (Odisha), Godavari and Krishna delta mangrove forest (Andhra Pradesh), Pichavaram mangrove forest (Tamilnadu) etc. |
| (b) | West coast mangrove habitats | Mangrove forest along Cochin estuary, mangrove forests along Zuari and Mandovi estuaries in Goa, Mangrove forest along Bhavnagar estuaries in Gujarat etc. |
| (c) | Andaman & Nicobar and Lakshadweep Island | Andamam & Nicobar Islands and Lakshadweep Atoll, comprising more than 200 and 37 islands, respectively, harbour island mangroves |

species of *Bruguiera* are present, comprising *B. cylindrica*, *B. gymnorrhiza* and *B. sexangula.* These genera have been over-exploited and regeneration is poor. *Kandelia candel*, a fourth genus of the family *Rhizophoraceae*, is very rare. *Avicennia alba*, *Avicennia marina* and *Avicennia officinalis* are members of the family *Avicenniaceae*, which are abundantly found in the Indian Sundarbans. These species represent a

**Table 6.8** Most common Indian mangrove flora (including mangrove associate species)

| Sl. No | Scientific names | Families |
|---|---|---|
| 1 | *Acanthus ilicifolius* L | Acanthaceae |
| 2 | *A. ebracteatus* Vahl | Acanthaceae |
| 3 | *A. volubilis* Wall | Acanthaceae |
| 4 | *Aegialitis rotundifolia* Roxb | Aegialitidaceae |
| 5 | *Sesuvium portulacastrum* L | Aizoaceae |
| 6 | *Crirtum defixum* Ker. Gawler | Amaryllidaceae |
| 7 | *Cerbera manghas* L | Apocynaceae |
| 8 | *C. odolum* Gaertner | Apocynaceae |
| 9 | *Cryptocoryne ciliata* Fish ex Schott | Araceae |
| 10 | *Nypa fruticans* Roxb | Arecaceae |
| 11 | *Phoenix paludosa* Wurmb | Arecaceae |
| 12 | *Sarcolobus globosus* Wall | Asclepiadaceae |
| 13 | *S. carinatas* Wall | Asclepiadaceae |
| 14 | *Hoya parasitica* Wall | Asclepiadaceae |
| 15 | *Pentatropis capensis* (L.f.) Bullock | Asclepiadaceae |
| 16 | *Avicennza alba* Blume | Avicenniaceae |
| 17 | *A. marina* (Forsk.) Vier | Avicenniaceae |
| 18 | *A. officinalis* L | Avicenniaceae |
| 19 | *Dolichandrone spathaceae* (L.f.) Schum | Bignoniaceae |
| 20 | *Heliotropium curassavicum* L | Boraginaceae |
| 21 | Caesalpinia bonuc L | Caesalpiniaceae |
| 22 | *C. crista* L | Caesalpiniaceae |
| 23 | *Suaeda nudiflora* Roxb | Chenopodiaceae |
| 24 | *S. mritima* Dumort | Chenopodiaceae |
| 25 | *S. monoica* Forssk | Chenopodiaceae |
| 26 | *Salicornia brachiata* Roxb | Chenopodiaceae |
| 27 | *Lumnitzera racemosa* Willd | Combretaceae |
| 28 | *L. littorea* (Jack.) Voigt | Combretaceae |
| 29 | *Ipomoea pes-caprae* L | Convolvulaceae |
| 30 | *Excoecaria agallocha* (L.) Sweet | Euphorbiaceae |
| 31 | *Barringtonia racemosa* (L.) Sprengal | Lecythidaceae |
| 32 | *Viscum orientale* Willd | Loranthaceae |
| 33 | *Dendrophthoe falcata* (L.f.) Etting | Loranthaceae |
| 34 | Hibiscus liliaceous L | Malvaceae |
| 35 | *Viscum orientale* Willd | Malvaceae |

(continued)

**Table 6.8** (continued)

| Sl. No | Scientific names | Families |
|---|---|---|
| 36 | *Thespesia populnea* (L.) Soland ex Correa | Malvaceae |
| 37 | *T. populnoides* (Roxb.) Kostel | Malvaceae |
| 38 | *Xylocarpus granetum* Koenig | Meliaceae |
| 39 | *X. moluccensis* (Lamk.) Roem | Meliaceae |
| 40 | *X. mekongensis* Pierre | Meliaceae |
| 41 | Aglaia cucullata Pellegrin | Meliaceae |
| 42 | *Cynometra ramiflora* L | Mimosaceae |
| 43 | *Aegiceras corniculatum* (L.) Blanco | Myrsinaceae |
| 44 | *Ardisia littoralis* SW | Myrsinaceae |
| 45 | *Pandanus foetidus* Roxb | Pandanaceae |
| 46 | *Derris scandens* Benth | Fabaceae |
| 47 | *D. trifoliata* Lour | Fabaceae |
| 48 | *D. indica* Bennett | Fabaceae |
| 49 | *Dalbergia spinosa* L | Fabaceae |
| 50 | *Porteresia coarctata* (Roxb.) Takeoka | Poaceae |
| 51 | *Aehiropus legopoides* (L.) Trin | Poaceae |
| 52 | *Myriostachya wightiana* Hk.f | Poaceae |
| 53 | *Urochandra setulosa* (Trin.) Hubbard | Poaceae |
| 54 | *Acrostichum aureum* L | Pteridaceae |
| 55 | *Rhizophora mucronata* Lamk | Rhizophoraceae |
| 56 | *R. apiculata* Blume | Rhizophoraceae |
| 57 | *R. stylosa* Griff | Rhizophoraceae |
| 58 | *R.lamarkii* Montr | Rhizophoraceae |
| 59 | *R. annamalayana* Kathir | Rhizophoraceae |
| 60 | *Bruguiera gymnorrhiza* (L.) Lamk | Rhizophoraceae |
| 61 | *B. sexangula* (Lour.) Poir | Rhizophoraceae |
| 62 | *B. parviflora* W. & A | Rhizophoraceae |
| 63 | *B. cylindrica* (L.) Blume | Rhizophoraceae |
| 64 | *Ceriops decandra* (Griff.) DingHou | Rhizophoraceae |
| 65 | *C. tagal* Robinson | Rhizophoraceae |
| 66 | *Kandelia candel* (L.) Druce | Rhizophoraceae |
| 67 | *Scyphiphora hydrophyllacae* Gaertn. F | Rubiaceae |
| 68 | *Hydrophylax maritima* L | Rubiaceae |

(continued)

**Table 6.8** (continued)

| Sl. No | Scientific names | Families |
|---|---|---|
| 69 | *Atalantia correa* M.Roem | Rutaceae |
| 70 | *Salvadora persica* L | Salvadoraceae |
| 71 | *Solarium trilobatum* L | Solanaceae |
| 72 | *Sonneratia apetala* Buch. Ham | Sonneratiaceae |
| 73 | *S. alba* Smith | Sonneratiaceae |
| 74 | *S. caseolaris* (L.) Engler | Sonneratiaceae |
| 75 | *S. griffithii* Kurz | Sonneratiaceae |
| 76 | *Heritiera fames* Buch. Ham | Sterculiaceae |
| 77 | *H. littoralis* Ait. Ex Dry | Sterculiaceae |
| 78 | *H. kanikensis* | Sterculiaceae |
| 79 | *Tamarix gallica* L | Tamaricaceae |
| 80 | *Tamarix dioica* Roxb | Tamaricaceae |
| 81 | *Brownlowia tersa* (L.) Koesterm | Tiliaceae |
| 82 | *Clerodendron inerme* (L.) Gaertn | Verbenaceae |

pioneer genus in mangrove plant succession and occur in association with the salt marsh grass *Porteresia coarctata*, in areas of accretion. *Avicennia* spp. also occurs in gregarious forms along the stretches of channels and creeks. *Sonneratia apetala* occurs in fringe areas in discontinuous strips while *S. caseolaris* occurs only in the western Sundarbans. These two species thrive luxuriantly in areas with high dilution factor. *Xylocarpus granatum* and *X. mekongensis*, which were once common species in the Sundarbans, are found scattered and have become scarce because of selective felling and poor regeneration. Stunted trees of *Heritiera fomes* are found scattered all over the forest. In the central Indian Sundarbans, which is extremely hypersaline in nature, the species is on the verge of extinction (Mitra 2013; Mitra and Zaman 2014, 2015, 2016). *Amoora cucullata* is also found in association with *Heritiera*, in freshwater conditions. *Aegialites rotundifolia*, representing a monoypic genus, is found in low-lying areas with *Acanthus ilicifolius*. Both *Amoora cucullata* and *Aegialites rotundifolia*, together with *Aegiceras corniculatum*, are present in very small numbers and are a cause for concern. *Aegiceras*, found on the banks of narrow creeks, are exploited for timber and fuel wood. *Lumnitzera racemosa* is sparsely distributed in areas of high salinity. *Nypa fruticans* is found in the creeks although this species has been over-exploited and regeneration is poor. *Brownlowia tersa* and *Cynometra ramiflora* are scarce. On high areas above tide level, thick patches of pure *Phoenix paludosa* occur and are increasing in number (Fig. 6.7). The species can thrive luxuriantly in high saline environment and offers a congenial shelter for Royal Bengal tiger.

The Bhitarkanika mangrove forest is the second largest mangrove forest in India. In April, 1975 mangroves of Bramhani and Baitarani delta of Kendrapara district have been declared as the Bhitarkanika Wild Life Sanctuary (BWLS), which covers an area of 145 km$^2$. Bhitarkanika is endowed with three major types of ecosystem

**Fig. 6.7** *Phoenix paludosa* patches in the Indian Sundarbans on high areas of the tide level

namely brackish water, marine and terrestrial—intricately mixed with each other. The community structure of mangrove vegetation is greatly influenced by dilution factor, tidal inflow, degree of inundation and salinity gradient. Several works have been carried out on the distribution of mangroves in the Bhitarkanika region (Banerjee and Das 1972; Mishra and Panigrahi 1987; Pattanaik and Choudhury 1989; Banerjee and Rao 1990; Dani et al. 1999; Mishra et al. 2005; Pattnaik et al. 2008; Upadhyay et al. 2008).

In Indian literatures most of the mangrove related studies focused on their taxonomic status. In course of time, the dimension of research shifted towards their ecosystem services. Since the last decade, mangrove related researches took a bent/turn towards climate change and carbon sequestration. In many countries, policy makers attempted to link mangrove plantation with carbon sequestration as a road map towards Clean Development Mechanism (CDM).

A survey was carried out at Bhitarkanika Wild Life Sanctuary (BWLS) by the present author during August 2016 to estimate stored carbon in the stem region of true mangrove and mangrove associate floral species. The stored carbon exhibited direct proportionality with stem biomass irrespective of the species. The stored carbon in the stem of the selected species varied from 0.03 t $ha^{-1}$ in *Thespesia populnea* to 11.32 t $ha^{-1}$ in *Avicennia officinalis* as shown in Fig. 6.8. The total stored carbon in the stems of the selected species is 61.97 t $ha^{-1}$, which synchronizes well with values (extrapolated through thumb rule) obtained from different mangrove regions of the

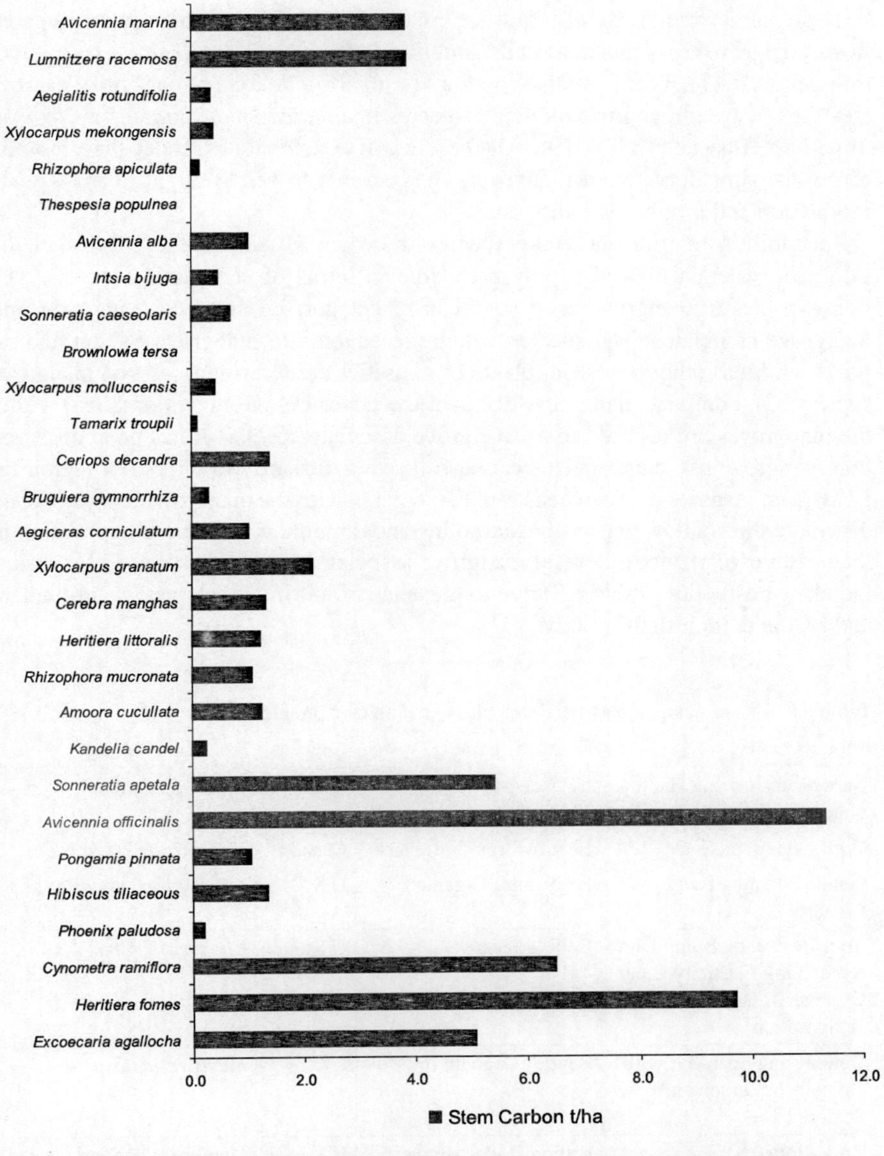


**Fig. 6.8** Stem carbon of mangrove trees in Bhitarkanika as recorded during August, 2016

World. The present study establishes the potential of BWLS mangrove flora in the domain of carbon sequestration, which is a new dimension in the ecosystem services of mangroves thriving in the deltaic complex of Brahmani and Baitarani Rivers.

It is interesting to note that few cutting edge researches are gradually taking a formative stage in the domain of ecosystem services of Indian Sundarban mangroves, which have direct linkage with the livelihood of coastal population. Fruit jelly, has been prepared from *S. apetala* fruit. Ice creams, cookies, breads have been prepared from mangrove associate seaweeds *Enteromorpha intestinalis* (Mitra et al. 2016; Pramanick 2017). There are also reports of preparing breakfast snacks like samosa and kachuri from mangrove associate species like *Suaeda maritima* and *Salicornia brachiata* (Biswas et al. 2019). All these are few innovations under the umbrella of provisioning services of mangroves, and are yet to see the light of mass scale production and proper marketing facility.

In addition to true mangrove species there are also associate species in the same ecosystem which also play great role in terms of ecosystem services. The floral species in mangrove ecosystem can be categorized into true mangroves and mangrove associates. The species which are adapted to mangrove habitat and do not extend into other terrestrial plant communities are referred to as true mangrove species. On contrary, plants that occur in the coastal environment and also within the mangroves are referred to as mangrove associate species. It has been observed that mangrove associate species can also thrive luxuriantly in terrestrial region far away from mangroves. Seaweeds often associated with the mangrove ecosystem and distributed in shallow water adjacent to intertidal mudflats have wide application in the health care services. Several mangrove associated seaweeds are used for manufacturing cosmetics (Table 6.9) due to presence of nutritional elements, antioxidant and iodine in their thallus body.

**Table 6.9** Use of mangrove associate species in health care products

| Product name | Company name | Country | Seaweed used |
|---|---|---|---|
| Super Moisturizing Hand Cream | Balance Me | UK | *Chondrus crispus* |
| Pacific Seaweed Soak | Beauty through Balance | Canada | *Macrocystis integrifolia* |
| Gentle Foaming Facial Cleanser | Be Natural Organic | U.S | *Ascophyllum nodosum* |
| Your Best Baby Bump Duo (containing Elasticity Belly Oil and All Day Moisture Body Lotion) | Belli Skincare | U.S | *Chondrus crispus* |
| Detoxifying Soap Bar with Grapefruit, Lemon and Seaweed | Bentley Organic | UK | *Fucus vesiculosus* |
| Brow Boost | Billion Dollar Brows | U.S | *Laminaria digitata* |
| Bliss Fabulous Every Day Eye cream | Bliss World LLC | U.S | *Laminaria digitata* |

(continued)

**Table 6.9** (continued)

| Product name | Company name | Country | Seaweed used |
|---|---|---|---|
| Beeswax and Royal Jelly Eye Cream | Burt's Bees | U.S | *Chondrus crispus* |
| Scottish Intensive Seaweed Conditioner | Diana Drummond | UK | *Laminaria digitata, Chondrus crispus, Ascophyllum nodosum, Ulva lactuca* |
| Sea Plus Renewal Night Cream | Alba Botanica | U.S | *Chondrus crispus, Laminaria digitata, Macrocystis pyrifera, Ulva compressa, Ulva lactuca* |

## 6.2 Respondent Analysis on Acidification of Sundarban Estuaries

Little quantitative information is available regarding the impact of acidification on the ecosystem services of estuaries. Very few people are aware of the phenomenon and its impact on the ecosystem. Considering this knowledge gap, the concept of acidification of Indian Sundarban estuaries and its subsequent impacts on biodiversity and livelihoods were brain stormed amongst the stakeholders associated with the mangrove ecosystem in the frame work of Indian Sundarbans during 2017–2019 through a set of questionnaires (*Vide* Table 6.16). The entire network of the study consisted of three stages (i) identification of respondents (policy maker, researcher, fisherman, agriculturist and local inhabitant) (ii) identification of major impacts of acidification (8 in numbers) and iii) evaluation of the respondent's response to construct the Acidification Impact Scale (AIS) through ranking (IR) and voting. Although impacts can be of various types, the present list captures the major impacts caused by lowering of aquatic pH in three sectors of Indian Sundarbans (western, central and eastern) and was ranked in terms of their importance by building an Acidification Assessment Matrix (AAM). However, as there is high probability of variation of this ranking with the category of respondents, therefore the views of the respondents were also considered (by inclusion of the % of voting along with their respective ranking factor) and finally Composite Acidification Impact Scale (CAIS) in context to Indian Sundarban estuaries was constructed as per the expression (Tables 6.10, 6.11, 6.12, 6.13, 6.14 and 6.15).

$$CAIS = AIS1 + AIS2 + AIS3 + AIS4 + AIS5$$

where, AIS = Impact Rank (IR) × % of Vote.

It is to be noted in this context that the sample size of respondents are variable e.g., for policy maker it is 65, but for other groups of respondents n = 90 for each group.

**Table 6.10** Impact of acidification on estuarine water with scaling in western Indian Sundarbans

| **Impact** | **Policy maker (respondent type 1)** | | |
|---|---|---|---|
| | **IR** | **% of vote** | **AIS1** |
| pH level of estuarine water | 7 | 23.1 | 161.7 |
| Pollution level | 7 | 22.2 | 155.4 |
| Fishes | 2 | 9.1 | 18.2 |
| Molluscs | 5 | 10.2 | 51.0 |
| Plankton | 4 | 9.7 | 38.8 |
| Mangrove flora | 3 | 6.2 | 18.6 |
| Livelihood | 6 | 11.3 | 67.8 |
| Socio-economic profile | 5 | 8.2 | 41.0 |
| **Impact** | **Researcher (respondent type 2)** | | |
| | **IR** | **% of vote** | **AIS2** |
| pH level of estuarine water | 7 | 23.7 | 165.9 |
| Pollution level | 7 | 23.0 | 161.0 |
| Fishes | 2 | 8.1 | 16.2 |
| Molluscs | 5 | 10.2 | 51.0 |
| Plankton | 4 | 9.1 | 36.4 |
| Mangrove flora | 3 | 7.4 | 22.2 |
| Livelihood | 6 | 11.5 | 69.0 |
| Socio-economic profile | 5 | 7.0 | 35.0 |
| **Impact** | **Fisherman (respondent type 3)** | | |
| | **IR** | **% of vote** | **AIS3** |
| pH level of estuarine water | 8 | 23.5 | 188.0 |
| Pollution level | 8 | 21.8 | 174.4 |
| Fishes | 1 | 9.7 | 9.7 |
| Molluscs | 8 | 10.0 | 80.0 |
| Plankton | 2 | 8.9 | 17.8 |
| Mangrove flora | 5 | 6.0 | 30.0 |
| Livelihood | 7 | 11.1 | 77.7 |
| Socio-economic profile | 6 | 9.0 | 54.0 |
| **Impact** | **Agriculturist (respondent type 4)** | | |
| | **IR** | **% of vote** | **AIS4** |
| pH level of estuarine water | 7 | 26.2 | 183.4 |
| Pollution level | 7 | 24.7 | 172.9 |
| Fishes | 2 | 8.9 | 17.8 |
| Molluscs | 5 | 11.7 | 58.5 |
| Plankton | 2 | 3.7 | 7.4 |

(continued)

**Table 6.10** (continued)

| **Impact** | **Policy maker (respondent type 1)** | | |
|---|---|---|---|
| | **IR** | **% of vote** | **AIS1** |
| Mangrove flora | 2 | 4.2 | 8.4 |
| Livelihood | 5 | 12.1 | 60.5 |
| Socio-economic profile | 6 | 8.5 | 51.0 |
| **Impact** | **Local inhabitant (respondent type 5)** | | |
| | **IR** | **% of vote** | **AIS5** |
| pH level of estuarine water | 5 | 29.3 | 146.5 |
| Pollution level | 8 | 21.5 | 172.0 |
| Fishes | 1 | 8.7 | 8.7 |
| Molluscs | 6 | 14.2 | 85.2 |
| Plankton | 3 | 5.6 | 16.8 |
| Mangrove flora | 6 | 3.2 | 19.2 |
| Livelihood | 7 | 10.3 | 72.1 |
| Socio-economic profile | 5 | 7.2 | 36.0 |

**Table 6.11** ANOVA results showing variations between impact of acidification and respondent's views on the impact in the western Indian Sundarbans

| Source of variation | SS | df | MS | F | P-value | $F_{crit}$ |
|---|---|---|---|---|---|---|
| Between impact | 138,973.3 | 7 | 19,853.33 | 161.5015 | 6.48E-21 | 2.35926 |
| Between respondents | 568.949 | 4 | 142.2372 | 1.157062 | 0.350728 | 2.714076 |
| Error | 3442.031 | 28 | 122.9297 | | | |
| Total | 142,984.3 | 39 | | | | |

Table 6.10 reflects the results of respondent analysis in the western sector of Indian Sundarbans, which is almost semi-urbanized with fish landing stations/harbours, tourism units and pilgrims. The mangroves in this sector have been mostly sacrificed for shrimp farms, which is one of the major livelihoods in this area.

The western sector of Indian Sundarbans is in the zone of 'noise' as there is the presence of industries, fish landing stations, tourism units, shrimp farms that discharge wastes of complex nature and causes the lowering of pH. Policy makers, researchers, fishermen, agriculturists and local inhabitants are aware of the phenomenon and imparted maximum weightage on lowering of the pH and pollution (Table 6.10). The sequence of CAIS, which is the combined opinion of all categories of stakeholders ranked the impact of acidification as pH level of estuarine water (845.5) > pollution level of the estuarine water (835.7) > livelihood (347.1) > molluscan community (325.7) > socio-economic profile (217.0) > plankton community (117.2) > mangrove flora (38.4) > fishes (70.6). The stakeholders also feel that

**Table 6.12** Impact of acidification on estuarine water with scaling in central Indian Sundarbans

| Impact | Policy maker (respondent type 1) | | |
|---|---|---|---|
| | **IR** | **% of vote** | **AIS1** |
| pH level of estuarine water | 7 | 36.3 | 254.1 |
| Pollution level | 8 | 19.3 | 154.4 |
| Fishes | 1 | 6.9 | 6.9 |
| Molluscs | 6 | 12.8 | 76.8 |
| Plankton | 4 | 4.2 | 16.8 |
| Mangrove flora | 3 | 1.8 | 5.4 |
| Livelihood | 6 | 11.2 | 67.2 |
| Socio-economic profile | 6 | 7.5 | 45.0 |
| **Impact** | **Researcher (respondent type 2)** | | |
| | **IR** | **% of vote** | **AIS2** |
| pH level of estuarine water | 8 | 22.9 | 183.2 |
| Pollution level | 7 | 20.7 | 144.9 |
| Fishes | 4 | 9.0 | 36.0 |
| Molluscs | 8 | 19.9 | 159.2 |
| Plankton | 1 | 4.6 | 4.6 |
| Mangrove flora | 2 | 3.8 | 7.6 |
| Livelihood | 8 | 10.1 | 80.8 |
| Socio-economic profile | 8 | 9.0 | 72.0 |
| **Impact** | **Fisherman (respondent type 3)** | | |
| | **IR** | **% of vote** | **AIS3** |
| pH level of estuarine water | 7 | 21.9 | 153.3 |
| Pollution level | 7 | 22.8 | 159.6 |
| Fishes | 3 | 9.1 | 27.3 |
| Molluscs | 8 | 17.2 | 137.6 |
| Plankton | 4 | 6.0 | 24.0 |
| Mangrove flora | 5 | 5.9 | 29.5 |
| Livelihood | 8 | 11.7 | 93.6 |
| Socio-economic profile | 8 | 5.4 | 43.2 |
| **Impact** | **Agriculturist (respondent type 4)** | | |
| | **IR** | **% of vote** | **AIS4** |
| pH level of estuarine water | 7 | 27.0 | 189.0 |
| Pollution level | 8 | 12.7 | 101.6 |
| Fishes | 5 | 9.2 | 46.0 |
| Molluscs | 8 | 15.8 | 126.4 |
| Plankton | 2 | 6.2 | 12.4 |

(continued)

**Table 6.12** (continued)

| Impact | Policy maker (respondent type 1) | | |
|---|---|---|---|
| | IR | % of vote | AIS1 |
| Mangrove flora | 2 | 8.0 | 16.0 |
| Livelihood | 8 | 12.0 | 96.0 |
| Socio-economic profile | 7 | 9.1 | 63.7 |
| **Impact** | **Local inhabitant (respondent type 5)** | | |
| | **IR** | **% of vote** | **AIS5** |
| pH level of estuarine water | 7 | 24.3 | 170.1 |
| Pollution level | 8 | 21.6 | 172.8 |
| Fishes | 3 | 8.2 | 24.6 |
| Molluscs | 8 | 16.1 | 128.8 |
| Plankton | 3 | 5.5 | 16.5 |
| Mangrove flora | 7 | 4.0 | 28.0 |
| Livelihood | 7 | 9.1 | 63.7 |
| Socio-economic profile | 8 | 11.2 | 89.6 |

**Table 6.13** ANOVA results showing variations between impact of acidification and respondent's views on the impact in the central Indian Sundarbans

| Source of variation | SS | df | MS | F | P-value | $F_{crit}$ |
|---|---|---|---|---|---|---|
| Between Impact | 148,517.1 | 7 | 21,216.72 | 37.0189 | 1.64E-12 | 2.35926 |
| Between Respondents | 383.129 | 4 | 95.78225 | 0.167121 | 0.953315 | 2.714076 |
| Error | 16,047.7 | 28 | 573.1323 | | | |
| Total | 164,947.9 | 39 | | | | |

the phenomenon of acidification will have substantial impact on livelihood as most of the livelihood of Indian Sundarbans is aqua-centric (Fig. 6.9).

ANOVA carried out with the data sets of Table 6.10 shows that there are no statistically significant differences between the views of stake holders on the impact of acidification in the western Indian Sundarbans (Table 6.11). However, the type of impact varied at 5% level of significance as $F_{obs}$ (161.5015) > $F_{crit}$ (2.35926).

The central Indian Sundarbans is noted for erosion and hypersaline ambient environment (due to blockage of fresh water on account of massive siltation (Fig. 6.10).

Tourism is one of the major livelihoods of the local inhabitants. The respondents' views towards the impact of acidification in this sector are presented in Table 6.12.

In the central Indian Sundarbans, the result of respondent analysis is bit different. With maximum weightage to pH level of the estuarine water, the adverse impact on the molluscan community is visible in this zone through thinning of the oyster shell and mortality and hence impact of molluscan community has been given considerable

**Table 6.14** Impact of acidification on estuarine water with scaling in eastern Indian Sundarbans

| Impact | Policy maker (respondent type 1) | | |
|---|---|---|---|
| | IR | % of vote | AIS1 |
| pH level of estuarine water | 7 | 19.7 | 137.9 |
| Pollution level | 7 | 13.7 | 95.9 |
| Fishes | 2 | 17.3 | 34.6 |
| Molluscs | 8 | 19.0 | 152.0 |
| Plankton | 4 | 5.2 | 20.8 |
| Mangrove flora | 1 | 6.6 | 6.6 |
| Livelihood | 6 | 11.0 | 66.0 |
| Socio-economic profile | 7 | 7.5 | 52.5 |
| **Impact** | **Researcher (respondent type 2)** | | |
| | **IR** | **% of vote** | **AIS2** |
| pH level of estuarine water | 7 | 44.5 | 311.5 |
| Pollution level | 7 | 12.8 | 89.6 |
| Fishes | 1 | 5.2 | 5.2 |
| Molluscs | 8 | 15.8 | 126.4 |
| Plankton | 4 | 1.2 | 4.8 |
| Mangrove flora | 1 | 1.8 | 1.8 |
| Livelihood | 7 | 11.2 | 78.4 |
| Socio-economic profile | 7 | 7.5 | 52.5 |
| **Impact** | **Fisherman (respondent type 3)** | | |
| | **IR** | **% of vote** | **AIS3** |
| pH level of estuarine water | 7 | 29.5 | 206.5 |
| Pollution level | 7 | 26.3 | 184.1 |
| Fishes | 4 | 7.1 | 28.4 |
| Molluscs | 7 | 13.9 | 97.3 |
| Plankton | 2 | 5.5 | 11.0 |
| Mangrove flora | 1 | 2.8 | 2.8 |
| Livelihood | 8 | 8.0 | 64.0 |
| Socio-economic profile | 8 | 6.9 | 55.2 |
| **Impact** | **Agriculturist (respondent type 4)** | | |
| | **IR** | **% of vote** | **AIS4** |
| pH level of estuarine water | 7 | 30.5 | 213.5 |
| Pollution level | 8 | 11.5 | 92.0 |
| Fishes | 3 | 4.3 | 12.9 |
| Molluscs | 5 | 11.5 | 57.5 |
| Plankton | 1 | 5.9 | 5.9 |

(continued)

**Table 6.14** (continued)

| Impact | Policy maker (respondent type 1) | | |
|---|---|---|---|
| | IR | % of vote | AIS1 |
| Mangrove flora | 1 | 2.7 | 2.7 |
| Livelihood | 4 | 24.4 | 97.6 |
| Socio-economic profile | 7 | 9.2 | 64.4 |
| **Impact** | **Local inhabitant (respondent type 5)** | | |
| | **IR** | **% of vote** | **AIS5** |
| pH level of estuarine water | 8 | 16.3 | 130.4 |
| Pollution level | 6 | 20.5 | 123.0 |
| Fishes | 2 | 6.9 | 13.8 |
| Molluscs | 5 | 19.3 | 96.5 |
| Plankton | 4 | 4.2 | 16.8 |
| Mangrove flora | 2 | 3.4 | 6.8 |
| Livelihood | 8 | 11.9 | 95.2 |
| Socio-economic profile | 5 | 17.5 | 87.5 |

**Table 6.15** ANOVA results showing variations between impact of acidification and respondent's views on the impact in the eastern Indian Sundarbans

| Source of variation | SS | df | MS | F | P-value | $F_{crit}$ |
|---|---|---|---|---|---|---|
| Between impact | 153,306.5 | 7 | 21,900.93 | 18.08322 | 7.62E-09 | 2.35926 |
| Between respondents | 1531.927 | 4 | 382.9816 | 0.316221 | 0.864691 | 2.714076 |
| Error | 33,911.32 | 28 | 1211.118 | | | |
| Total | 188,749.7 | 39 | | | | |

score. It is to be noted that many local inhabitants of central Indian Sundarbans are engaged in lime generation from molluscs (Fig. 6.11) and hence people are quite aware of that situation. The CAIS values in this region varies as per the order pH level of estuarine water (947.7) > adverse impact on molluscan population (733.3) > pollution level of the estuarine water (628.8) > livelihood (401.3) > socio-economic profile (313.5) > fishes (140.8) > mangrove flora (86.5) > plankton (74.0) (Table 6.12).

We have also carried out ANOVA with the data set of Table 6.12 and observed that there are no significant differences in opinion between the stakeholders of the central sector of Indian Sundarbans in context to impact of acidification as $F_{obs}$ is less than $F_{crit}$ (Table 6.13). However, the type of impact varied at 5% level of significance as $F_{obs}$ (37.0189) > $F_{crit}$ (2.35926).

The eastern Indian Sundarban is a pristine zone with Reserve Forest and minimum anthropogenic pressure. The mangrove forest is luxuriant in this sector (Fig. 6.12).

**Table 6.16** Questionnaires to evaluate the impact of acidification through respondent's analysis

| | |
|---|---|
| 1 | What is your name and profession? |
| 2 | Have you heard the term acidification? What is the normal pH of estuarine water around you? |
| 3 | What are the major effects of acidification? |
| 4 | How acidification impacts fishes in Sundarban estuaries? Rank the impact with the score from 1 to 8, where 1 stands for very least impact and 8 stands for highest degree of impact |
| 5 | How the molluscan community is helpful in context of livelihood? |
| 6 | How acidification impacts oysters (kasturi in local language), snails and other molluscs in Sundarban estuaries? Rank the impact with the score from 1 to 8, where 1 stands for very least impact and 8 stands for highest degree of impact |
| 7 | Do you have any idea on the bioremediation potential of mangroves? |
| 8 | What are the sources of pollutants in the estuaries of Indian Sundarbans? |
| 9 | Do you think that lowering of pH is related to release of industrial and domestic wastes in the estuaries of Sundarbans? |
| 10 | Do you have any idea about climate change and Green House Gas (GHG) emission? |
| 11 | Which is the major GHG in your area and what are its sources?Do you believe that excess carbon dioxide emission causes acidification of Sundarban estuaries? |
| 12 | Do you believe that excess carbon dioxide emission causes acidification of Sundarban estuaries? |
| 13 | How mangroves are useful in minimizing the GHG level in Sundarban atmosphere? |
| 14 | What is your major source of income? Does it get affected with acidification? |
| 15 | Do you have any idea about alternative livelihood and its relation with acidification? |

The respondent analysis, however, highlighted the CAIS values as per the order pH of the estuarine water (999.8) > pollution of the estuarine water (529.7) > livelihood (401.2) > socio-economic profile (312.1) > adverse impact on fish (94.9) > plankton (59.3) > mangrove (20.7). The least weightage of acidification on mangrove as provided by the stakeholders confirm that the stakeholders are quite aware of the resilience of mangrove vegetation to acidification (Table 6.14).

Results of ANOVA with the data sets of Table 6.14 exhibit no significant variations between stakeholders' views, but the impact of acidification differed greatly at 5% level of significance (Table 6.15).

## 6.3 Status of Heavy Metals in Context to Acidification of Sundarban Estuaries

Conservative wastes are non-biodegradable in nature and include heavy metals like Zn, Cu, Pb, Cr, Hg etc. Unlike common domestic wastes having organic matter, conservative wastes are not degraded by microbes, and hence remain in the ambient media almost permanently. We are concerned of this long residence time and hence

**Fig. 6.9** Aqua-centric livelihood in Sundarbans

in this section we have tried to focus on the effect of acidification on heavy metals considering Indian Sundarbans as our case study zone. This Indo-Gangetic plain is the biggest alluvial tract of the World, and this alluvial fill is essentially of Quaternary age. The lower Ganga delta in the undivided Sundarbans was built up in the seaboard by the alluvium fine clay, silt and sand particles, carried down from the Great Himalaya and Chotanagpur hills. In this silt and clay transport the two rivers Ganga and Brahmaputra and their numerous tributaries has taken active part. The stretch of coastal quaternaries exhibits varied geomorphological signatures (like sand dunes, beach ridge, intertidal clayey/sandy flats, tidal shoals etc.) evolved out of dynamic and varied interactions of marine agencies like waves, tides and littoral currents, combined with fluvial and aeolin components. These silt, clay and sand particles have been deposited step by step or layer by layer in the sea board and due to rapid formation have also resulted in uneven thickness and width or surface of the horizons in the estuarine mouths, facing the Bay of Bengal. The macrotidal (tidal range > 4 m) Sundarbans is characterized by embroidery of tidal creeks, encompassing the islands and offshore linear tidal shoals, aligned perpendicular to the shore line and separated by swales.

In the present day scenario, the Hooghly-Matla estuary has became the most anthropogenically stressed ecosystem due to presence of multifarious industries that discharge wastes in to aquatic system mostly without any treatment. Presence of heavy metals in the ambient media of Sundarbans has been reported by a number of

**Fig. 6.10** Erosion of inter-tidal mudflats—A major threat to mangrove vegetation

researchers and extensive studies have been carried out by our team since last few decades (Mitra et al. 1994, 2000, 2013; Trivedi et al. 1995; Mitra 1998; Bhattacharyya et al. 2001; Das et al. 2005; Barua et al. 2011; Ray Chaudhuri et al. 2014; Zaman et al. 2014).

Three stations namely Diamond Harbour (Station 1), Namkhana (Station 2) and Ajmalmari (Station 3) were selected for the present programme, which are located in South 24 Parganas district of West Bengal. The latitude and longitude of the three sampling stations are presented in Table 6.17.

Surface water pH was measured by using a portable pH meter (sensitivity = ±0.02). Analysis of dissolved heavy metals was done with samples collected during high tide condition from all the three stations. Before analysis, each water sample was collected and stored in clean TARSON bottles and was filtered through a 0.45 μm Millipore membrane. The filtrate was treated with diethyl dithiocarbamate and extracted in carbon tetrachloride (Chakraborti et al. 1987). The extracted was evaporated to dryness and the residue was mineralized with 0.1 ml of concentrated nitric acid. Analytical blank was prepared and treated with the same reagents. Analyses were done in triplicate by direct aspiration into AAS (Perkin-Elmer Model: 3030) equipped with a HGA-500 graphite furnace atomizer and a deuterium background corrector. The accuracy of the dissolved heavy metal determinations is indicated by good agreement between our values and reported for certified reference seawater materials (CASS 2) (Table 6.18).

**Fig. 6.11** Collected molluscan shell for lime generation

In order to analyse biologically available heavy metals in sediments, samples from surface (1 cm depth) were collected by scrapping using a pre-cleaned and acid washed plastic scale and immediately kept in clean polythene bags, which were sealed. The samples were washed with metal free double distilled water and dried in an oven at 105 °C for 5–6 h, freed from visible shells or shell fragments, ground to powder in a mortar and stored in acid washed polythene bags. Analyses of biologically available metals were done after re-drying the samples, from which 1 gm was taken and digested with 0.5 (N) HCl as per the standard procedure outlined by Malo (1977). The resulting solutions were then stored in polythene containers for analysis. The solutions were finally aspirated in the flame Atomic Absorption Spectrophotometer (Perkin Elmer: Model 3030) for the determination of metal concentrations. No detectable trace metals were found in the reagent blank. Analysis of the NIES Sargasso sample was carried out to assure the quality of the data (Table 6.19).

The pH ranged from 8.11 (during 2014) to 8.32 (during 1984) at Diamond Harbour during HT (Fig. 6.13).

The pH ranged from 8.27 (during 2013–2014) to 8.33 (during 1984–1986; 1988–1989) at Namkhana during HT (Fig. 6.14).

The pH ranged from 8.29 (during 2006–2007; 2010–2014) to 8.32 (during 1984–1989; 1992–1993) at Ajmalmari during HT (Fig. 6.15).

The surface water pH is an important environmental parameter on which the life of any aquatic organism depends. It showed more or less a constant value throughout

**Fig. 6.12** Luxuriant mangrove forest in eastern Indian Sundarbans

**Table 6.17** Stations with coordinates

| Stations | Geographical locations | |
|---|---|---|
| | Latitude | Longitude |
| Diamond Harbour | 22° 11′ 4.2″ N | 88° 11′ 22.2″ E |
| Namkhana | 21° 45′ 53.7″ N | 88° 13′ 51.5″ E |
| Ajmalmari | 21° 49′ 42.9″ N | 88° 37′ 13.7″ E |

**Table 6.18** Analysis of reference material for near shore seawater (CASS 2)

| Element | Certified value ($\mu g\ l^{-1}$) | Laboratory results ($\mu g\ l^{-1}$) |
|---|---|---|
| Zn | $1.97 \pm 0.12$ | $2.01 \pm 0.14$ |
| Cu | $0.675 \pm 0.039$ | $0.786 \pm 0.058$ |
| Pb | $0.019 \pm 0.006$ | $0.029 \pm 0.009$ |

**Table 6.19** Analysis of reference material (NIES Sargasso sample) for sediments obtained from the National Institute of Environmental Studies, Japan

| Element | Certified value ($\mu g\ g^{-1}$) | Laboratory results ($\mu g\ g^{-1}$) |
|---|---|---|
| Zn | 28.6 | 26.2 |
| Cu | 14.9 | 13.7 |
| Pb | 2.4 | 2.9 |

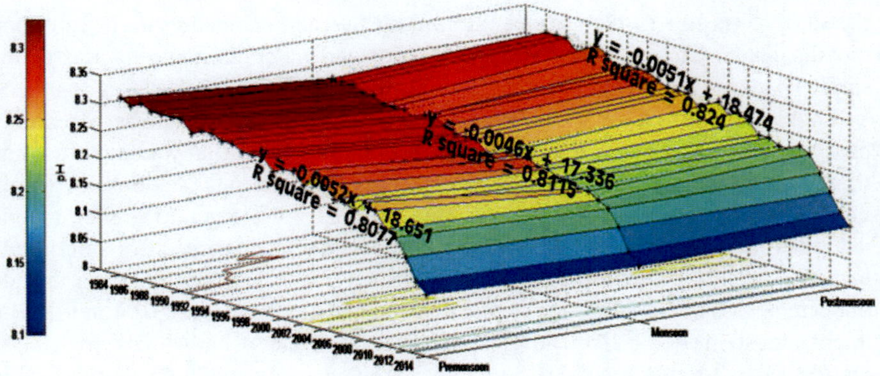


Fig. 6.13 Temporal variation of surface water pH in Diamond Harbour (Stn. 1)

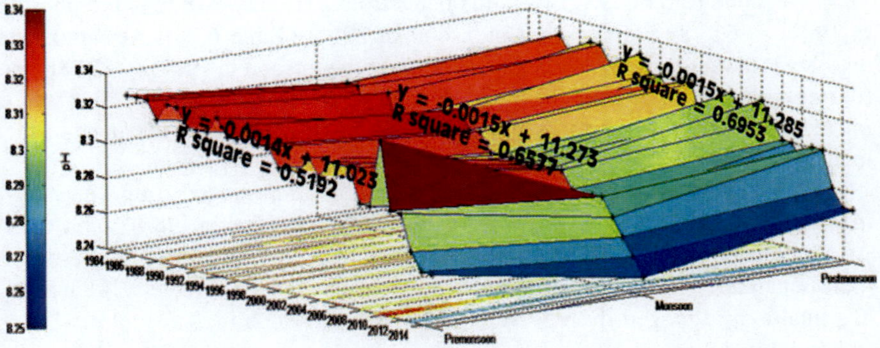


Fig. 6.14 Temporal variation of surface water pH in Namkhana (Stn. 2)

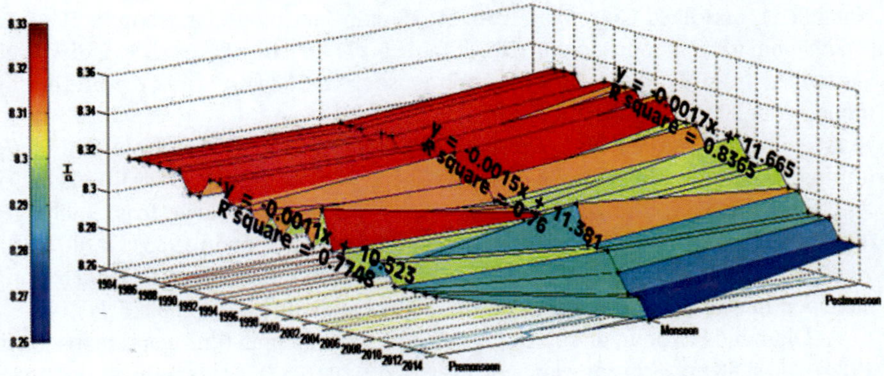


Fig. 6.15 Temporal variation of surface water pH in Ajmalmari (Stn. 3)

the study period in all the stations with minor fluctuations mainly in the monsoon months, when, dilution by run-off, precipitation and water discharge suppressed the surface water pH in all the sampling stations.

Looking on the time series data of 31 years, the trend of acidification becomes quiet pronounced. Acidification of marine and estuarine water is usually a direct function of rise of carbon dioxide in the atmosphere. In all the selected stations, an almost uniform decrease in surface water pH was observed. At Diamond Harbour, the pH value has decreased at the rate of 0.081 unit/decade in premonsoon, 0.074 unit/decade in monsoon and 0.077 unit/decade in postmonsoon (Fig. 6.13). At Namkhana, the rate of decrease is 0.024 unit/decade in premonsoon and monsoon and 0.019 unit/decade in postmonsoon (Fig. 6.14). The decrease of surface water pH at Ajmalmari is 0.012 unit/decade in premonsoon, 0.023 unit/decade in monsoon and 0.019 unit/decade in postmonsoon (Fig. 6.15). This decrease is quite close to the observations recorded in several oceans and estuaries.

At Diamond Harbour, dissolved Zn ranged from 201.00 ppm (during premonsoon in 1984) to 626.89 ppm (during monsoon in 2014) (Table 6.20). At Namkhana, dissolved Zn ranged from 187.55 ppm (during premonsoon in 1984) to 527.88 ppm (during monsoon in 2014) (Table 6.21), whereas at Ajmalmari, it was observed to range from 104.85 ppm (during premonsoon in 1984) to 286.61 ppm (during monsoon in 2014) (Table 6.22).

At Diamond Harbour, the sediment Zn concentration ranged from 18.99 μg/g (during monsoon in 2014) to 94.98 μg/g (during premonsoon in 1984) (Table 6.23). At Namkhana, the sediment Zn concentration ranged from 16.54 μg/g (during monsoon in 2014) to 92.75 μg/g (during premonsoon in 1984) (Table 6.24), whereas at Ajmalmari, it ranged from 12.98 μg/g (during monsoon in 2014) to 64.71 μg/g (during premonsoon in 1984) (Table 6.25).

At Diamond Harbour, dissolved Cu ranged from 63.25 ppm (during premonsoon in 1984) to 319.37 ppm (during monsoon in 2014) (Table 6.26), whereas at Namkhana, dissolved Cu ranged from 50.95 ppm (during premonsoon in 1984) to 297.98 ppm (during monsoon in 2014) (Table 6.27). In case of Ajmalmari, dissolved Cu ranged from 39.68 ppm (during premonsoon in 1984) to 140.12 ppm (during monsoon in 2014) (Table 6.28).

At Diamond Harbour, the sediment Cu concentration ranged from 9.86 μg/g (during monsoon in 2014) to 39.85 μg/g (during premonsoon in 1984) (Table 6.29). At Namkhana, the sediment Cu concentration was found to range from 7.93 μg/g (during monsoon in 2014) to 36.23 μg/g (during premonsoon in 1985) (Table 6.30), whereas at Ajmalmari, the sediment Cu concentration ranged from 1.08 μg/g (during monsoon in 2013) to 15.55 μg/g (during premonsoon in 1984) (Table 6.31).

At Diamond Harbour, dissolved Pb ranged from 9.54 ppm (during premonsoon in 1984) to 108.88 ppm (during monsoon in 2014) (Table 6.32). At Namkhana, dissolved Pb ranged from 8.59 ppm (during premonsoon in 1984) to 89.42 ppm (during 2014) (Table 6.33). In case of Ajmalmari, it was observed to range from 4.65 ppm (during premonsoon in 1984) to 24.83 ppm (during monsoon in 2014) (Table 6.34).

At Diamond Harbour, the sediment Pb concentration ranged from 0.93 μg/g (during monsoon in 2014) to 24.09 μg/g (during premonsoon in 2009) (Table

**Table 6.20** Average yearly variation of dissolved Zn (in ppm) in Diamond Harbour during 1984–2014

| Year | Premonsoon | Monsoon | Postmonsoon |
|---|---|---|---|
| 1984 | 201 | 263.64 | 250.64 |
| 1985 | 212.55 | 274.68 | 261.68 |
| 1986 | 232.05 | 315.54 | 302.54 |
| 1987 | 300.34 | 349.37 | 336.37 |
| 1988 | 315.68 | 354.05 | 341.05 |
| 1989 | 322.98 | 359.22 | 346.22 |
| 1990 | 305.78 | 362.85 | 349.85 |
| 1991 | 331.89 | 363.31 | 350.31 |
| 1992 | 340.06 | 375.83 | 362.83 |
| 1993 | 351.17 | 402.33 | 389.33 |
| 1994 | 378.9 | 414.55 | 401.55 |
| 1995 | 389.99 | 448.46 | 435.46 |
| 1996 | 395.52 | 453.84 | 440.84 |
| 1997 | 399.27 | 466.14 | 453.14 |
| 1998 | 406.57 | 410.84 | 397.84 |
| 1999 | 398.99 | 414.66 | 401.66 |
| 2000 | 410.29 | 430.35 | 417.35 |
| 2001 | 413.59 | 441.22 | 428.22 |
| 2002 | 402.44 | 444.55 | 431.55 |
| 2003 | 413.78 | 447.71 | 434.71 |
| 2004 | 417.78 | 457.06 | 444.06 |
| 2005 | 423.89 | 427.75 | 414.75 |
| 2006 | 434.56 | 475.22 | 462.22 |
| 2007 | 436.89 | 453.85 | 440.85 |
| 2008 | 450 | 496.91 | 483.91 |
| 2009 | 365.21 | 403.22 | 390.22 |
| 2010 | 468.89 | 491.44 | 478.44 |
| 2011 | 456.22 | 499.31 | 486.31 |
| 2012 | 471.89 | 515.21 | 502.21 |
| 2013 | 489.02 | 525.75 | 512.75 |
| 2014 | 491.54 | 626.89 | 601.89 |

6.35). At Namkhana, the sediment Pb concentration ranged from 0.89 μg/g (during monsoon in 2014) to 21.12 μg/g (during premonsoon in 1984) (Table 6.36). At Ajmalmari, it was observed to range from 0.44 μg/g (during monsoon in 2013) to 10.79 μg/g (during premonsoon in 1984) (Table 6.37).

The distribution of dissolved metals followed the order Zn > Cu > Pb in all the three sampling stations and exhibited a significant seasonal pattern with high values during monsoon and lowest during premonsoon. In the present study the significant

**Table 6.21** Average yearly variation of dissolved Zn (in ppm) in Namkhana during 1984–2014

| Year | Premonsoon | Monsoon | Postmonsoon |
|---|---|---|---|
| 1984 | 187.55 | 247.27 | 234.29 |
| 1985 | 191.9 | 255.53 | 242.55 |
| 1986 | 195.98 | 306.84 | 293.86 |
| 1987 | 203.67 | 312.93 | 299.95 |
| 1988 | 213.89 | 316.08 | 303.1 |
| 1989 | 227.8 | 328.74 | 315.76 |
| 1990 | 231.11 | 346.83 | 333.85 |
| 1991 | 246.67 | 356.2 | 343.22 |
| 1992 | 264.89 | 359.86 | 346.88 |
| 1993 | 276.9 | 404.83 | 391.85 |
| 1994 | 283.59 | 410.97 | 397.99 |
| 1995 | 293.77 | 418.7 | 405.72 |
| 1996 | 299.62 | 434.8 | 421.82 |
| 1997 | 312.5 | 449.82 | 436.84 |
| 1998 | 345.87 | 456.08 | 443.1 |
| 1999 | 333.51 | 436.27 | 423.29 |
| 2000 | 382.05 | 415.2 | 402.22 |
| 2001 | 393.44 | 432.83 | 419.85 |
| 2002 | 399.75 | 434.87 | 421.89 |
| 2003 | 401.44 | 442.93 | 429.95 |
| 2004 | 412.52 | 449.64 | 436.66 |
| 2005 | 416.56 | 443.2 | 430.22 |
| 2006 | 418 | 463.83 | 450.85 |
| 2007 | 405.23 | 465.19 | 452.21 |
| 2008 | 426.91 | 482.98 | 470 |
| 2009 | 302.21 | 381.82 | 368.84 |
| 2010 | 417.88 | 478.76 | 465.78 |
| 2011 | 456 | 483.96 | 470.98 |
| 2012 | 461.09 | 511.32 | 498.34 |
| 2013 | 470.59 | 514.88 | 501.9 |
| 2014 | 489.05 | 527.88 | 513.66 |

temporal variation of heavy metals in all the stations may be attributed to huge run-off from the adjacent landmasses during the monsoon and also due to chemical speciation (Mitra and Choudhury 1998; Chakraborty et al. 2006). During monsoon, the dilution factor (df) of all the sampling stations in the coastal and estuarine zone of West Bengal increases many folds that results in the decrease of salinity and pH. The lowering of pH might facilitate the dissolution of the precipitated form of metal and increase the amount of metallic ions in solution. The increase of Cu concentrations

**Table 6.22** Average yearly variation of dissolved Zn (in ppm) in Ajmalmari during 1984–2014

| Year | Premonsoon | Monsoon | Postmonsoon |
|---|---|---|---|
| 1984 | 104.85 | 136.55 | 123.66 |
| 1985 | 107.32 | 139.82 | 133.84 |
| 1986 | 111.66 | 141.2 | 138.43 |
| 1987 | 117.54 | 143.65 | 140.66 |
| 1988 | 131.65 | 144.78 | 142.44 |
| 1989 | 140.82 | 148.32 | 147.18 |
| 1990 | 142.33 | 153.55 | 146.45 |
| 1991 | 147.94 | 168.21 | 151.94 |
| 1992 | 149.2 | 170.12 | 156.88 |
| 1993 | 152 | 172.42 | 161.43 |
| 1994 | 154.75 | 175.21 | 167.39 |
| 1995 | 158.66 | 177.39 | 172.85 |
| 1996 | 162.32 | 184.46 | 180.05 |
| 1997 | 166.81 | 195.35 | 182.66 |
| 1998 | 168.45 | 199.85 | 185.91 |
| 1999 | 170.3 | 204.61 | 189.23 |
| 2000 | 175 | 211.32 | 194.21 |
| 2001 | 177.21 | 212.02 | 202.54 |
| 2002 | 179.06 | 220.4 | 217.36 |
| 2003 | 181.25 | 226.77 | 221.41 |
| 2004 | 185.5 | 235.75 | 229.5 |
| 2005 | 187.22 | 241.82 | 235.89 |
| 2006 | 189.17 | 244.09 | 240.62 |
| 2007 | 193.45 | 252.15 | 248.22 |
| 2008 | 195.62 | 254.1 | 251.33 |
| 2009 | 121.42 | 201.29 | 159.85 |
| 2010 | 198.63 | 260.05 | 257.21 |
| 2011 | 199.13 | 267.25 | 259.5 |
| 2012 | 201.35 | 266.5 | 261.22 |
| 2013 | 207.42 | 270.13 | 265 |
| 2014 | 213.44 | 286.61 | 280.02 |

with the decrease of salinity was also reported by several workers while working in the Porto Novo (India) waters (Sundararaj and Krishnamurthy 1972). Again the maximum run-off occurs during monsoon season (UNESCO 2012).

The distribution of biologically available heavy metals in sediment followed the order: Zn > Cu > Pb in all the three selected sampling stations and exhibited unique seasonal pattern with high values during premonsoon and lowest during monsoon.

**Table 6.23** Average yearly variation of sediment Zn (in ppm dry wt) in Diamond Harbour during 1984–2014

| Year | Premonsoon | Monsoon | Postmonsoon |
|---|---|---|---|
| 1984 | 94.98 | 76.5 | 83.44 |
| 1985 | 94.65 | 74.32 | 80.61 |
| 1986 | 93.56 | 71.29 | 78.23 |
| 1987 | 93.5 | 73.62 | 73.55 |
| 1988 | 93.49 | 64.31 | 65.42 |
| 1989 | 91.88 | 60.22 | 67.61 |
| 1990 | 91.22 | 55.75 | 60.21 |
| 1991 | 90.77 | 51.49 | 62.15 |
| 1992 | 89.85 | 49.63 | 55.5 |
| 1993 | 83.85 | 46.4 | 49.66 |
| 1994 | 82.82 | 38.32 | 40.41 |
| 1995 | 81.65 | 35.55 | 40.66 |
| 1996 | 76.51 | 31.68 | 39.32 |
| 1997 | 76.29 | 30.32 | 39.49 |
| 1998 | 73.22 | 28.73 | 35.66 |
| 1999 | 71.84 | 26.55 | 36.82 |
| 2000 | 71.22 | 25.08 | 35.75 |
| 2001 | 70.44 | 24.81 | 29.8 |
| 2002 | 70.39 | 23.95 | 30.02 |
| 2003 | 69.84 | 23.36 | 30.32 |
| 2004 | 69.22 | 27.52 | 31.31 |
| 2005 | 68.59 | 26 | 31.25 |
| 2006 | 67.81 | 21.15 | 27.21 |
| 2007 | 66.85 | 20.08 | 29.95 |
| 2008 | 64.79 | 20 | 29.66 |
| 2009 | 83.88 | 44.55 | 65.88 |
| 2010 | 61.21 | 27.56 | 29.21 |
| 2011 | 57.85 | 21.39 | 22.45 |
| 2012 | 56.02 | 19.73 | 21.41 |
| 2013 | 51.9 | 19.05 | 21.32 |
| 2014 | 46.44 | 18.99 | 20.66 |

The increase of metallic ions in the aquatic phase resulted in the decrease of metal concentration in the sediment bed, which is a pure case of compartmentation of heavy metals in the estuarine systems. Estuarine bottom sediments accumulate metals and affect the near bottom water layer due to re-suspension or dissolution processes (Li et al. 1984; Khan et al. 1998).

The significant negative correlations between dissolved metals and biologically available metals (especially in case of Zn) in all the three stations (for Diamond

**Table 6.24** Average yearly variation of sediment Zn (in ppm dry wt) in Namkhana during 1984–2014

| Year | Premonsoon | monsoon | Postmonsoon |
|---|---|---|---|
| 1984 | 92.75 | 73.21 | 80.77 |
| 1985 | 92.08 | 73.05 | 79.43 |
| 1986 | 90.5 | 69.83 | 76 |
| 1987 | 89.22 | 66.55 | 72.13 |
| 1988 | 87.99 | 63.89 | 71.02 |
| 1989 | 86.22 | 60.02 | 63.29 |
| 1990 | 85.66 | 52.43 | 62.88 |
| 1991 | 80.39 | 50.99 | 60 |
| 1992 | 79.23 | 46.44 | 59.43 |
| 1993 | 78.15 | 45.89 | 51.84 |
| 1994 | 76 | 37.36 | 47.35 |
| 1995 | 74.82 | 33.28 | 39.83 |
| 1996 | 74.21 | 30.41 | 39 |
| 1997 | 73.85 | 29.49 | 38.64 |
| 1998 | 71.22 | 27.32 | 33.24 |
| 1999 | 70.33 | 25.16 | 31.85 |
| 2000 | 69.44 | 25 | 34.32 |
| 2001 | 69.21 | 23.97 | 28.78 |
| 2002 | 68.72 | 22.02 | 29.1 |
| 2003 | 64.88 | 21.58 | 29.36 |
| 2004 | 64.55 | 24.68 | 27.41 |
| 2005 | 64.21 | 25.51 | 30.31 |
| 2006 | 62.88 | 19.83 | 24.85 |
| 2007 | 62.29 | 19.01 | 23.32 |
| 2008 | 60.29 | 17.99 | 27.88 |
| 2009 | 59.73 | 39.36 | 27.48 |
| 2010 | 54.55 | 25.62 | 28.32 |
| 2011 | 50.33 | 20.38 | 20.49 |
| 2012 | 49.88 | 18.7 | 21 |
| 2013 | 42.44 | 18.62 | 20.15 |
| 2014 | 40.39 | 16.54 | 20.05 |

Harbour, dissolved Zn × sediment Zn = −0.4887, $p < 0.05$; for Namkhana, dissolved Zn × sediment Zn = −0.6352, $p < 0.01$; for Ajmalmari, dissolved Zn × sediment Zn = −0.7152, $p < 0.01$) stand as proof of the compartmentation process. But in case of other two metals the correlations between dissolved concentration and biologically available concentration are at 10% level of signiicance.

Finally the present results conclude that the process of acidification largely regulates the concentration of conservative pollutants (primarily heavy metals) in the

**Table 6.25** Average yearly variation of sediment Zn (in ppm dry wt) in Ajmalmari during 1984–2014

| Year | Premonsoon | Monsoon | Postmonsoon |
|---|---|---|---|
| 1984 | 64.71 | 52.85 | 60.41 |
| 1985 | 62.31 | 50.4 | 59.83 |
| 1986 | 61.35 | 50.22 | 55.41 |
| 1987 | 60.08 | 49.65 | 53.22 |
| 1988 | 59.15 | 46.32 | 54.44 |
| 1989 | 57.16 | 50.05 | 52 |
| 1990 | 55.72 | 43.35 | 49.71 |
| 1991 | 53.83 | 41.85 | 46.83 |
| 1992 | 51.65 | 40.61 | 46.01 |
| 1993 | 51.65 | 40.85 | 44.43 |
| 1994 | 50.17 | 33.78 | 44 |
| 1995 | 50.15 | 32.72 | 38.39 |
| 1996 | 50.12 | 29.36 | 37.85 |
| 1997 | 49.12 | 26.45 | 33.24 |
| 1998 | 49.1 | 25.38 | 32.24 |
| 1999 | 48.75 | 24.32 | 29.85 |
| 2000 | 48.32 | 21.41 | 26.32 |
| 2001 | 47.54 | 20.08 | 24.49 |
| 2002 | 46.93 | 19.9 | 25.02 |
| 2003 | 46.71 | 17.39 | 24.36 |
| 2004 | 45.88 | 17 | 23.88 |
| 2005 | 43.85 | 16.82 | 23.41 |
| 2006 | 43.42 | 16.71 | 21.47 |
| 2007 | 43.34 | 15.99 | 20.33 |
| 2008 | 51.66 | 15.31 | 19.71 |
| 2009 | 40.32 | 15 | 17.83 |
| 2010 | 35.81 | 13.98 | 16.55 |
| 2011 | 33.44 | 14.02 | 16.32 |
| 2012 | 31.88 | 13.65 | 17.05 |
| 2013 | 29.66 | 13.34 | 15.08 |
| 2014 | 26.44 | 12.98 | 14.96 |

water and sediments. Depending on the aquatic pH, bioaccumulation and subsequent toxic effects may occur in plankton, nekton or benthic community. Hence during the planning process of cleaning the lower Gangetic delta region focus and proper weightage should be given on the role of acidification and/or aquatic pH as this parameter seems to be a major driver in the domain of aquatic health of Indian Sundarbans.

**Table 6.26** Average yearly variation of dissolved Cu (in ppm) in Diamond Harbour during 1984–2014

| Year | Premonsoon | Monsoon | Postmonsoon |
|---|---|---|---|
| 1984 | 63.25 | 70.08 | 68.33 |
| 1985 | 67.82 | 72.44 | 70.22 |
| 1986 | 69.33 | 79.32 | 73.85 |
| 1987 | 73.21 | 81.55 | 79.61 |
| 1988 | 76.85 | 87.32 | 80.1 |
| 1989 | 79.83 | 88.1 | 82.15 |
| 1990 | 82.91 | 93.41 | 87.63 |
| 1991 | 93.44 | 116.38 | 98.41 |
| 1992 | 97.88 | 123.5 | 102.86 |
| 1993 | 113.23 | 130.88 | 117.55 |
| 1994 | 121.55 | 139.41 | 128.3 |
| 1995 | 111.86 | 142.88 | 138.55 |
| 1996 | 137.44 | 153.65 | 141.62 |
| 1997 | 142.65 | 168.22 | 150.02 |
| 1998 | 158.44 | 180.93 | 160.11 |
| 1999 | 173.85 | 199.99 | 175.33 |
| 2000 | 189.93 | 208.85 | 191.44 |
| 2001 | 195.21 | 214.32 | 197.88 |
| 2002 | 200.24 | 231.45 | 205.66 |
| 2003 | 211.42 | 242.66 | 217.85 |
| 2004 | 217.33 | 251.75 | 219.22 |
| 2005 | 220.64 | 260.08 | 224.75 |
| 2006 | 223.54 | 265.73 | 226.82 |
| 2007 | 214.54 | 268.82 | 217.34 |
| 2008 | 241.34 | 271.35 | 252.55 |
| 2009 | 137.85 | 283.1 | 255.75 |
| 2010 | 253.65 | 290.09 | 258.32 |
| 2011 | 261.73 | 295.45 | 265.71 |
| 2012 | 273.54 | 297.98 | 280.31 |
| 2013 | 281.99 | 314.2 | 285.93 |
| 2014 | 299.50 | 319.37 | 301.32 |

## 6.4 Take Home Messages

A. Ecosystem comprising of biotic and abiotic components provide a wide range of services to mankind. In addition to meet the daily needs of human beings, the floral community of ecosystem is the source of several life saving drugs. Disasters like super cyclones, landslides, tsunamis etc. are often checked by

**Table 6.27** Average yearly variation of dissolved Cu (in ppm) in Namkhana during 1984–2014

| Year | Premonsoon | Monsoon | Postmonsoon |
|---|---|---|---|
| 1984 | 50.95 | 59.75 | 52.85 |
| 1985 | 52.68 | 61.78 | 54.71 |
| 1986 | 58.43 | 63.75 | 60.2 |
| 1987 | 69.21 | 75.88 | 71.55 |
| 1988 | 71.44 | 79.23 | 75.23 |
| 1989 | 74.44 | 85.44 | 76.09 |
| 1990 | 75.79 | 89.41 | 79.82 |
| 1991 | 80.96 | 93.22 | 82.45 |
| 1992 | 85.93 | 102.85 | 87.88 |
| 1993 | 99.81 | 109.31 | 101.66 |
| 1994 | 105.71 | 116.42 | 107.22 |
| 1995 | 107.83 | 120.85 | 111.35 |
| 1996 | 120.69 | 127.39 | 123.79 |
| 1997 | 117.55 | 129.88 | 125.68 |
| 1998 | 140.39 | 151.33 | 143.42 |
| 1999 | 148.72 | 158 | 149.85 |
| 2000 | 151.65 | 163.45 | 156.23 |
| 2001 | 165.39 | 175.22 | 166.38 |
| 2002 | 188.54 | 191.83 | 189.95 |
| 2003 | 195.25 | 205.78 | 197.05 |
| 2004 | 205.71 | 213.44 | 206.82 |
| 2005 | 212.86 | 224.78 | 217.85 |
| 2006 | 218.54 | 234.88 | 219.32 |
| 2007 | 220.79 | 242.71 | 222.43 |
| 2008 | 238.21 | 258.05 | 240.8 |
| 2009 | 132 | 262.31 | 242.45 |
| 2010 | 248.76 | 271.86 | 253.85 |
| 2011 | 255.55 | 283.22 | 257.39 |
| 2012 | 260.08 | 291.01 | 263.03 |
| 2013 | 264 | 295 | 265.43 |
| 2014 | 270.06 | 297.98 | 271.88 |

ecosystem and plays great role in reducing their intensities. However, the services are mostly unmarked and therefore the concept is not known to many. Today the concept of ecosystem services has become an important model for linking the functioning of ecosystems to societal benefits. However, little quantitative information is available regarding the impact of acidification on the ecosystem services of estuaries due to which a sound management policy could

**Table 6.28** Average yearly variation of dissolved Cu (in ppm) in Ajmalmari during 1984–2014

| Year | Premonsoon | Monsoon | Postmonsoon |
|---|---|---|---|
| 1984 | 39.68 | 55.38 | 44.75 |
| 1985 | 40.22 | 57.84 | 46.82 |
| 1986 | 41.31 | 59.93 | 47.33 |
| 1987 | 43.15 | 62.08 | 49.44 |
| 1988 | 46.32 | 65.77 | 48.14 |
| 1989 | 48.11 | 69.81 | 51.33 |
| 1990 | 49.05 | 72.44 | 53.45 |
| 1991 | 52.14 | 73.1 | 57.61 |
| 1992 | 54.63 | 74.98 | 59.22 |
| 1993 | 56.15 | 79.1 | 63.15 |
| 1994 | 59.81 | 82.4 | 65.77 |
| 1995 | 61.41 | 87.3 | 67.89 |
| 1996 | 63.02 | 89.28 | 71.42 |
| 1997 | 65.07 | 94.46 | 73.21 |
| 1998 | 68.72 | 97.1 | 75.15 |
| 1999 | 70.15 | 99.2 | 77.2 |
| 2000 | 72.43 | 103.14 | 78.14 |
| 2001 | 76.81 | 107.51 | 81.21 |
| 2002 | 78.41 | 111.62 | 82.65 |
| 2003 | 79.21 | 114.41 | 84.27 |
| 2004 | 79.99 | 116.08 | 80.51 |
| 2005 | 82.69 | 117.31 | 83.66 |
| 2006 | 84.65 | 121.55 | 87.1 |
| 2007 | 87.91 | 123.78 | 89.21 |
| 2008 | 87.99 | 127.08 | 88.55 |
| 2009 | 43.91 | 129.36 | 91.63 |
| 2010 | 86.66 | 131.05 | 87.41 |
| 2011 | 87.2 | 132.13 | 89.93 |
| 2012 | 89.21 | 135.62 | 91.66 |
| 2013 | 94.05 | 137 | 97.14 |
| 2014 | 96.12 | 140.12 | 98.23 |

not be undertaken for Estuarine Acidification (EA) preferably for Sundarban estuaries.

B. Considering the knowledge gap in the domain of acidification of Indian Sundarban estuaries and its subsequent impacts on biodiversity and livelihoods, a brain storming was carried out involving the stakeholders associated with the mangrove ecosystem in the frame work of Indian Sundarbans during 2017–2019 through a set of questionnaires. The entire network of the study consisted

**Table 6.29** Average yearly variation of sediment Cu (in ppm dry wt) in Diamond Harbour during 1984–2014

| Year | Premonsoon | Monsoon | Postmonsoon |
|---|---|---|---|
| 1984 | 39.85 | 32.16 | 37.82 |
| 1985 | 35.55 | 29.44 | 33.43 |
| 1986 | 36.24 | 29 | 33.94 |
| 1987 | 37.83 | 26.84 | 36.41 |
| 1988 | 30.22 | 24.32 | 29.88 |
| 1989 | 31.44 | 23.98 | 30.41 |
| 1990 | 34.31 | 23 | 33.2 |
| 1991 | 31.15 | 24.16 | 29.95 |
| 1992 | 33.62 | 22.89 | 32.86 |
| 1993 | 29.23 | 22.14 | 27.46 |
| 1994 | 30.05 | 21.95 | 27.89 |
| 1995 | 27.14 | 19.77 | 26.44 |
| 1996 | 26.23 | 19.21 | 26.05 |
| 1997 | 25.75 | 18.65 | 25.15 |
| 1998 | 25 | 18.34 | 24.84 |
| 1999 | 24.13 | 17.49 | 23.66 |
| 2000 | 24.02 | 16.32 | 23.88 |
| 2001 | 23.19 | 16.21 | 22.49 |
| 2002 | 22.48 | 15.84 | 21.83 |
| 2003 | 25.78 | 15.21 | 24.32 |
| 2004 | 29.2 | 14.66 | 26.41 |
| 2005 | 21.15 | 13.21 | 20.72 |
| 2006 | 22 | 13.05 | 20.44 |
| 2007 | 19.99 | 12.88 | 20.91 |
| 2008 | 19.05 | 12.23 | 18.94 |
| 2009 | 31.44 | 11.64 | 17.49 |
| 2010 | 22.39 | 11.21 | 21.36 |
| 2011 | 20.56 | 10.94 | 19.22 |
| 2012 | 18.98 | 10.63 | 17.85 |
| 2013 | 17.64 | 10.05 | 16.43 |
| 2014 | 14.55 | 9.86 | 13.22 |

of three stages (i) identification of respondents (policy maker, researcher, fisherman, agriculturist and local inhabitant) (ii) identification of major impacts of acidification (8 in numbers) and (iii) evaluation of the respondent's response to construct the Acidification Impact Scale (AIS) through ranking and voting. It is observed from the results of this exercise that all the stakeholders are aware of the deterioration of estuarine ecosystem services, but they are confused about the exact causes of such impacts. Fisherman, agriculturist and local inhabitant

**Table 6.30** Average yearly variation of sediment Cu (in ppm dry wt) in Namkhana during 1984–2014

| Year | Premonsoon | Monsoon | Postmonsoon |
|---|---|---|---|
| 1984 | 36.23 | 28.49 | 31.1 |
| 1985 | 33.41 | 26.22 | 32.49 |
| 1986 | 34.82 | 27.14 | 31.15 |
| 1987 | 35.39 | 24.34 | 30.02 |
| 1988 | 29.14 | 23 | 29.11 |
| 1989 | 28.66 | 22.8 | 28.44 |
| 1990 | 29.81 | 22.41 | 28.35 |
| 1991 | 28.99 | 21.36 | 27.31 |
| 1992 | 24.82 | 21.08 | 22.65 |
| 1993 | 24.96 | 21 | 21.49 |
| 1994 | 24.33 | 19.45 | 20.14 |
| 1995 | 23.85 | 18.71 | 23 |
| 1996 | 23.06 | 19 | 22.09 |
| 1997 | 23.12 | 17.48 | 21 |
| 1998 | 22.94 | 17 | 20.41 |
| 1999 | 22.13 | 16.15 | 20.69 |
| 2000 | 21.95 | 15.99 | 21.45 |
| 2001 | 21.03 | 15.34 | 20.6 |
| 2002 | 20.69 | 14.98 | 19.21 |
| 2003 | 20.94 | 14.64 | 18.38 |
| 2004 | 20.12 | 13.32 | 17.95 |
| 2005 | 19.74 | 13.08 | 18.33 |
| 2006 | 19.44 | 12.31 | 16.39 |
| 2007 | 18.31 | 11.66 | 20.66 |
| 2008 | 18.49 | 11.41 | 17.72 |
| 2009 | 23.48 | 10.45 | 15.85 |
| 2010 | 16.42 | 9.62 | 14.68 |
| 2011 | 15.19 | 9.48 | 14.93 |
| 2012 | 15.36 | 9.81 | 13.68 |
| 2013 | 16.08 | 8.85 | 12.05 |
| 2014 | 13.12 | 7.93 | 11.89 |

imparted weightage on the adverse impacts on shelled organisms, molluscs, and all of them also opined on the alteration of aquatic pH level and pollution of Sundarban estuaries, but the very term EA is not very transparent to them.

C. Conservative wastes are non-biodegradable in nature and include heavy metals like Zn, Cu, Pb, Cr, Hg etc. Unlike common domestic wastes having organic matter, conservative wastes are not degraded by microbes, and hence remain in the ambient media almost permanently. We are concerned of this long residence

**Table 6.31** Average yearly variation of sediment Cu (in ppm dry wt) in Ajmalmari during 1984–2014

| Year | Premonsoon | Monsoon | Postmonsoon |
|---|---|---|---|
| 1984 | 15.55 | 10.28 | 10.79 |
| 1985 | 14.83 | 9.52 | 10.43 |
| 1986 | 13.36 | 9.93 | 10.02 |
| 1987 | 13 | 8.83 | 9.33 |
| 1988 | 12.84 | 9.19 | 9.69 |
| 1989 | 11.65 | 8.56 | 9.06 |
| 1990 | 11.39 | 7.81 | 8.41 |
| 1991 | 10.85 | 8.22 | 8.82 |
| 1992 | 9.84 | 7.55 | 7.88 |
| 1993 | 8.63 | 2.38 | 8.15 |
| 1994 | 9.01 | 6.39 | 7.33 |
| 1995 | 7.89 | 6.06 | 6.66 |
| 1996 | 9.44 | 6.73 | 7.99 |
| 1997 | 8.1 | 5.28 | 2.54 |
| 1998 | 7.65 | 2.04 | 5.88 |
| 1999 | 8.75 | 7.41 | 8.01 |
| 2000 | 7.02 | 5.37 | 5.97 |
| 2001 | 6.69 | 3.58 | 4.43 |
| 2002 | 6.6 | 1.93 | 4.08 |
| 2003 | 5.44 | 2.84 | 3.34 |
| 2004 | 4.88 | 3.05 | 3.55 |
| 2005 | 4.93 | 2.27 | 2.77 |
| 2006 | 5.14 | 2.73 | 3.06 |
| 2007 | 4.78 | 2.56 | 3.33 |
| 2008 | 4.85 | 1.63 | 2.13 |
| 2009 | 8.32 | 1.52 | 2.02 |
| 2010 | 3.98 | 1.48 | 1.98 |
| 2011 | 3.04 | 1.52 | 1.57 |
| 2012 | 2.96 | 1.43 | 1.33 |
| 2013 | 2.08 | 1.08 | 1.46 |
| 2014 | 2.13 | 1.19 | 1.25 |

time and hence in Sect. 6.3 we have tried to focus on the effect of acidification on heavy metals considering Indian Sundarbans as our case study zone. The results of our study with a data bank of 31 years conclude that the process of acidification largely regulates the concentration of conservative pollutants (primarily heavy metals) in the water and sediments of Indian Sundarban estuaries. Depending on the aquatic pH, bioaccumulation and subsequent toxic effects may occur in plankton, nekton or benthic community. Hence during the

**Table 6.32** Average yearly variation of dissolved Pb (in ppm) in Diamond Harbour during 1984–2014

| Year | Premonsoon | Monsoon | Postmonsoon |
|---|---|---|---|
| 1984 | 9.54 | 14.73 | 11.48 |
| 1985 | 11.23 | 17.85 | 13.65 |
| 1986 | 13.41 | 21.32 | 15.83 |
| 1987 | 15.66 | 27.65 | 19.23 |
| 1988 | 17.85 | 30.33 | 20.44 |
| 1989 | 14.93 | 34.88 | 23.85 |
| 1990 | 19.21 | 39.21 | 26.33 |
| 1991 | 19.64 | 42.65 | 26.98 |
| 1992 | 20.95 | 47.83 | 28.22 |
| 1993 | 20.89 | 49.95 | 31.35 |
| 1994 | 22.54 | 52.16 | 35.46 |
| 1995 | 23.31 | 57.65 | 38.32 |
| 1996 | 27.73 | 59.99 | 40.05 |
| 1997 | 28.15 | 60.23 | 41.78 |
| 1998 | 28.33 | 64.79 | 46.35 |
| 1999 | 29.71 | 68.66 | 44.82 |
| 2000 | 30.62 | 70.23 | 49.33 |
| 2001 | 32.05 | 74.39 | 50.05 |
| 2002 | 33.89 | 76.82 | 51.78 |
| 2003 | 34.73 | 79.99 | 54.32 |
| 2004 | 35.11 | 80.23 | 55.69 |
| 2005 | 35.85 | 84.88 | 57.31 |
| 2006 | 37.34 | 87.31 | 56.43 |
| 2007 | 37.39 | 89.02 | 59.78 |
| 2008 | 38.32 | 93.41 | 61.28 |
| 2009 | 23.21 | 96.46 | 63.43 |
| 2010 | 38.43 | 97.23 | 67.88 |
| 2011 | 40.6 | 99.1 | 68.01 |
| 2012 | 41.44 | 101.61 | 72.44 |
| 2013 | 43.34 | 107.35 | 75.18 |
| 2014 | 47.98 | 108.88 | 77.28 |

planning process of cleaning the lower Gangetic delta region focus and proper weightage should be given on the role of acidification and/or aquatic pH as this parameter seems to be a major driver in the domain of aquatic health of Indian Sundarbans.

**Table 6.33** Average yearly variation of dissolved Pb (in ppm) in Namkhana during 1984–2014

| Year | Premonsoon | Monsoon | Postmonsoon |
|---|---|---|---|
| 1984 | 8.59 | 12.44 | 10.69 |
| 1985 | 9.21 | 13.31 | 11.31 |
| 1986 | 10.08 | 13.65 | 12.18 |
| 1987 | 10.99 | 15.83 | 13.09 |
| 1988 | 11.44 | 17.31 | 13.54 |
| 1989 | 12.66 | 17.99 | 14.76 |
| 1990 | 13.85 | 19.22 | 15.95 |
| 1991 | 14.79 | 20.05 | 16.89 |
| 1992 | 15.66 | 21.46 | 17.76 |
| 1993 | 17.94 | 23.98 | 20.04 |
| 1994 | 18.55 | 24.73 | 20.65 |
| 1995 | 19.21 | 27.62 | 21.31 |
| 1996 | 21.33 | 28.31 | 23.43 |
| 1997 | 24.79 | 29.35 | 26.89 |
| 1998 | 23.83 | 30.44 | 25.93 |
| 1999 | 25.97 | 35.63 | 28.07 |
| 2000 | 27.88 | 41.38 | 29.98 |
| 2001 | 28.22 | 44.75 | 30.32 |
| 2002 | 30.65 | 46.82 | 32.75 |
| 2003 | 32.41 | 50.35 | 34.51 |
| 2004 | 30.79 | 53.44 | 32.89 |
| 2005 | 33.81 | 57.81 | 39.91 |
| 2006 | 35.42 | 60.05 | 41.52 |
| 2007 | 36.21 | 64.66 | 38.31 |
| 2008 | 38 | 70.91 | 40.1 |
| 2009 | 19.49 | 77.41 | 21.59 |
| 2010 | 35.48 | 78.35 | 37.58 |
| 2011 | 37.79 | 80.21 | 43.89 |
| 2012 | 40.83 | 84.68 | 45.93 |
| 2013 | 42 | 87.38 | 44.1 |
| 2014 | 45.15 | 89.42 | 48.37 |

**Table 6.34** Average yearly variation of dissolved Pb (in ppm) in Ajmalmari during 1984–2014

| Year | Premonsoon | Monsoon | Postmonsoon |
|---|---|---|---|
| 1984 | 4.65 | 9.12 | 7.39 |
| 1985 | 4.82 | 9.75 | 8.1 |
| 1986 | 5.13 | 9.83 | 8.15 |
| 1987 | 5.15 | 10.21 | 8.06 |
| 1988 | 6 | 10.72 | 9.11 |
| 1989 | 6.73 | 10.54 | 9.15 |
| 1990 | 6.85 | 10.98 | 9.34 |
| 1991 | 7.12 | 11.2 | 9.76 |
| 1992 | 7.66 | 11.44 | 10.1 |
| 1993 | 8.15 | 12.65 | 10.71 |
| 1994 | 8.31 | 13.1 | 11.02 |
| 1995 | 9 | 13.44 | 11.66 |
| 1996 | 9.54 | 13.93 | 11.44 |
| 1997 | 9.63 | 14.75 | 12.08 |
| 1998 | 9.88 | 15.01 | 12.95 |
| 1999 | 10.01 | 15.63 | 12.99 |
| 2000 | 10.53 | 16.17 | 13.33 |
| 2001 | 10.77 | 16.7 | 13.55 |
| 2002 | 11.21 | 17 | 13.78 |
| 2003 | 11.78 | 17.44 | 13.83 |
| 2004 | 12.02 | 18.23 | 13.97 |
| 2005 | 12.14 | 19.01 | 14.02 |
| 2006 | 12.99 | 19.39 | 15.11 |
| 2007 | 13.61 | 20.55 | 15.44 |
| 2008 | 13.45 | 22.05 | 15.92 |
| 2009 | 8.16 | 21.38 | 16.08 |
| 2010 | 14.32 | 22.88 | 16.05 |
| 2011 | 15.15 | 23.67 | 16.37 |
| 2012 | 15.73 | 24.04 | 17.15 |
| 2013 | 16.02 | 24.78 | 17.39 |
| 2014 | 17.14 | 24.83 | 20.56 |

**Table 6.35** Average yearly variation of sediment Pb (in ppm dry wt) in Diamond Harbour during 1984–2014

| Year | Premonsoon | Monsoon | Postmonsoon |
|---|---|---|---|
| 1984 | 23.19 | 15.65 | 22.69 |
| 1985 | 21.21 | 14.32 | 18.61 |
| 1986 | 19.93 | 14.01 | 17.94 |
| 1987 | 21.44 | 13.98 | 16.04 |
| 1988 | 16.64 | 12.64 | 15.43 |
| 1989 | 15.33 | 13.05 | 14.73 |
| 1990 | 19.05 | 12.01 | 18.45 |
| 1991 | 15.02 | 11.64 | 14.42 |
| 1992 | 18.09 | 9.83 | 17.49 |
| 1993 | 14.88 | 8.44 | 14.28 |
| 1994 | 14.96 | 8.12 | 14.36 |
| 1995 | 13.94 | 7.75 | 13.34 |
| 1996 | 11.54 | 6.62 | 7.6 |
| 1997 | 9.2 | 6.01 | 7.94 |
| 1998 | 13.18 | 5.98 | 12.58 |
| 1999 | 9.98 | 5.44 | 9.52 |
| 2000 | 10.02 | 5.23 | 9.48 |
| 2001 | 14.33 | 5.01 | 13.73 |
| 2002 | 11.2 | 4.86 | 10.7 |
| 2003 | 9.81 | 3.85 | 9.22 |
| 2004 | 9.72 | 3.63 | 9.31 |
| 2005 | 8.56 | 3.54 | 7.83 |
| 2006 | 9.33 | 3.3 | 7.16 |
| 2007 | 10.66 | 3.21 | 8.06 |
| 2008 | 9.21 | 3.09 | 8.71 |
| 2009 | 14.09 | 2.89 | 7.59 |
| 2010 | 7.05 | 1.93 | 6.55 |
| 2011 | 6.29 | 1.44 | 5.79 |
| 2012 | 5.83 | 1.01 | 5.33 |
| 2013 | 5.75 | 1.06 | 5.25 |
| 2014 | 5.01 | 0.93 | 4.88 |

**Table 6.36** Average yearly variation of sediment Pb (in ppm dry wt) in Namkhana during 1984–2014

| Year | Premonsoon | Monsoon | Postmonsoon |
|---|---|---|---|
| 1984 | 21.12 | 13.34 | 19.87 |
| 1985 | 20.47 | 13.01 | 17.52 |
| 1986 | 17.49 | 12.86 | 16.89 |
| 1987 | 16.83 | 11.55 | 16.23 |
| 1988 | 16 | 12.04 | 15.4 |
| 1989 | 15 | 12 | 14.4 |
| 1990 | 14.19 | 11.54 | 13.59 |
| 1991 | 13.6 | 11.21 | 13 |
| 1992 | 13.44 | 8.85 | 12.84 |
| 1993 | 12.49 | 7.79 | 11.89 |
| 1994 | 12.05 | 8 | 11.45 |
| 1995 | 9.97 | 7.23 | 9.47 |
| 1996 | 11.37 | 5.61 | 10.87 |
| 1997 | 8.66 | 5 | 8.16 |
| 1998 | 9.08 | 4.96 | 8.84 |
| 1999 | 9.34 | 4.33 | 8.58 |
| 2000 | 9.11 | 4.02 | 8.61 |
| 2001 | 10.41 | 3.98 | 9.91 |
| 2002 | 8.13 | 3.34 | 7.63 |
| 2003 | 7.5 | 3.01 | 6.9 |
| 2004 | 7.05 | 3.15 | 6.55 |
| 2005 | 11.2 | 3.21 | 7 |
| 2006 | 6.85 | 2.97 | 5.7 |
| 2007 | 8.95 | 2.68 | 8.45 |
| 2008 | 6.2 | 2.44 | 6.35 |
| 2009 | 5.32 | 2.33 | 4.82 |
| 2010 | 4.88 | 2.31 | 4.38 |
| 2011 | 4.79 | 1.85 | 4.29 |
| 2012 | 3.96 | 1.4 | 3.46 |
| 2013 | 3.88 | 0.98 | 3.38 |
| 2014 | 3.2 | 0.89 | 2.98 |

**Table 6.37** Average yearly variation of sediment Pb (in ppm dry wt) in Ajmalmari during 1984–2014

| Year | Premonsoon | Monsoon | Postmonsoon |
|---|---|---|---|
| 1984 | 10.79 | 9.41 | 10.28 |
| 1985 | 10.43 | 9.32 | 9.52 |
| 1986 | 10.02 | 8.85 | 9.93 |
| 1987 | 9.33 | 7.79 | 8.83 |
| 1988 | 9.69 | 7.02 | 9.19 |
| 1989 | 9.06 | 7 | 8.56 |
| 1990 | 8.41 | 7.33 | 7.81 |
| 1991 | 8.82 | 6.89 | 8.22 |
| 1992 | 8.71 | 6.54 | 7.55 |
| 1993 | 8.15 | 4.28 | 4.38 |
| 1994 | 7.33 | 3.35 | 7.39 |
| 1995 | 6.66 | 5.82 | 6.06 |
| 1996 | 7.99 | 5.79 | 6.73 |
| 1997 | 5.54 | 4.85 | 5.28 |
| 1998 | 5.88 | 4.89 | 5.04 |
| 1999 | 8.01 | 2.98 | 7.41 |
| 2000 | 5.97 | 2.63 | 5.37 |
| 2001 | 5.43 | 2.54 | 3.58 |
| 2002 | 4.08 | 2.33 | 2.93 |
| 2003 | 3.34 | 1.85 | 2.84 |
| 2004 | 3.55 | 2.02 | 3.05 |
| 2005 | 5.77 | 2.19 | 2.27 |
| 2006 | 5.06 | 2.17 | 5.73 |
| 2007 | 5.33 | 3.35 | 3.56 |
| 2008 | 2.13 | 2.08 | 2.63 |
| 2009 | 2.02 | 1.54 | 1.62 |
| 2010 | 1.98 | 1.31 | 1.48 |
| 2011 | 1.57 | 0.96 | 1.07 |
| 2012 | 1.33 | 0.62 | 0.83 |
| 2013 | 1.11 | 0.48 | 0.68 |
| 2014 | 1.05 | 0.44 | 0.59 |

## References

Banerjee LK, Das GC (1972) New distributional records from Orissa coast. Bull Bot Surv India 14(1–4):184–186

Banerjee LK, Rao TA (1990) Mangroves of Orissa Coast and their ecology. Bishen Singh Mahendra Pal Singh, Dehra Dun, India, 118

Barbier EB (1991) An approach to economic evaluation of tropical wetlands: with examples from Guatemala and Nicaragua. In: Girvan NP, Simons D (eds) Caribbean ecology and economics, Caribbean conservation association, St. Michael, Barbados, pp 207–231

Barbier EB, Acreman M, Knowler D (1997) Economic valuation of wetlands: a guide for policy makers and planners. Gland, Switzerland, Ramsar Convention Bureau

Barua P, Mitra A, Banerjee K, Chowdhury MSN (2011) Seasonal variation of heavy metals accumulation in water and oyster (*Saccostrea cucullata*) inhabiting central and western sector of Indian Sundarbans. Environ Res J 5(3):121–130

Bhattacharyya DP, Mahapatra B, Hadar KC, Banerjee K, Mitra A (2001) Spectrophotometric analysis of heavy metals in coastal waters of West Bengal. Indian J Phys 75B(2):147–149

Biswas S, Pal N, Fazli P, Zaman S, Mitra A (2019) Samosa and Kachuri from mangrove associate species: an innovative utilization of coastal flora. Int J Res Appl Sci Eng Tech 7(1):73–79

Brander LM, Raymond JG, Florax M, Vermat JE (2006) The empirics of wetland valuation: a comprehensive summary and a meta-analysis of the literature. Environ Resour Econ 33:223–250 https://doi.org/10.1007/s10640-005-3104-4,CopyrightSpringer

Brauman KA, Daily GC, Duarte TK, Mooney HA (2007) The nature and value of ecosystem services: an overview high-lighting hydrologic services. Annu Rev Environ Resour 32:67–98

Brouwer R (2000) Environmental value transfer: state of the art and future prospects. Ecol Econo 32:137–152

Brscic K (2006) The impact of agrotourism on agricultural production. J Cent Euro Agri 7(3)

Carpenter SR, Caraco NF, Correll DL, Howarth RW, Sharpley AN, Smith VH (1998) Nonpoint pollution of surface waters with phosphorus and nitrogen. Ecol App 8(3):559–568

Chakraborti D, Adams F, Van Mol W, Irgolic JK (1987) Determination of trace metals in natural waters at nanogram per litre levels by electrothermal atomic absorption spectrophotometry after extraction with sodium diethyldithiocarbamate. Anal Chem Acta 196:23–31

Chakraborty P, Gopalpillai Y, Murimboh J, Fasfous II, Chakrabarti CL (2006) Kinetic speciation of nickel in mining and municipal effluents. Anal Bioanal Chem 386:1803–1813

Chan KMA, Shaw MR, Cameron DR, Underwood EC, Daily GC (2006) Conservation planning for ecosystem ser-vices. PLoS Biol 4:2138–2152

Chaudhuri AB, Choudhury A (1994) Mangroves of the Sundarbans, 1:247 (IUCN, Bangkok, Thailand)

Clark WC, Jones DD, Holling CS (1979) Lessons for ecological policy design: a case study of ecosystem management. Ecol Model 7:1–53

Dani CS, Kar CS, Behura BK (1999) Bhitarkanika—a unique mangrove eco-system. Nature and Wildlife Conservation Society of Orissa, Bhubaneswar, India, pp 30–43

Das S, Mitra A, Banerjee K, Mukherjee D, Bhattacharyya DP (2005) Heavy metal pollution in marine and estuarine environment: the case of North and South 24 Parganas district of coastal west Bengal. Indian J Environ Ecoplan 10(1):19–25

del-Val E, Armesto JJ, Barbosa O, Christie DA, Gutiérrez AG, Jones CG, Marquet PA, Weathers KC (2006) Rain forest islands in the chilean semiarid region: fog-dependency. Ecosyst Persistence Tree Regeneration Ecosys 9:598–608

Enfors E, Gordon L, Peterson GD, Bossio D (2008) Making investments in dryland development work: participatory scenario planning in the Makanya catchment, Tanzania. Ecol Soc 13(2):42

Engel S, Iskandarani M, del Pilar UM (2005) Improved water supply in the Ghanaian Volta basin: who uses it and who participates in community decision-making? Int Food Policy Res Inst EPT Discuss Paper 129:1–61

Foley JA, DeFries R, Asner GP, Barford C, Bonan G, Carpenter SR et al (2005) Global consequences of land use. Science 309:570–574

Hughes TP, Bellwood DR, Folke C, Steneck RS, Wilson J (2005) New paradigms for supporting the resilience of marine ecosystems. Trends Ecol Evol 20:380–386

Kareiva P, Watts S, McDonald R, Boucher T (2007) Domesticated nature: shaping landscapes and ecosystems for human welfare. Sci 316:1866–1869

Khan YSA, Hossain MS, Hossain SMGA, Halimuzzaman AHM (1998) An environmental assessment of trace metals in the Ganges–Brahmaputra–Meghna estuary. J Remote Sens Environ 2:103–117

Kramer RA, Richter DD, Pattanayak S, Sharma NP (1997) Ecological and economic analysis of watershed protection in Eastern Madagascar. J Environ Manage 49(3):277–295

Kristensen E (2008) Mangrove crabs as ecosystem engineers, with emphasis on sediment processes. J Sea Res 59(1–2):30–43

Li YH, Burkhardt M, Teraoka H (1984) Desorption and coagulation of trace elements during estuarine mixing. Geo Cosmo Acta 48:2011–2019

Malo BA (1977) Partial extraction of metals from aquatic sediments. Environ Sci Technol 11:277–288

Millennium Ecosystem Assessment (2003) Ecosystems and human well-being. Washington (DC), Island Press

Millennium Ecosystem Assessment (2005) Ecosystems and human well-being: synthesis. A report of the millennium ecosystem assessment. World Resources Institute and Island Press, Washington, DC

Mishra PK, Sahu JR, Upadhyay VP (2005) Species diversity in Bhitarkanika mangrove ecosystem in Orissa, India. Lyonia 8:73–81

Mishra SC, Panigrahi G (1987) Studies on the mangrove flora of Orissa with particular reference to Rhizophoraceae. R. Br. J Eco Taxo Bot 11:121–132

Mitra A (1998) Status of coastal pollution in West Bengal with special reference to heavy metals. J Indian Ocean Stud 5(2):135–138

Mitra A (2013) Sensitivity of mangrove ecosystem to changing climate. Springer. https://doi.org/10.1007/978; 81-322- 1509-7:323

Mitra A, Choudhury A (1998) Impact of heavy metals on ichthyoplankton richness around Nayachar Island in Hooghly estuary. Riv Behavi Cont 25:80–83

Mitra A, Zaman S (2014) Carbon sequestration by coastal floral community. Published by The Energy and Resources Institute (TERI) TERI Press, India

Mitra A, Zaman S (2015) Blue carbon reservoir of the blue planet. Published by Springer. ISBN 978-81-322-2106–7 (Springer. DOI https://doi.org/10.1007/978-81-322-2107-4)

Mitra A, Zaman S (2016) Basics of marine and estuarine ecology. Springer. ISBN 978-81-322-2705-2

Mitra A, Trivedi S, Chaudhuri A, Gupta A, Chakraborty K, Sanyal T, Choudhury, A (1994) Comparative study of heavy metal accumulation by different mangrove plants. In: Proceedings of national convention on environment of India—challenges for the 21st century. Institution of Public Health Engineers, Calcutta (India), T-III/9–T-III/14

Mitra A, Choudhury A, Mitra S, Mukherjee S (2000) Heavy metal concentration in Indian coastal fishes. Res J Chem Environ 4(4):35–37

Mitra A, Banerjee K, Bhattacharyya DP (2004) The other face of mangroves. Published by Department of Environment, Government of West Bengal

Mitra A, Zaman S, Bhattacharyya SB (2013) Heavy metal pollution in the lower Gangetic ecosystem. In: Abedin MA, Habiba U, Shaw R (eds) Water insecurity: a social dilemma. Published by Emerald Group Publishing Limited (ISBN: 9781781908822)

Mitra A, Sundaresan J, Bera D, Zaman S (2016) Seaweed resource utilization: an adaptation to climate change. Published by CSIR-National Institute of Science Communication And Information Resources (NISCAIR), New Delhi, ISBN 978-81-7236-349-9:120

Monfreda C, Ramankutty N, Foley JA (2008) Farming the planet: 2. Geographic distribution of crop areas, yields, physio-logical types, and net primary production in the year 2000. Global Biogeochem Cycles 22. https://doi.org/10.1029/2007GB002947

Moorthy P (1995) Effects of UV-B radiation on mangrove environment: physiological responses of Rhizophora apiculata Blume. Ph.D. thesis, Annamalai University, India, p 130

Pattanaik SN, Choudhury BP (1989) Present status and future development of mangrove vegetation in Bhitarkanika in Indo-US Workshop on Wetlands, Mangroves and Biosphere Reserve, 89–95, Government of India, Ministry of Environments and Forests, New Delhi

Pattnaik C, Redy CS, Dhal NK, Das R (2008) Utilisation of mangrove forest in Bhitarkanika Wild Life Sanctuary, Orissa. Indian J Tradit Know 7(4):598–603

Peterson GD, Beard D, Beisner B, Bennett E, Carpenter S, Cumming G et al (2003) Assessing future ecosystem services: a case study of the northern highland lake district, Wisconsin. Conserve Ecol 7(3):1

Pramanick P (2017) Study on the ecology and biochemical aspects of the major seaweeds of Indian Sundarbans with few edible products. Ph.D Thesis, p 153

Ramankutty N, Evan AT, Monfreda C, Foley JA (2008) Farming the planet: 1. Geographic distribution of global agricultural lands in the year 2000. Global Biogeochem Cycles 22, doi:https://doi.org/10.1029/2007 GB002952

Ray Chaudhuri T, Fazli P, Zaman S, Pramanick P, Bose R, Mitra A (2014) Impact of acidification on heavy metals in Hooghly Estuary. J Harmon Res Appl Sci 2(2):91–97

Ricketts TH, Regetz J, Steffan-Dewenter I, Cunningham SA, Kremen C, Bogdanski A, Gemmill-Herren B, Greenleaf SS, Klein AM, Mayfield MM, Morandin LA, Ochieng A, Potts SG, Viana BF (2008) Landscape effects on crop pollination services: are there general patterns? Ecol Lett 11:499–515

Rodriguez JP, Beard TD, Bennett EM, Cumming GS, Cork SJ, Agard J et al (2006) Trade-offs across space, time, and ecosystem services. Ecol Soc 11(1):28

Sundararaj V, Krishnamurthy K (1972) Studies on phytoplankton pigments in Porto Novo waters (India). I. Mangrove J of Exp Mar Biol Ecol 14(3):275–284

Thangam TS (1990) Studies on marine plants for mosquito control. Ph.D. thesis, Annamalai University, India, p 68

Trivedi S, Mitra A, Bag M, Ghosh I, Choudhury A (1995) Heavy metal concentration in mud skipper *Boleophthalmus boddaerti* of Nayachar Island, India. Indian J Environ Health NEERI 37(1):42–45

Twilley RR, Chen RH, Hargis T (1992). Carbon sinks in mangroves and their implications to carbon budget of tropical coastal ecosystems. In: Natural sinks of $CO_2$. Springer, Netherlands

Upadhyay VP, Mishra PK, Sahu JR (2008) Distribution of mangrove species within Bhitarkanika National Park in Orissa, India. Trees Life J 3:1–5

UNESCO (United Nations Educational, Scientific and Cultural Organization) (2012) World's groundwater resources are suffering from poor governance. UNESCO Natural Sciences Sector News. Paris, UNESCO

Van Bochove J, Sullivan E, Nakamura T (2014) The importance of mangroves to people: a call to action. United Nations Environment Programme

Wolf EC, Cooper DJ, Hobbs NT (2007) Hydrologic regime and herbivory stabilize an alternative state in Yellowstone National Park. Ecol Appl 17:1572–1587

Woodward R, Wui Y (2001) The economic value of wetland services: a meta-analysis. Ecol Econ 257–270

Zaman S, De UK, Pramanick P, Thakur S, Amin G, Fazli P, Mitra A (2014) Forecasting heavy metal level on the basis of acidification trend in the major estuarine system of Indian Sundarbans. J Energy Environ Carbon Cred 4(2):1–9

Zedler JB (2003) Wetlands at your service: reducing impacts of agriculture at the watershed scale. Front Ecol Environ 1:65–72

## *Internet References*

http://cices.eu/
http://www.fao.org/gpa/sediments/habitat.htm
http://www.reefrelief.org/Documents/mangrove.html

# Chapter 7
# Regulatory Role of Blue Carbon in Estuarine Acidification

**Abstract** Blue carbon comprising the coastal vegetation (mangroves, marsh grass, sea grass etc.) exerts a regulatory effect on the pH of the estuarine water as these halophytes absorb carbon dioxide for photosynthesis. The key biochemical actor in the process is Ribulose-1,5-Biphosphate Carboxylase Oxygenase (RuBisCo), present in the halophytes that fixes the carbon and push shift the pH towards alkaline value. The present chapter critically addresses the role of blue carbon in maintaining an equilibrium condition of estuarine water in context to pH.

## 7.1 Mangroves

Mangrove forest area accounts for 0.7% of the total tropical forest in the world. Most mangroves are concentrated in Asia (42%), followed by Africa (20%), North and Central America (15%), Oceania (12%) and South America (15%). The entire mangrove spectrum of the world encompasses 31 families under which 48 genera and 96 species are included as documented in the site http://www.bsienvis.nic.in/Database/IndianMangroves_3941.aspx#WORLD DISTRIBUTION OF MANGROVES (Annexure 7). According to a study conducted by a team of U.S. Forest Service and university scientists coastal mangrove forests store more carbon than almost any other forest on Earth. Their findings are published online in the journal *Nature Geoscience* (www.nature.com/naturegeoscience.com). The potential of mangrove forest to store such large amounts of carbon may be attributed, in part, to the deep organic-rich soils on which it thrives (Fig. 7.1).

The soil depth mostly increases with time as the silt of the overlying aquatic phase deposit on the existing soil bed (intertidal mudflats). This happens because the intertidal mudflats suffer total submergence by silty water during high tide (Fig. 7.2).

Mangrove-sediment/substratum stores carbon on an average five times larger than those typically observed in temperate, boreal and tropical terrestrial forests, on a per-unit-area basis. The mangrove forest's complex root systems, which anchor the plants into underwater sediment, slow down incoming tidal waters allowing organic and inorganic material to settle on the sediment surface. Low oxygen conditions slow decay rates, resulting in much of the carbon accumulating in the soil. In fact,

 
A. Mitra and S. Zaman, *Estuarine Acidification*,
https://doi.org/10.1007/978-3-030-84792-0_7

**Fig. 7.1** Organic carbon—rich soil (substratum) of mangrove ecosystem

**Fig. 7.2** Complete submergence of intertidal mudflats during high tide

mangroves have more carbon in their soil alone than most tropical forests have in all their biomass and soil combined.

The capacity of mangroves, sea grasses, and salt marshes to sequester carbon dioxide from the atmosphere is becoming increasingly recognized at an international level. Of all the biological carbon, also termed as 'green carbon', captured in the world, over half (55%) is captured by mangroves, sea grasses, salt marshes, and other marine living organisms, which are also known more specifically as 'blue carbon'. Mangroves, salt marshes, and sea grasses form much of the earth's blue carbon sinks (http://www.recoftc.org/site/resources/Mangroves-more-Carbon-Rich-and-Important-for-Climate-Change.php).

It has been reported that mangroves are among the most carbon-dense forests in the tropics (sample-wide mean: 1023 MgC $ha^{-1} \pm 88$ s:e:m:) and exceptionally high compared to mean carbon storage of the world's major forest domains (Donato et al. 2011). Estuarine sites contained a mean of 1074 MgC $ha^{-1}$ ($\pm$171 s:e:m:); oceanic sites contained $990 \pm 96$ MgC $ha^{-1}$. AGC pools are sizeable (mean 159 MgC $ha^{-1}$, maximum 435 MgC $ha^{-1}$), but Below Ground Storage in soils dominate, accounting for 71–98% and 49–90% of total storage in estuarine and oceanic sites, respectively. High per-hectare carbon storage coupled with a pan-tropical distribution (total area ~14 million ha; FAO, 2007; Bouillon et al. 2009), suggests mangroves are a globally important surface carbon reserve.

Mangroves exhibit almost all the salient features of plants that follow the route of $C_3$ photosynthesis. However, it is interesting to note that although mangroves are considered under the umbrella of $C_3$ photosynthetic plants, they can also be compared with 'seaweed' as they are able to grow in submerged and high saline environment. There are four important evidences to prove that mangrove plants use $C_3$ photosynthetic biochemistry. These evidences are highlighted in points.

(1) Leaves of mangroves lack bundle sheath cells, which is the characteristic of plants employing $C_4$ photosynthesis.
(2) Carbon isotope composition of mangrove leaves offers a direct evidence of the presence mangroves under the category of $C_3$ photosynthetic plants. It has been documented by Smith and Epstein (1971) that the carbon isotope composition of leaves shows a range of $\delta^{13}C$ values from −23.2% to −34.3% as reported for $C_3$ plants.
(3) The carbon dioxide compensation point at 25 °C and saturated light intensity ranges from 45 to 60 $\mu l\,l^{-1}$, consistent with the photorespiratory activity typical of $C_3$ photosynthesis.
(4) In mangroves, the photosynthetic rates are maximal at leaf temperature less than 35 °C and generally become light saturated at intensities ranging from 30 to 50% of full sunlight.

Finally, there is no convincing evidence in mangroves on transformation of environmentally induced from $C_3$ to either $C_4$ or CAM photosynthetic biochemistry. The gas exchange in mangroves is strongly regulated by the amount and properties of the enzyme ribulose-1, 5-biphosphate carboxylase/oxygenase (RuBisCO), which catalyzes the combination of carbon dioxide with the acceptor molecule ribulose-1, 5-biphosphate (Fig. 7.3).

The ability of mangroves to scrub carbon from the ambient media is the key to regulate the process of acidification. Basically mangroves absorb carbon dioxide from the atmosphere and the water column and store it as carbon in their vegetative

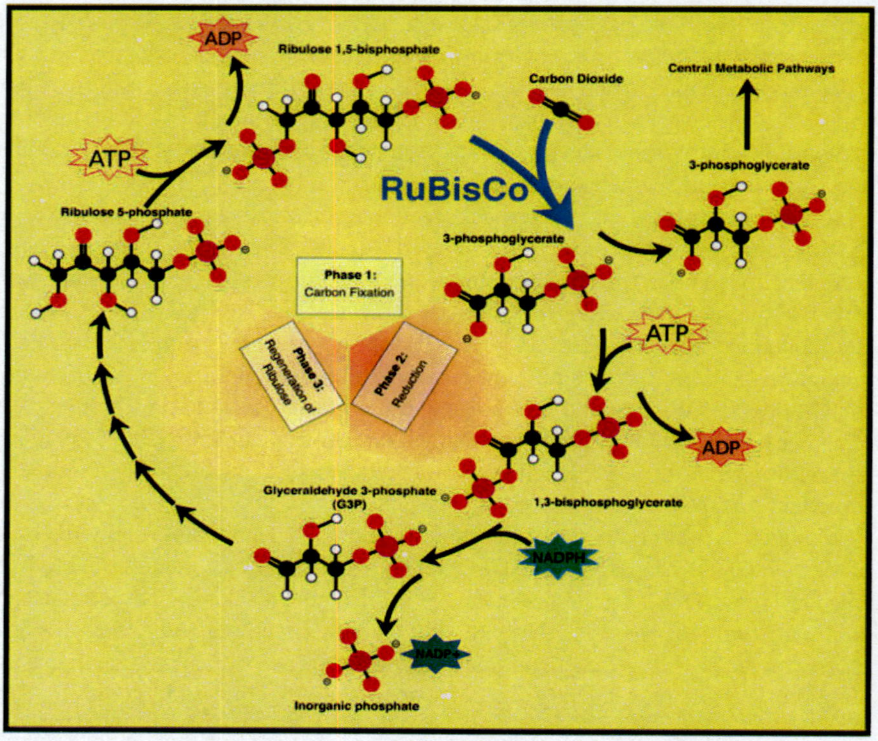


**Fig. 7.3** RuBisCo action in mangrove flora (*Source* Mitra and Zaman 2016)

and reproductive parts. The principal carbon-fixing enzyme in plants is ribulose 1, 5-biphosphate carboxylase/oxygenase (RuBisCO). The magnitude of the mangrove vegetation pool thus exerts a regulatory role on the pH of the estuarine water as these halophytes absorb carbon dioxide for photosynthesis and shift the equilibrium towards alkalinity.

We have made a detailed study to assess the inter-relationship between mangrove vegetation cover-cum-biomass and ambient aquatic pH during 2019 considering six sampling stations in eastern sector of Indian Sundarbans (Table 7.1).

**Table 7.1** Sampling stations with coordinates in western Indian Sundarbans

| Sampling stations | Latitude | Longitude |
|---|---|---|
| Arbesi (Stn. 1) | 22° 12′ 07.44″ N | 89° 00′ 03.15″ E |
| Jhilla (Stn. 2) | 22° 10′ 20.55″ N | 88° 56′ 04.13″ E |
| Harinbhanga (Stn. 3) | 21° 59′ 17.79″ N | 88° 59′ 05.32″ E |
| Khatuajhuri (Stn. 4) | 22° 03′ 09.52″ N | 89° 01′ 03.12″ E |
| Chamta (Stn. 5) | 21° 53′ 05.91″ N | 88° 57′ 26.52″ E |
| Chankhali (Stn. 6) | 21° 51′ 31.73″ N | 89° 00′ 50.92″ E |

We observed high pH reduction in areas with low mangrove pool confirming the positive role of mangroves in retarding the pace of acidification (Table 7.2).

The polynomial presented in Fig. 7.4 (considering aquatic pH as dependent variable and forest area as independent variable for six selected stations in eastern Indian Sundarbans) shows the positive role of mangrove vegetation in retarding the pace of EA.

**Table 7.2** Mangrove pool (tonnes), % pH reduction in the eastern Indian Sundarbans

| Station | Total biomass of dominate species (tonnes/ha) Date type: primary | | Forest area (ha) Date type: primary | Mangrove pool (tonnes) Date type: primary | % pH reduction/decade Date type: primary |
|---|---|---|---|---|---|
| Arbesi (Stn. 1) | *S. apetala* | 67.12 | 7200 | 1,971,000.0 | 1.08 |
| | *A. alba* | 62.2 | | | |
| | *E. agallocha* | 21.79 | | | |
| | *A. officinalis* | 63.56 | | | |
| | *A. marina* | 59.08 | | | |
| | | *273.75* | | | |
| Jhilla (Stn. 2) | *S. apetala* | 65.33 | 8387 | 2,309,863.67 | 0.96 |
| | *A. alba* | 63.88 | | | |
| | *E. agallocha* | 21.97 | | | |
| | *A. officinalis* | 64.13 | | | |
| | *A. marina* | 60.1 | | | |
| | | *275.41* | | | |
| Harinbhanga (Stn. 3) | *S. apetala* | 66.21 | 7465 | 1,971,357.20 | 0.72 |
| | *A. alba* | 59.31 | | | |
| | *E. agallocha* | 21.84 | | | |
| | *A. officinalis* | 62.17 | | | |
| | *A. marina* | 54.55 | | | |
| | | *264.08* | | | |
| Uajhuri (Stn. 4) | *S. apetala* | 66.18 | 8666 | 24,022,648.5 | 0.69 |
| | *A. alba* | 64.77 | | | |
| | *E. agallocha* | 21.58 | | | |
| | *A. officinalis* | 64.97 | | | |
| | *A. marina* | 59.75 | | | |
| | | *277.25* | | | |
| Chamta (Stn. 5) | *S. apetala* | 64.99 | 12,955 | 3,641,002.75 | 0.72 |
| | *A. alba* | 66.09 | | | |

(continued)

**Table 7.2** (continued)

| Station | Total biomass of dominate species (tonnes/ha) Date type: primary | | Forest area (ha) Date type: primary | Mangrove pool (tonnes) Date type: primary | % pH reduction/decade Date type: primary |
|---|---|---|---|---|---|
| | *E.agallocha* | 22.02 | | | |
| | *A. officinalis* | 67.94 | | | |
| | *A. marina* | 60.01 | | | |
| | | *281.05* | | | |
| Chandkhali (Stn. 6) | *S. apetala* | 62.14 | 9635 | 2,536,895.5 | 0.36 |
| | *A. alba* | 60.54 | | | |
| | *E.agallocha* | 22.87 | | | |
| | *A. officinalis* | 62.58 | | | |
| | *A. marina* | 55.17 | | | |
| | | *263.3* | | | |

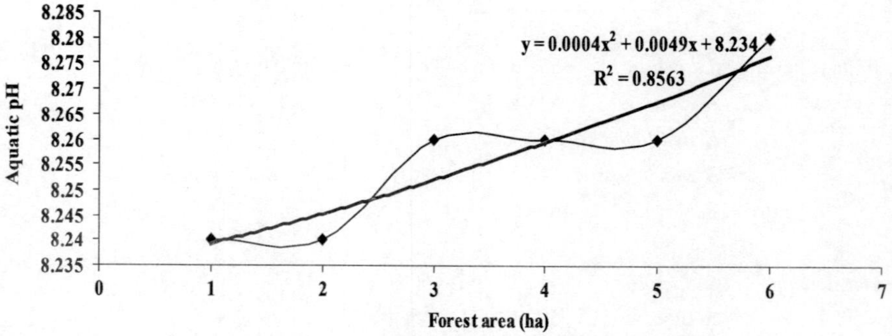


**Fig. 7.4** Inter-relationships between forest area and aquatic pH in eastern Indian Sundarbans

## 7.2 Sea Grass and Saltmarsh Grass

The burning of fossil fuels like coal, oil etc. results in the release of carbon dioxide in to the atmosphere, which triggers the process climate change at local, regional and global scales. The atmospheric carbon dioxide is absorbed into the aquatic phase of oceans, seas, bays and estuaries where chemical reactions with the seawater produce carbonic acid that poses an adverse impact on the marine life particularly to organisms with calcareous shell. Sea grass beds provide several ecosystem services like providing food and shelter for marine organisms, erosion control, bioremediation etc. They also absorb carbon dioxide to run the process of photosynthesis and thus can buffer the pH of the oceans and estuaries thereby help neutralize the adverse effects caused by acidification.

The present authors assessed the stored carbon in major species of sea grass sampled from the Gulf of Mannar region during 2014 and documented the stored carbon both in AGB and BGB as highlighted in Tables 7.3 and 7.4 respectively. Three study sites namely Koswari (08° 52.34′N, 78° 13.04′E; Stn. 1), Kariyachalli (08° 57.36′N, 78° 14.42′E; Stn. 2) and Vilanguchalli (08° 56.22′N, 78° 15.59′E; Stn. 3) and three species (*Cymodocea serrulata, Thalassia hemprichii, Halophila ovalis*) were selected for estimation of stored carbon. Simultaneously pH values of the ambient aquatic phase were also estimated to evaluate the buffering potential of sea grass. It is observed that pH values are high in sites having more total biomass of sea grass (Table 7.5). These results confirm that sea grass meadows offer the buffering ecosystem service to prevent acidification of the aquatic phase.

In the selected quadrate, the order of abundance of the sea grass species was *Cymodocea serrulata* > *Thalassia hemprichii* > *Halophila ovalis*. Pronounced variation was observed between AGB and BGB in all the three species. The ratio of AGB and BGB ranged from 1:1.25 to 1:1.30. The Above Ground Carbon (AGC) and

**Table 7.3** Species-wise above ground biomass (AGB) and above ground carbon (AGC) per unit area in Gulf of Mannar

| Species | AGB (g dry wt. $m^{-2}$) | | | AGC (g dry wt. $m^{-2}$) | | |
|---|---|---|---|---|---|---|
| | Stn 1 | Stn 2 | Stn 3 | Stn 1 | Stn 2 | Stn 3 |
|  *Cymodocea serrulata* | 115.23 (49.6%) | 109.60 (49.7%) | 101.85 (49.9%) | 57.15 | 54.47 | 50.82 |
|  *Thalassia hemprichii* | 56.98 (48.5%) | 49.79 (48.9%) | 38.64 (47.9%) | 27.63 | 24.35 | 18.51 |
|  *Halophila ovalis* | 28.87 (41.23%) | 25.19 (43.05%) | 19.49 (45.16%) | 11.90 | 10.84 | 8.80 |

**Table 7.4** Species-wise below ground biomass (BGB) and below ground carbon (BGC) per unit area in Gulf of Mannar

| Species | BGB (g dry wt. $m^{-2}$) | | | BGC (g dry wt. $m^{-2}$) | | |
|---|---|---|---|---|---|---|
| | Stn 1 | Stn 2 | Stn 3 | Stn 1 | Stn 2 | Stn 3 |
|  *Cymodocea serrulata* | 144.04 (48.4%) | 140.29 (48.6%) | 132.41 (49.1%) | 69.72 | 68.18 | 65.01 |
|  *Thalassia hemprichii* | 71.79 (47.9%) | 63.23 (48.2%) | 49.84 (48.5%) | 34.39 | 30.48 | 24.17 |
|  *Halophila ovalis* | 36.31 (39.69%) | 33.14 (38.43%) | 24.75 (38.11%) | 14.41 | 12.74 | 9.43 |

Below Ground Carbon (BGC) also exhibited significant variation with highest value in *Cymodocea serrulata* followed by *Thalassia hemprichii* and *Halophila ovalis.*

The coincidences of high biomass of sea grass with high pH of the surrounding water is a litmus test to prove the positive role of the sea grass species in preventing the pace of lowering the aquatic pH.

Soil Organic Carbon (SOC) was also monitored simultaneously to evaluate the amount of stored carbon in the underlying soil of the intertidal region in the study area. SOC values ranged from 0.89% (in the underlying soil of *Thalassia hemprichii* in station 3) to 1.98% (in the underlying soil of *Cymodocea serrulata* in station 1) (Table 7.6). ANOVA results also confirm significant variations ($p < 0.01$) in SOC between stations and also between underlying substrata of the respective species (Table 7.7).

Very scanty references are available on the mechanisms of carbon fixation in sea grasses and questions concerning the kinetics of the active forms of carbon utilized and possible photosynthetic pathways. Steemann Nielsen (1975) postulated that the exogenous form of carbon assimilated by sea grasses is $HCO_3^-$. This may

**Table 7.5** Species-wise total biomass (TB) and total carbon (TC) per unit area in the Gulf of Mannar along with pH of the surrounding aquatic phase

| Species | TB (g dry wt. $m^{-2}$) | | | TC (g dry wt. $m^{-2}$) | | | pH value | | |
|---|---|---|---|---|---|---|---|---|---|
| | Stn 1 | Stn 2 | Stn 3 | Stn 1 | Stn 2 | Stn 3 | | | |
|  *Cymodocea serrulata* | 259.27 | 249.89 | 234.26 | 126.87 | 122.65 | 115.83 | 8.34 | 8.32 | 8.32 |
|  *Thalassia hemprichii* | 128.77 | 111.02 | 88.48 | 62.02 | 54.83 | 42.68 | 8.30 | 8.30 | 8.29 |
|  *Halophila ovalis* | 65.18 | 58.33 | 44.24 | 26.31 | 23.58 | 18.23 | 8.30 | 8.29 | 8.27 |

**Table 7.6** Soil organic carbon (SOC) level associated with the sea grass species

| Species | SOC (%) | | |
|---|---|---|---|
| | Stn 1 | Stn 2 | Stn 3 |
| *Cymodocea serrulata* | 1.98 | 1.36 | 1.09 |
| *Thalassia hemprichii* | 1.07 | 0.98 | 0.89 |
| *Halophila ovalis* | 1.13 | 0.97 | 1.01 |

**Table 7.7** ANOVA results showing variations of AGB, BGB, AGC and BGC of Seagrass between species and stations in Gulf of Mannar

| Variables | $F_{cal}$ | $F_{crit}$ |
|---|---|---|
| *AGB* | | |
| Between species | 1104.33 | 6.94 |
| Between station | 27.76 | 6.94 |
| *BGB* | | |
| Between species | 1029.25 | 6.94 |
| Between station | 19.57 | 6.94 |
| *AGC* | | |
| Between species | 643.76 | 6.94 |
| Between station | 12.50 | 6.94 |
| *BGC* | | |
| Between species | 999.93 | 6.94 |
| Between station | 13.96 | 6.94 |
| *SOC* | | |
| Between underlying substratum of species | 4.58 | 6.94 |
| Between station | 2.61 | 6.94 |

be attributed to extremely low concentration of carbon dioxide in seawater compared with $HCO_3^-$ (about 200 times less) and the slow rate at which carbon dioxide could be re-supplied to the plant from other forms of carbon. The complete dependency of few sea grass species on $HCO_3^-$ was demonstrated by Beer et al. (1977). These species are *Halophila stipulacea, Thalassodendron ciliatum, Halodule uninervis* and *Syringodium isoetifolium.* It is of interest whether sea grass species belong to the $C_3$ or $C_4$ group of plants because many species are tropical, and hence exposed to high temperatures, especially in shallow water. This is one of the conditions which might have contributed to the evolution of $C_4$ plants. While certain features such as the $\delta^{13}C$ values measured by Parker (1964), Smith and Epstein (1971), Doohan and Newcomb (1976) and Benedict and Scott (1976) suggest that various sea grass species belong to the $C_4$ type, other features such as the high rates of photorespiration (Hough 1976) and photosynthetic inhibition by high oxygen concentrations (Downton et al. 1976) are typically associated with the $C_3$ pathway. Results of a more appropriate test for the determination of $C_3$ or $C_4$ photosynthesis, i.e. short-time $\delta^{14}C$ experiments, have been reported previously only by Benedict and Scott (1976). They show that *Thalassia testudinurn* is a $C_4$ plant. In 1979 Beer and Waisel conducted short-time $\delta^{14}C$ pulse-chase experiments on two other sea grass species. Both species show a typical $C_3$ pattern with phosphoglyceric acid being the first major stable substance labelled. The label is subsequently transferred to phosphorylated sugars and phosphate esters, and later to saccharose or aspartic acid. Although saccharose and aspartic acid were not separated in the scraping and counting of the chromatograms, two distinct spots could be identified on the X-ray film, the saccharose spot usually of higher density than the aspartic acid. In any case, neither malic nor aspartic acids were primary products of $^{14}C$ fixation showing that these sea grass species are not $C_4$ plants. Although $^{14}C$ pulse-chase experiments should ideally result in the same total activity of label in all treatments, the relative activity of fixed $^{14}C$ increased with increasing chase times. This is due to absorption of $H^{14}CO_3^-$ which continues to be metabolized after the plant has been transferred to a ~4C-free medium. Therefore, phosphoglyceric acid levels in 15 and 45 s chase periods are somewhat overestimated while main labelled products are underestimated. The regain of spotted ~4C activity was about 60%, including the activity of the whole plates. Some activity was lost to the running solvent in the tank and about 10% was lost as an evenly distributed smear over the whole plate. The two sea grass species investigated are unlike *Thalassia testudinum* which was shown earlier to be a $C_4$ plant (Benedict and Scott 1976). They are also somewhat different from *Halophila stipulacea and Thalassodendron ciliatum* plants which, although mostly resembling $C_3$ plants, show a relatively higher labelling into malic and aspartic acids (Beer unpublished). Sea grasses do not exhibit the structural differentiation of photosynthetic tissue (Kranz anatomy) which is usually found in $C_4$ plants and photosynthesis seems to be carried out mainly in the chloroplast-rich epidermis. $C_4$ photosynthesis as in *Thalassia testudinurn* is therefore not associated with the Kranz anatomy (Benedict and Scott 1976). It is likely that variations in photosynthetic pathways found in sea grasses are linked to the levels of photosynthetic enzymes as well as $HCO_3^-$ and/or carbon dioxide rather than to differences in anatomy. In some freshwater angiosperms, the ratio of phosphoenolpyruvate carboxylase to ribulosebiphosphate carboxylase differs as a response to seasonal changes of the environment (Bowes et al. 1978). If this is also true for sea grasses, one could expect to find variations in photosynthetic pathways both between and within species as a response to growth conditions. Also, nothing would hinder both pathways being more or less active simultaneously. The main exogenous form of carbon utilized by six sea grass species investigated so far is $HCO_3^-$. This agrees well with the high $HCO_3^-$ compared with carbon dioxide concentrations found in seawater. The former concentration (about 2.5 mM) is saturating for photosynthesis in these plants. Whether $HCO_3^-$ is fixed directly by phosphoenolpyruvate carboxylase or transformed to carbon dioxide and fixed via ribulosebiphosphate carboxylase probably depends on the levels of these and other enzymes as well as which form of carbon is available. $^{14}C$ pulse-chase experiments have shown that at least one sea grass species is a $C_4$-plant while others are $C_3$ plants. Experiments based on other parameters sometimes show contradictory or intermediary photosynthetic pathways. The variations in photosynthetic pathways reflect an ability of sea grasses to adapt to various ecological conditions.

**Salt marsh** communities are widely distributed on the shoreward side of mud flats in temperate and subarctic regions of the world. The dominant plant life in this community on many North American shores is cordgrass (*Spartina* sp.) that has moved to the shallow intertidal area. The cordgrass *Spartina alterniflora* dominates the marshlands of the Atlantic coast, while *Spartina foliosa* is dominant along the coast of California. On the Gulf Coast of Florida, the tallest and most marsh grass is *Juncus* sp., which is found in higher elevations (higher up on shore). The species found most seaward is tall cordgrass, *Spartina alterniflora.*

The estuaries of Indian sub-continent sustains an important saltmarsh grass species namely *Porteresia coarctata* (Fig. 7.5), which has a great role in stabilizing the islands. Their long network of roots binds the soil particles very intricately and retards the process of erosion. The growth, biomass and subsequent carbon storage capacity of the species is controlled by the ambient aquatic salinity. The species exhibits a luxuriant growth in relatively low saline regions (salinity range within 5–20 psu) with subsequently higher carbon content, but at salinity around 30 psu, the growth is stunted.

A time series analysis (2001–2012) was conducted by the present authors on the stored carbon of the species sampled from the east and west coasts of India. Interestingly the species sampled from the Hooghly estuary in the east coast exhibited higher biomass and carbon compared to those sampled from the Mandovi estuary in the west coast. The entire data bank of biomass and stored carbon in *P. coarctata* is presented Figs. 7.6, 7.7, 7.8 and 7.9.

In Indian Sundarbans, *Porteresia coarctata* is widely available in the intertidal zone surrounding almost all the islands. This species provides different categories of ecosystem services like provisioning services (as fodder), regulating services (bioremediation, carbon sequestration, erosion checking etc.) and supporting services

**Fig. 7.5** Saltmarsh grass *Porteresia coarctata* in Indian Sundarbans

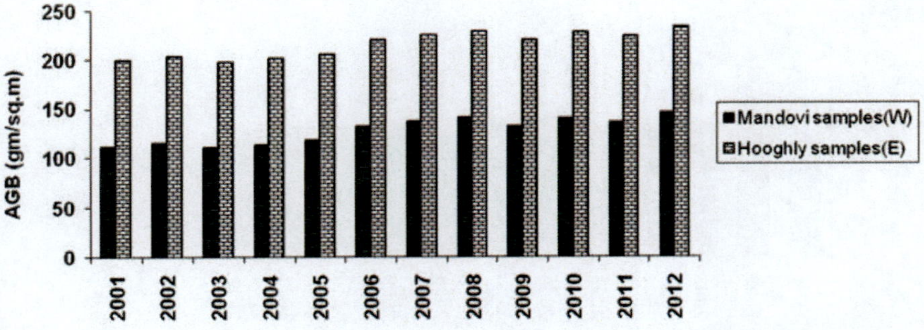


**Fig. 7.6** AGB in *P. coarctata* samples from the Hooghly and Mandovi estuary during 2001–2012

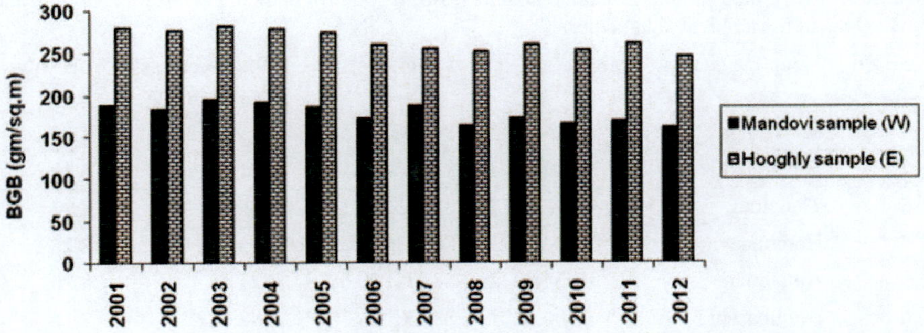


**Fig. 7.7** BGB in *P. coarctata* samples from the Hooghly and Mandovi estuary during 2001–2012

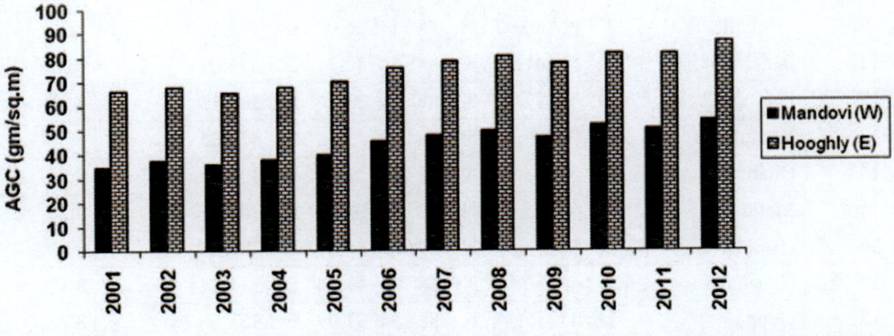


**Fig. 7.8** AGC in *P. coarctata* samples from the Hooghly and Mandovi estuary during 2001–2012

(nursery and breeding ground of several fish species). In this section, we have documented the role of the species in regulating the pH of the water body surrounding the islands. The monitoring of the parameters was conducted during March, 2019 in 24 selected stations in Indian Sundarbans (Table 7.8 and Fig. 7.10).

We estimated the total biomass (sum of Above Ground Biomass and Below Ground Biomass) of the species by the standard quadrant method. Simultaneously pH of the surrounding water around the salt marsh bed was monitored in all the selected

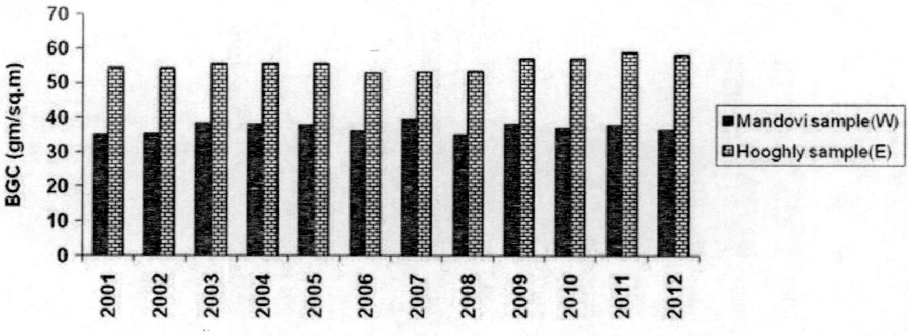


**Fig. 7.9** BGC in *P. coarctata* samples from the Hooghly and Mandovi estuary during 2001–2012

**Table 7.8** Sampling stations in Indian Sundarbans with Total Biomass of *Porteresia coarctata* and pH values of the ambient aquatic phase

| S. No. | Sampling station | Latitude | Longitude | Total biomass (g dry wt. $m^{-2}$) | pH value |
|---|---|---|---|---|---|
| 1 | Muriganga | 21° 38′ 25.86″ N | 88° 08′ 53.55″ E | 289.54 | 8.05 |
| 2 | Saptamukhi | 21° 36′ 02.49″ N | 88° 23′ 47.18″ E | 301.49 | 8.10 |
| 3 | Thakuran | 21° 49′ 43.17″ N | 88° 33′ 21.57″ E | 335.46 | 8.22 |
| 4 | Herobhanga | 21° 59′ 34.32″ N | 88° 41′ 46.52″ E | 393.45 | 8.28 |
| 5 | Ajmalmari | 21° 51′ 34.72″ N | 88° 39′ 00.68″ E | 327.45 | 8.20 |
| 6 | Dhulibasani | 21° 47′ 06.62″ N | 88° 33′ 48.20″ E | 348.65 | 8.26 |
| 7 | Chulkathi | 21° 41′ 53.62″ N | 88° 34′ 10.31″ E | 354.70 | 8.27 |
| 8 | Arbesi | 22° 11′ 43.14″ N | 89° 01′ 09.04″ E | 401.68 | 8.33 |
| 9 | Jhilla | 22° 09′ 51.53″ N | 88° 57′ 57.07″ E | 385.29 | 8.31 |
| 10 | Pirkhali | 22° 06′ 00.97″ N | 88° 51′ 06.04″ E | 373.25 | 8.31 |
| 11 | Panchmukhani | 21° 59′ 41.58″ N | 88° 54′ 14.71″ E | 391.02 | 8.32 |
| 12 | Harinbhanga | 21° 57′ 17.85″ N | 88° 59′ 33.24″ E | 363.13 | 8.30 |
| 13 | Khatuajhuri | 22° 03′ 06.55″ N | 89° 01′ 05.33″ E | 379.85 | 8.30 |
| 14 | Chamta | 21° 53′ 18.56″ N | 88° 57′ 11.40″ E | 360.38 | 8.29 |
| 15 | Matla | 21° 53′ 15.30″ N | 88° 44′ 08.74″ E | 303.45 | 8.15 |
| 16 | Chandkhali | 21° 51′ 13.59″ N | 89° 00′ 44.68″ E | 329.92 | 8.19 |
| 17 | Goashaba | 21° 43′ 50.64″ N | 88° 46′ 41.44″ E | 311.48 | 8.18 |
| 18 | Gona | 21° 41′ 15.44″ N | 88° 54′ 31.09″ E | 353.66 | 8.28 |
| 19 | Chhotahardi | 21° 44′ 42.24″ N | 88° 44′ 17.79″ E | 343.85 | 8.27 |
| 20 | Baghmara | 21° 39′ 04.45″ N | 89° 04′ 40.59″ E | 356.96 | 8.29 |
| 21 | Mayadwip | 21° 35′ 50.23″ N | 88° 47′ 09.95″ E | 390.40 | 8.30 |
| 22 | Jambu Island | 21° 35′ 42.03″ N | 88° 10′ 22.76″ E | 289.43 | 7.99 |
| 23 | Lothian | 21° 38′ 21.20″ N | 88° 20′ 29.32″ E | 356.29 | 8.32 |
| 24 | Sagar Island | 21° 38′ 51.55″ N | 88° 02′ 20.97″ E | 213.45 | 7.98 |

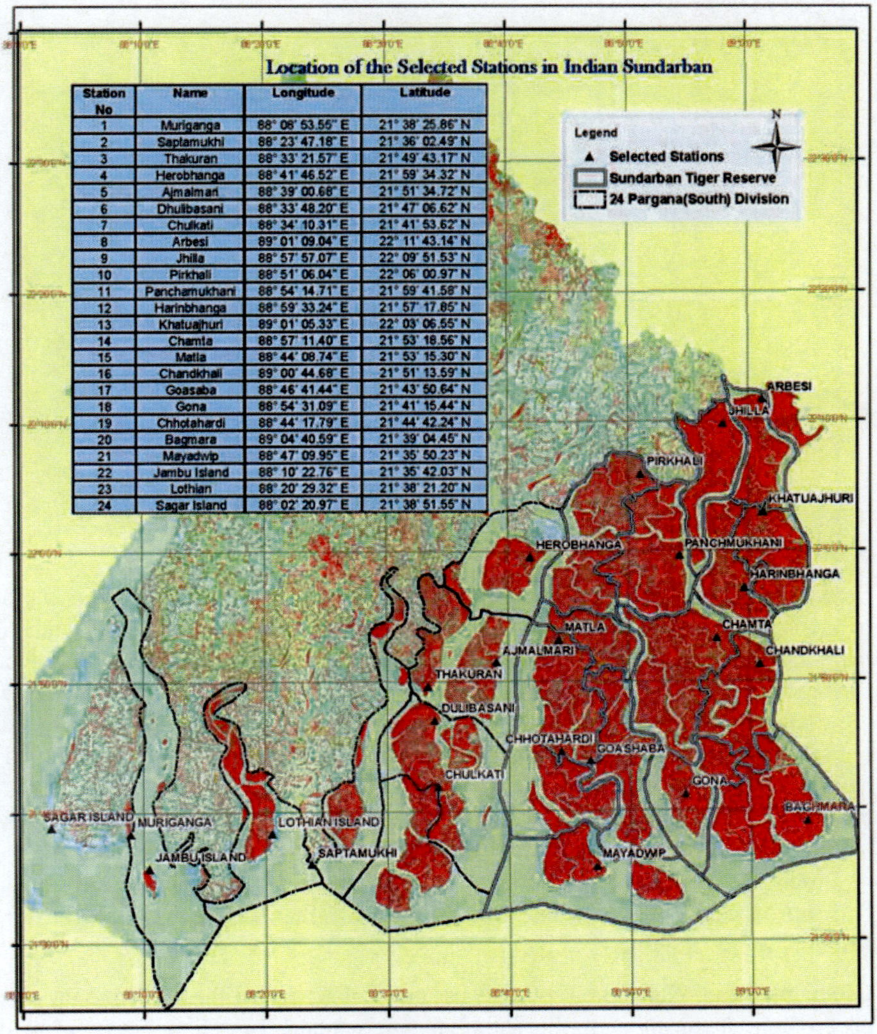


**Fig. 7.10** Location of 24 sampling stations in Indian Sundarbans

24 stations. For each observational station, at least three samples were collected from the surface during low tide condition within 500 m of each other. We observed a significant positive correlation between total biomass of the species and aquatic pH ($r = 0.9204$; $p < 0.01$). The inter-relationship between these variables is also highlighted by the scatter plot (Fig. 7.11), which confirms the regulatory role of the salt marsh grass species, *P. coarcatata* in buffering the water bodies surrounding the vegetation. The present result strongly suggests the conservation and massive plantation of the species to retard the rate of acidification in the coastal and estuarine regions.

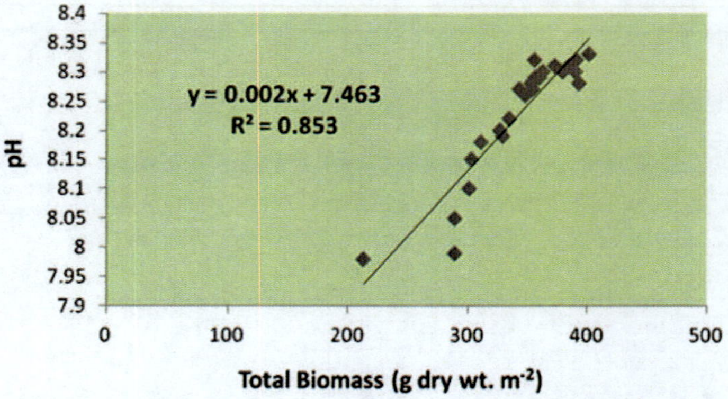


**Fig. 7.11** Scatterplot showing the positive relationship between total biomass of *Porteresia coarctata* and aquatic pH

## 7.3 Seaweeds

Seaweeds or marine macroalgae are the group of plants that thrive either in marine or brackish water environment. They are macroalgae and contain photosynthetic pigments. Like the terrestrial producer community, the seaweeds can prepare their own food with the help of sunlight and nutrients present in the seawater. Seaweeds are found in the coastal region between high tide to low tide and in the sub-tidal region up to a depth where 0.01% photosynthetic light is available. They require hard substratum for their growth, which may be tree trunks (preferably mangroves) that get submerged during the high tide or even brick or boulders that are often laid to enter the island from the adjacent bays or estuaries.

Based on the colour of the pigmentation, the seaweeds are broadly classified into various categories like Chlorophyceae (green), Phaeophyceae (brown), Rhodophyceae (red) etc.

**Chlorophyceae**: Algae are typically green in colour due to the predominant presence of the green pigment, the chlorophyll *a*, contained in chloroplasts. Most of the green algae are found occurring in the littoral zones of the marine environment and produce motile reproductive bodies e.g., *Codium* sp., *Enteromorpha intestinalis, Enteromorpha compressa, Ulva lactuca, Chaetomorpha* sp., *Cladophora* sp. etc.

**Phaeophyceae**: The brown algae are distinguished by their colour, which varies from olive green through light golden to a deep shade of brown. Motile reproductive cells are commonly found in the brown algae e.g., Kelps like *Postelsia* sp., *Nereocystis* sp., *Macrocystis* sp., and *Laminaria* sp., *Fucus* sp., *Padina gymnospora, and Rosenvingea intricata.* Since brown algae live primarily in shallow waters or on shoreline rocks, they exhibit adaptations that protect them from the constant pounding of the waves. The body of brown algae is very flexible, which allow them to bend or orient with the wave action. Unlike red and green algae, brown algae usually do not reproduce by fragmentation because the tissues are too highly specialized to regenerate new parts. A major exception to this rule is *Sargassum* sp. Large masses of this seaweed are found floating in the Atlantic Ocean, where they tend to accumulate in an area known as the Sargasso Sea. These free-floating clumps of sargassum weed form a complex, three-dimensional habitat that sustains a diverse group of marine organisms.

**Rhodophyceae**: The seaweeds under this category are recognized by their bright pink colour caused by the biloprotein pigments, r-phycoerythrin and r-phycocyanin. The fresh water forms, however, are of bluish green in colour. Red algae produce large amount of polysaccharides around their cells. Several of these polysaccharides are commercially important. The majority of the marine forms occur from low tide marks to greater depths up to 100 m beneath the surface of the sea. e.g., *Porphyra* sp., *Chondrus crispus, Catenella repens* etc.

A list of common seaweeds (belonging to three different classes) found in the tropical regions is highlighted in Table 7.9.

**Ecosystem Services of Seaweeds**

Today seaweeds are used for various commercial purposes. Agar and algin extracted from the seaweeds have a variety of industrial applications. Apart from consuming the seaweed as food, they are used as fodder, manure, medicine and also as ingredients of several health products as discussed here under different headings.

**As Food**: In many countries like Malaysia, Indonesia, Korea, and Australia seaweeds are used for human consumption. These are used in the preparation of salads, soups, jellies and vinegar. The common species utilized for the preparation of food are *Porphyra* sp., *Gracilaria* sp., *Laurencia* sp., *Caulerpa* sp., *Sargassum* sp. etc. These algae are rich in Vitamin B.

**As Fodder**: Some important species of seaweeds that are used in cattle feed are *Ulva* sp., *Enteromorpha* sp., *Gracilaria* sp., *Padina* sp., *Sargassum* sp. etc. Poultry feed is also made in certain areas by mixing seaweeds with trash fishes.

**As Manure**: Seaweeds have unique capacity to bioaccumulate trace elements from ambient media and therefore they are rich sources of macro and micro-elements required for the growth of plants. Seaweeds like *Sargassum* sp. and *Gracilaria* sp. are used as manure for coconut plantation in India, particularly in the south Indian states.

**As Medicine**: Seaweeds have wide applications in the sphere of medicine. *Sargassum* sp. and *Turbinaria* sp. are good sources of alginates, which are known to prolong the period of activity of certain drugs. *Hypnea* sp. and *Acanthophora* sp. are the major sources of carrageenans, which have been found to be useful in ulcer therapy.

**As Industrial Raw Material**: Agar–agar is a gelatinous colloidal carbohydrate present in the cell wall of algae and is extracted from some members of Rhodophyceae. In India, agar has wide use in the preparation of food, ice creams, jellies, soups, bacteriological samples and cosmetics. Being a non-toxic substance

**Table 7.9** Common seaweed species of marine and estuarine ecosystems

| Scientific name | Systematic position | Description | Figure |
|---|---|---|---|
| *Caulerpa racemosa* | Division—Chlorophyta<br>Order—Bryopsidales<br>Family—Caulerpaceae | Plants with a horizontal rhizome like stem anchored by rhizoids. The erect branches bear clusters of small branchlets with swollen tips |  |
| *Caulerpa scalpelliformis* | Division—Chlorophyta<br>Order—Bryopsidales<br>Family—Caulerpaceae | Flat lanceolate blades arise from the stolen anchored in sand |  |
| *Caulerpa serrulata* | Division—Chlorophyta<br>Order—Bryopsidales<br>Family—Caulerpaceae | Spirally twisted blades with toothed edges (serrulate) arise from the stolon anchored in sand by the root like rhizoides |  |

(continued)

**Table 7.9** (continued)

| Scientific name | Systematic position | Description | Figure |
|---|---|---|---|
| *Chaetomorpha linum* | Division—Chlorophyta<br>Order—Cladphorales<br>Family—cladophoraceae | *Chaetomorpha* means 'hair-shaped' and the thick, unbranched filaments form bright green mats in mangrove channels | 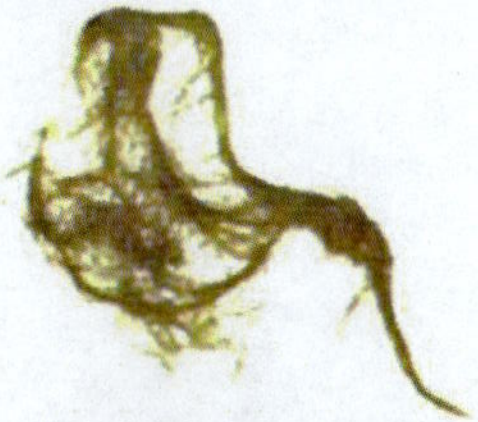 |
| *Codium arabicum* | Division—Chlorophyta<br>Order—Bryopsidales<br>Family—Codiaceae | This species forms convoluted flattened cushions |  |
| *Codium dwarkense* | Division—Chlorophyta<br>Order—Bryopsidales<br>Family—Codiaceae | The cylindrical, spongy branches are primarily dichotomous (with regularly forked branches) with some side branches |  |

(continued)

**Table 7.9** (continued)

| Scientific name | Systematic position | Description | Figure |
|---|---|---|---|
| *Codium indicum* | Division—Chlorophyta<br>Order—Bryopsidales<br>Family—Codiaceae | Plants are spongy with irregularly forked cylindrical branches attached by a disc-shaped holdfast to rocks. *Codium* plants vary in shape from this branched type to thick crusts to flat blades | 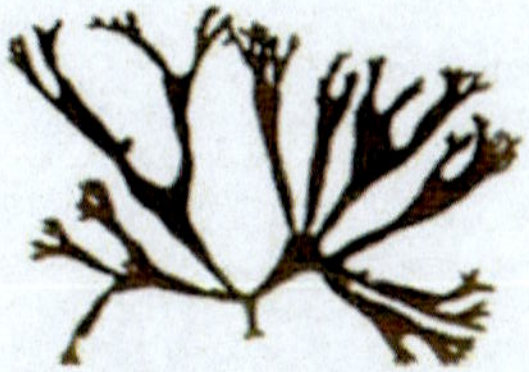 |
| *Enteromorpha flexuosa* | Division—Chlorophyta<br>Order—Ulvales<br>Family—Ulvaceae | *Enteromorpha* forms thin, hollow tubes which are generally small but may reach over 10 cm in length in good conditions. These fronds are mostly unbranched and flattened |  |
| *Ulva fasciata* | Division—Chlorophyta<br>Order—Ulvales<br>Family—Ulvaceae | The blade is green divided into long and narrow strap-like lobes reaching to 15 cm on Arabian sea coasts. The margins are irregularly toothed |  |

(continued)

**Table 7.9** (continued)

| Scientific name | Systematic position | Description | Figure |
|---|---|---|---|
| *Ulva grandis* | Division—Chlorophyta<br>Order—Ulvales<br>Family—Ulvaceae | The blade is dark green, expanded to more than a meter long and grows extensively along the Mirbat peninsula on the Arabian Sea coasts |  |
| *Dictyota pardalis* | Division—Phaeophyta<br>Order—Dictyotales<br>Family—Dictyotaceae | This species may be up to 20 cm high with branches often growing in a twisting or spiral manner. The lower branches may be covered with small proliferations |  |
| *Hormophysa cuneiformis* | Division—Phaeophyta<br>Order—Fucales<br>Family—Cystoseiraceae | Plants to 40 cm, stiff with wide, flat margins to the branches. The margins are toothed |  |

(continued)

**Table 7.9** (continued)

| Scientific name | Systematic position | Description | Figure |
|---|---|---|---|
| *Jolyna laminarioides* | Division—Phaeophyta<br>Order—Scytosiphonales<br>Family—Scytosiphonaceae | The strap-shaped blades are usually unbranched and reach up to 30 cm | 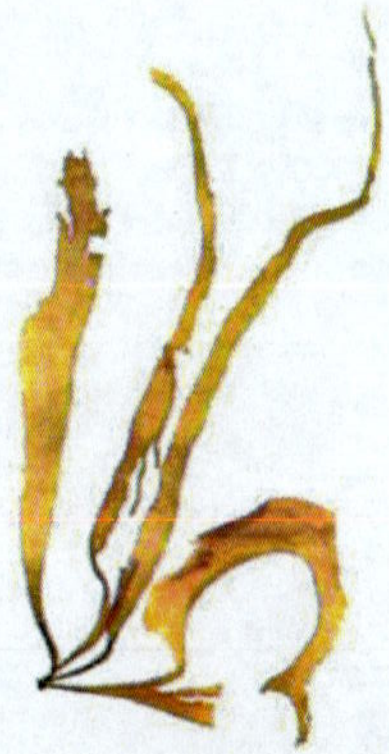 |
| *Padina boergesenii* | Division—Phaeophyta<br>Order—Dictyotales<br>Family—Dictyotaceae | *Padina* has fan-like blades which are attached by a discoid holdfast to rocks. This species has a double row of hairs in concentric bands |  |

(continued)

**Table 7.9** (continued)

| Scientific name | Systematic position | Description | Figure |
|---|---|---|---|
| *Sargassum ilicifolium* | Division—Phaeophyta<br>Order—Fucales<br>Family—Sargassaceae | These are long main stem bearing many leaves which are usually toothed with a distinct midrib. The spherical air bladders float the photosynthetic leaves up into the lighted surface waters |  |
| *Stoechospermum marginatum* | Division—Phaeophyta<br>Order—Dictyotales<br>Family—Dictyotaceae | This brown seaweed, which may reach 40 cm in height, has regularly dichotomously branched blades. These are flat, stiff and strap-like with in-rolled tips. Fertile plants have characteristic marginal dark lines crowded with sporangia |  |
| *Turbinaria conodes* | Division—Phaeophyta<br>Order—Fucales<br>Family—Sargassaceae | This stiff, erect plant has distinctive turban-like blade attached to a central stem fixed to rocks by a spreading stolon. Each blade is roughly triangular with marginal teeth and a lobed centre |  |

(continued)

**Table 7.9** (continued)

| Scientific name | Systematic position | Description | Figure |
|---|---|---|---|
| *Amphiroa anceps* | Division—Rhodophyta<br>Order—Corallinales<br>Family—Corallinaceae | This calcified, coralline red alga shows regular dichotomous branching with narrow fork angles in upper parts of the plant. There is clear articulation with segments alternating with non-calcified 'joints'. The plant is very brittle when dry and then easily falls to pieces |  |
| *Euptilota fergusonii* | Division—Rhodophyta<br>Order—Cerramiales<br>Family—Ceramiaceae | This bushy, dark red alga grows to 15 cm. the delicate-looking branchlets end I very fine tips with 2–3 spiny outgrowths |  |
| *Galaxaura marginata* | Division—Rhodophyta<br>Order—Nemaliales<br>Family—Galaxauraceae | This bushy, mounded seaweed grows to 10 cm or more with lightly calcified pinkish blades loosely entangled. The dichotomously branched blades show faint cross banding near the tips and have thickened margins |  |

(continued)

**Table 7.9** (continued)

| Scientific name | Systematic position | Description | Figure |
|---|---|---|---|
| *Galaxaura obtusata* | Division—Rhodophyta<br>Order—Nemaliales<br>Family—Galaxauraceae | This rose-red plant may reach 10 cm. The cylindrical branches are repeatedly forked |  |
| *Gelidium latifolium* | Division—Rhodophyta<br>Order—Geliadiales<br>Family—Gelideaceae | This identification is not confirmed but the plant has a dense appearance with many fine branches |  |
| *Gracilaria canaliculata* | Division—Rhodphyta<br>Order—Gracilariales<br>Family—Gracilarriaceae | These tough, fleshy plants reach a length of 7 cm tall with interwining branches throughout. The branches are cylindrical in cross section |  |
| *Haliptilon subulatum* | Division—Rhodophyta<br>Order—Cryptonemiales<br>Family—Corallinaceae | This attractive coralline alga has brittle, heavily calcified pink segments connected by flexible joints. There is a regular arrangement of main segments like inverted triangles with cylindrical side branches |  |

(continued)

**Table 7.9** (continued)

| Scientific name | Systematic position | Description | Figure |
|---|---|---|---|
| *Halymenia porphyriformis* | Division—Rhodophyta<br>Order—Cryptonemiales<br>Family—Halymeniaceae | This striking seaweed is rosy-red with a flat, lobed leaf-like lamina with a few perforations. It reaches 20 cm or more and is attached to rocks by a small disc-shaped holdfast |  |
| *Hypnea boergesenii* | Division—Rhodophyta<br>Order—Gigartinales<br>Family—Hypnaceae | This species grows up to 10—15 cm |  |
| *Hypnea bryoides* | Division—Rhodophyta<br>Order—Gigartinales<br>Family—Hypnaceae | This striking plant grows to 20 cm with tough, wiry branches forming dense clumps from a holdfast. The branches have a characteristics 'hairy' appearance from a dense covering of tiny spiny outgrowths |  |

(continued)

**Table 7.9** (continued)

| Scientific name | Systematic position | Description | Figure |
|---|---|---|---|
| *Martensia elegans* | Division—Rhodophyta<br>Order—Ceramiales<br>Family—Delesseriaceae | This remarkable elegant plant grows to 10 cm. The net-like blades are divided into zones with every other zone forming a network of elongated holes resembling a cross section of wood in a tree |  |
| *Porphyra* sp. | Division—Rhodophyta<br>Order—Bangiales<br>Family—Bangiaceae | This alga attaches itself to the rocks by multicellular rhizoidal attachments, usually disc-shaped |  |
| *Portieria hornemanni* | Division—Rhodophyta<br>Order—Cryptonemiales<br>Family—Rhizophyllidaceae | This conspicuous, bright red alga grows up to 8 cm in tufts, It branches repeatedly to form a delicate, lacy appearance |  |

(continued)

**Table 7.9** (continued)

| Scientific name | Systematic position | Description | Figure |
|---|---|---|---|
| *Sarcodia montagneana* | Division—Rhodophyta<br>Order—Gigartinales<br>Family—Sarcodiaceae | This plant occurs as dense clusters of thick, flat triangular blades to 8 cm high. The blades are irregularly or dichotomously divided with distinctive rounded projections along the margins |  |
| *Sarconema filiforme* | Division—Rhodophyta<br>Order—Gigartinales<br>Family—Solieriaceae | The dichotomously branched tips taper to fine points. It is coloured yellow–brown in exposed areas but dark red in shaded sites |  |
| *Sarconema scinaioides* | Division—Rhodophyta<br>Order—Gigartinales<br>Family—Solieriaceae | The plant is up to 15 cm tall, tough and rubbery with dichotomous branching throughout. The branches are cylindrical with short tips and are thicker than other species of *Sarconema* |  |
| *Sebdenia flabellata* | Division—Rhodophyta<br>Order—Gigartinales<br>Family—Solieriaceae | The dichotomous branches are cylindrical and very gelatinous. Large specimens found late in the season (November/December) are filled with jelly |  |

agar is also used for the preparation of wine, beer, hand lotions etc. Algin and alginates have extensive uses in the preparation of various medicines, cosmetics, paper products, textile products, paints, milk products etc.

**As biological treatment of wastes**: Seaweeds can also be utilized for the biological treatment of industrial wastes by utilising their unique bioaccumulation capacity. It has been found that *Catenella repens*, *Ulva* sp., and *Enteromorpha intestinalis* are unique absorber of Zn, Cu, Fe etc. from the ambient aquatic medium. These algae, if cultured in the treatment plant of coastal industries can trap substantial amount of metals from the released industrial wastes.

**In coral reef building**: Many macroalgae make important contributions to the construction of the coral reef structure or reef skeleton by depositing calcium carbonate ($CaCO_3$ or limestone). Crustose Calcareous Algae (CCA, especially the Order Corallinales: e.g. *Porolithon* sp) are considered important in building and cementing the carbonate framework of coral reefs. CCA bind adjacent substrata and provide a calcified tissue barrier against erosion. This process is particularly important on many reef crests of the Great Barrier Reef, where Crustose Coralline Algae are dominant. Estimates of calcification rates of CCA at Lizard Island indicate significant annual deposition rates (1–10.3 kg $CaCO_3$ $m^{-2}$ $yr^{-1}$). Non-coralline CCA (e.g. *Peyssonnelia*) is also important in deep areas (between 80 and 120 m) at the edge of the continental platform in the Southern Reef, where they form large algal frameworks several meters high.

**Seaweeds as biofuels**: Prospecting macroalgae (seaweeds) as feed stocks for bioconversion into biofuels and commodity chemical compounds is limited primarily because of tractable microorganisms that can metabolize alginate polysaccharides. Today researchers have discovered a 36-kilo–base pair DNA fragment from *Vibrio splendidusen* coding enzymes for alginate transport and metabolism. The genomic integration of this ensemble, together with an engineered system for extracellular alginate depolymerization, generated a microbial platform that can simultaneously degrade, uptake and metabolize alginate. When further engineered for ethanol synthesis, this platform enables bioethanol production directly from macroalgae via a consolidated process, achieving a titer of 4.7% volume/volume and a yield of 0.281 weight ethanol/weight dry macroalgae (equivalent to ~80% of the maximum theoretical yield from the sugar composition in macroalgae).

**Carbon sequestration**: Unlike other blue carbon sectors (mangroves, sea grasses, and salt marshes), kelp forests and seaweed beds do not have such sedimentary substrata. Instead, their carbon-rich biomass detaches and is broken down in food chains by organisms that range in scale from grazing animals to pelagic and seabed bacteria. Knowledge on the scale of conversion of inorganic carbon into biomass, its subsequent sinking to the seabed and its sequestration over thousands of years form the basis of understanding the oceans as a potential sink for increasing levels of atmospheric carbon dioxide ($CO_2$). The other modes of fate of seaweed biomass depend on natural processes. The seaweed can be consumed by herbivores, whose fecal matters sink to the bottom and may remain there for a while. Moreover, distal

**Table 7.10** Sampling stations with coordinates and salient features

| Station | Coordinates | Salient features |
|---|---|---|
| Nayachar Island (Stn. 1 in the western sector) | 21° 45′ 24″ N; 88° 15′ 24″ E | It is located in the Hooghly estuary and faces the Haldia port-cum-industrial complex that houses a variety of industrial units |
| Sagar South (Stn. 2 in the western sector) | 21° 39′ 04″ N; 88° 01′ 47″ E | Situated at the confluence of the River Hooghly and the Bay of Bengal on the western sector of Indian Sundarbans |
| Gosaba (Stn. 3 in the central sector) | 22° 15′ 45″ N; 88° 39′ 46″ E | Located in the Matla Riverine stretch in the central sector of Indian Sundarbans |
| Annpur in Satjelia Island (Stn. 4 in the central sector) | 22° 11′ 52″ N; 88° 50′ 43″ E | Located in the central sector of Indian Sundarbans. Noted for its wilderness and mangrove diversity; selected as control zone |

portions of the fronds disintegrate during the summer season and those fragments enter the detritus food chain. Exudation as a dissolved organic material can be a critical loss. Therefore, some of the seaweed carbon will return to the water column and be either recaptured during photosynthesis or eventually returned to the atmosphere. However, a significant fraction of the algal carbon can be sequestered on the sea floor for a long period, perhaps centuries depending on location, currents etc.

Compared to other vegetation, the carbon sequestration potential of seaweeds in estuarine and deltaic environments, are however poorly understood. In this section, the seasonal and spatial variations in biomass production and carbon content of three major species of seaweeds (*E. intestinalis, U. lactuca* and *C. repens*) inhabiting two different sectors of lower Gangetic delta (western and central) with contrasting salinity have been estimated during 2019 (Table 7.10).

Seasonal samplings for biomass and carbon estimation of seaweed species (*E. intestinalis, U. lactuca and C. repens*) were carried out at low tides during May, 2019 (premonsoon), September, 2019 (monsoon) and November, 2019 (postmonsoon) from the intertidal mudflats. Samples of seaweed species were scrapped and handpicked from sluice gates, mangrove trunk and concrete jetties. Immediately after collection, the thallus of each species was thoroughly washed separately in the ambient seawater, as well as with double distilled water, to remove adhering debris and sediments. Altogether 10 quadrants (area = 1 $m^2$) were sampled form each of the selected stations for each species, randomly mixed and weighed accurately in an electronic balance (IRD Balance; Model No. 290). The biomass was expressed in g $m^{-2}$.

For carbon estimation, seaweed samples were separately dried in a hot air oven (60 °C) for 72 h until a constant weight was obtained. Dried samples were ground to a fine powder. Direct estimation of percent carbon in the thallus body of each seaweed species for each season and for each sampled locations was separately carried out by Vario MACRO elementar make CHN analyzer, after grinding and random mixing the oven dried seaweed samples. At about 990 °C, the seaweed sample was mineralized separately. Formation of carbon monoxide is possible at this temperature even in the presence of excess oxygen. The complete oxidation is reached through a tungsten trioxide catalyst which is passed by the gaseous reaction products. The samples were finally analyzed through CHN mode, which is the most universal of the analysis modes because of the combination of the reagent design and the optimize combustion control parameters and expressed in percentage.

The biomass of *E. intestinalis* ranged from 2844.59 gm $m^{-2}$ (at Stn. 3, during premonsoon 2019) to 3185.99 gm $m^{-2}$ (at Stn. 1, during postmonsoon 2019). The carbon content exhibited lowest value at Stn. 3 (955.85 g $m^{-2}$ during monsoon 2019) and highest at Stn. 2 (1163.93 g $m^{-2}$ during premonsoon 2019). It is interesting to note that the mean order of biomass of *E. intestinalis* is postmonsoon (3098.68 g $m^{-2}$) > monsoon (3015.23 g $m^{-2}$) > premonsoon (2994.55 g $m^{-2}$) (Table 7.11). However, the carbon content varied as per the order premonsoon (1312.43 g $m^{-2}$) > postmonsoon (1185.03 g $m^{-2}$) > monsoon (1094.44 g $m^{-2}$) (Table 7.12). The mean seasonal values of both biomass and stored carbon are highlighted in Fig. 7.12.

The biomass of the species exhibited an almost uniform value in both the sectors and three seasons ($p < 0.01$) as revealed through ANOVA (Table 7.13), but the carbon content showed a significant spatial and seasonal variations ($p < 0.01$) (Table 7.14). The highest carbon content in premonsoon may be attributed to congenial temperature and solar radiation in the study area that have profound influence on the photosynthetic rate.

**Table 7.11** Average seasonal variations in biomass or standing stock (gm $m^{-2}$) of seaweeds with standard deviations; the values are the average of four selected stations during 2019

| Species | Season | | |
|---|---|---|---|
| | Premonsoon | Monsoon | Postmonsoon |
| *Enteromorpha intestinalis* | 2994.55 ± 26.85 | 3015.23 ± 30.63 | 3098.68 ± 33.25 |
| *Ulva lactuca* | 499.01 ± 9.01 | 251.66 ± 7.36 | 173.11 ± 5.54 |
| *Catenella repens* | 232.05 ± 4.52 | 69.05 ± 2.81 | 133.83 ± 1.73 |

**Table 7.12** Average seasonal variations in carbon content (gm $m^{-2}$) of seaweeds with standard deviations; the values are the average of four selected stations during 2019

| Species | Season | | |
|---|---|---|---|
| | Premonsoon | Monsoon | Postmonsoon |
| *Enteromorpha intestinalis* | 1312.43 ± 2.85 | 1094.44 ± 1.93 | 1185.03 ± 2.03 |
| *Ulva lactuca* | 157.08 ± 1.88 | 73.31 ± 1.29 | 49.33 ± 1.10 |
| *Catenella repens* | 57.21 ± 0.93 | 15.04 ± 0.70 | 29.32 ± 0.62 |

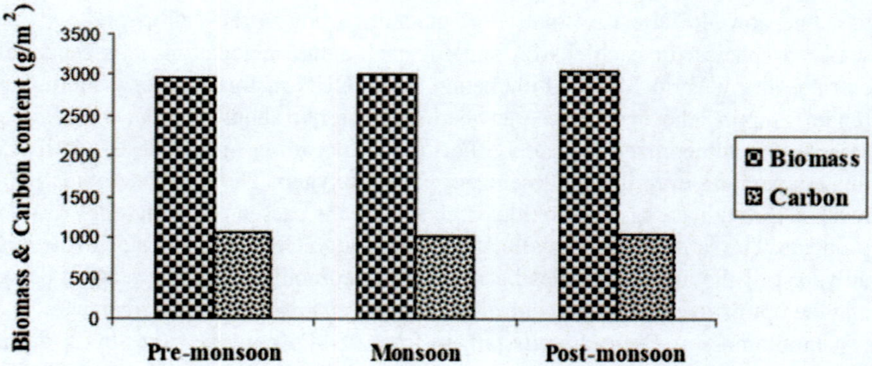


**Fig. 7.12** Mean seasonal variations in biomass and carbon content of *Enteromorpha intestinalis* during 2019

**Table 7.13** Results of ANOVA for seaweed standing stock (biomass) in Indian Sundarbans during premonsoon, monsoon and postmonsoon seasons considering the average values of four selected stations during 2019

| Variable | $F_{calculated}$ | $F_{critical}$ |
|---|---|---|
| *Enteromorpha intestinalis* | | |
| Between sectors | 1.59 | 18.51 |
| Between stations | 7.77 | 4.76 |
| Between seasons | 2.03 | 4.14 |
| *Ulva lactuca* | | |
| Between sectors | 1.12 | 18.51 |
| Between stations | 3.73 | 4.76 |
| Between seasons | 10.32 | 4.14 |
| *Catenella repens* | | |
| Between sectors | 7.38 | 18.51 |
| Between stations | 6.89 | 4.76 |
| Between seasons | 14.70 | 4.14 |

In case of *Ulva lactuca*, the biomass ranged from 95.66 g $m^{-2}$ (at Stn. 1, during monsoon 2019) to 793.22 g $m^{-2}$ (at Stn. 2, during premonsoon 2019). The carbon content showed the lowest value at Stn. 1 during monsoon 2019 (25.11gm $m^{-2}$) and highest at Stn. 2 during premonsoon 2019 (251.02 g $m^{-2}$). The mean order of biomass is premonsoon (499.01 g $m^{-2}$) > monsoon (251.66 g $m^{-2}$) > postmonsoon (173.11 g $m^{-2}$) (Table 7.11), and the carbon content exhibited a similar trend as per the order premonsoon (157.08 g $m^{-2}$) > monsoon (73.31 g $m^{-2}$) > postmonsoon (49.33 g $m^{-2}$) (Table 7.12 and Fig. 7.13). The significant seasonal variations in the values of biomass and carbon in *U. lactuca* are the results of synergistic effect of temperature, solar radiation and salinity.

The biomass of *C. repens* collected from the selected stations ranged from 42.11 (at Stn. 1, during postmonsoon 2019) to 315.00 g $m^{-2}$ (at Stn. 2, during premonsoon

**Table 7.14** Results of ANOVA for carbon content in seaweeds in unit area in Indian Sundarbans during premonsoon, monsoon and postmonsoon seasons considering the average values of four selected stations during 2019

| Variable | $F_{calculated}$ | $F_{critical}$ |
|---|---|---|
| *Enteromorpha intestinalis* | | |
| Between sectors | 25.15 | 18.51 |
| Between stations | 38.31 | 4.76 |
| Between seasons | 4.64 | 4.14 |
| *Ulva lactuca* | | |
| Between sectors | 0.84 | 18.51 |
| Between stations | 3.54 | 4.76 |
| Between seasons | 10.34 | 4.14 |
| *Catenella repens* | | |
| Between sectors | 5.38 | 18.51 |
| Between stations | 5.86 | 4.76 |
| Between seasons | 13.23 | 4.14 |

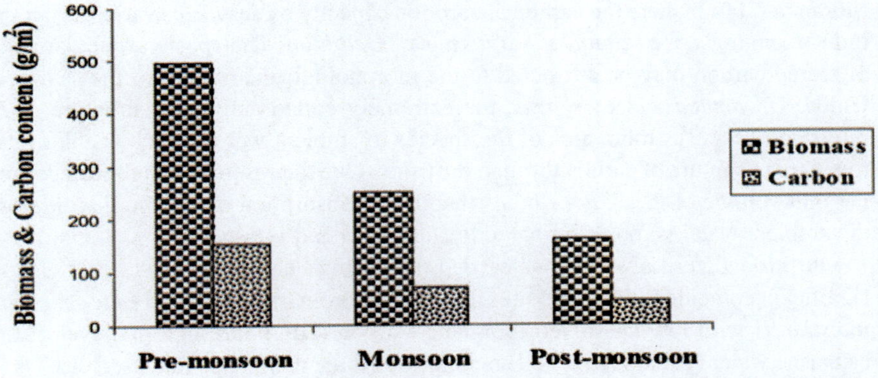


**Fig. 7.13** Mean seasonal variations in biomass and carbon content of *Ulva lactuca* during 2019

2019). The mean order of biomass is premonsoon (232.05 g $m^{-2}$) > postmonsoon (133.83 g $m^{-2}$) > monsoon (69.05 g $m^{-2}$) for *C. repens* (Table 7.11). In the thallus body of the species, the values of stored carbon ranged from 9.01 (at Stn.1, during postmonsoon 2019) to 76.22 g $m^{-2}$ (at Stn. 2, during premonsoon 2019). The carbon content in this species varied as per the order premonsoon (57.21 g $m^{-2}$) > post-monsoon (29.32 g $m^{-2}$) > monsoon (15.04 g $m^{-2}$) (Table 7.12). The mean seasonal values of both biomass and stored carbon in *C. repens* are highlighted in Fig. 7.14.

ANOVA results (Tables 7.13 and 7.14) also confirm significant spatial and seasonal variations in the biomass and carbon content of the species ($p < 0.01$).

The present study indicates that carbon storage in seaweed species is species-specific in nature. The highest value is observed in *E. intestinalis* (average 1042.77 g $m^{-2}$) followed by *U. lactuca* (average 89.79 g $m^{-2}$) and *C. repens* (average 32.64 g $m^{-2}$). Similar observation was also documented through a study done by

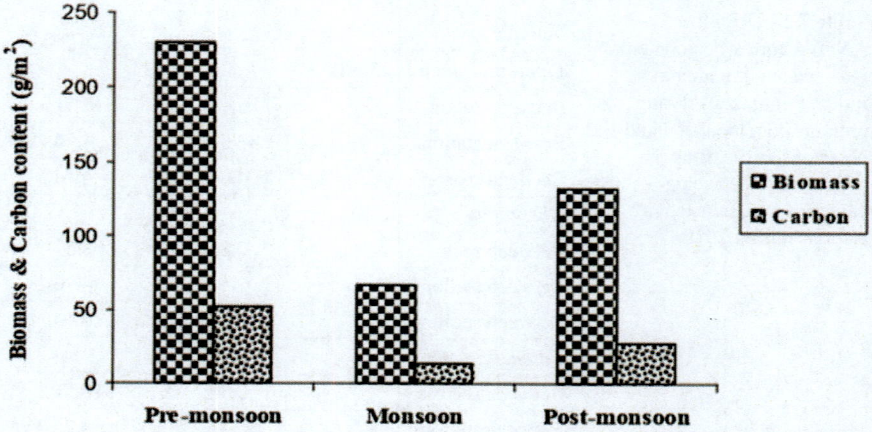


**Fig. 7.14** Mean seasonal variations in biomass and carbon content of *Catenella repens* during 2019

Muraoka (2004) where the carbon absorption capacity by seaweeds varied as per the order *Laminaria* > *Ecklonia* > *Sargassum* > *Gelidium.* The species-wise variation of stored carbon may be attributed to the morphological structure of the seaweed. Unlike *U. lactuca* and *C. repens*, the extremely coiled and spiral structure of *E. intestinalis* exposes more area of the species to ambient water, which enables relatively more capture of carbon through diffusion. Due to presence of high surface area per unit volume of the *E. intestinalis* thallus, the absorption of carbon dioxide from the ambient water is more compared to *U. lactuca* and *C. repens.*

Our ground zero observations carried out during February, 2020 clearly exhibit significant coincidence of those sites having high biomass of seaweed *Enteromorpha intestinalis* with low Dissolved Inorganic Carbon (DIC) and high pH level of the estuarine water (Table 7.15). DIC encompasses three main constituents: free $CO_2$ (a gas), the bicarbonate ion ($HCO_3^-$), and the carbonate ion ($CO_3^{2-}$). Bicarbonate and carbonate constitute the major buffers in most natural waters and account for most of the acid neutralizing capacity. Free $CO_2$ is the most dynamic of the constituents of DIC and is the dominant acid in most natural waters. The ratio of $CO_2$ to $HCO_3^-$ and $CO_3^{2-}$ acts as the major driver of pH in most natural waters.

It is observed from the present field experiment that the TB of *E. intestinalis* ranged between 0.879 kg dry wt. $m^{-2}$ (at Sagar Island) to 1.505 kg dry wt. $m^{-2}$ (at Arbesi) (Fig. 7.15). DIC ranged from 1738 μmol/L (at Chulkathi) to 2132 μmol/L (at Sagar Island) (Fig. 7.16). The pH value, which is the primary proxy to estuarine acidification ranged from 7.98 (at Sagar Island) to 8.33 (at Arbesi) (Fig. 7.17).

The gas exchange in seaweed is strongly regulated by ribulose-1,5-biphosphate carboxylase/oxygenase (RuBisCO), which catalyzes the combination of carbon dioxide with the acceptor molecule ribulose-1,5-biphosphate. It is to be noted in this context that (aqueous dissolved carbon dioxide; $CO_2$ (aq)) represents only 0.5–1% of the total DIC pool, while $HCO_3^-$ represents approximately 90% of the DIC

**Table 7.15** Sampling stations in Indian Sundarbans with Total Biomass (TB) of *Enteromorpha intestinalis*, DIC and pH values of the ambient aquatic phase during February, 2020

| Stn. No. | Sampling station | Coordinates | Total biomass (kg dry wt. $m^{-2}$) | DIC (µmol/L) | pH value |
|---|---|---|---|---|---|
| 1 | Muriganga | 21° 38′ 25.86″ N; 88° 08′ 53.55″ E | 1.005 | 1921 | 8.05 |
| 2 | Saptamukhi | 21° 36′ 02.49″ N; 88° 23′ 47.18″ E | 1.239 | 1863 | 8.10 |
| 3 | Thakuran | 21° 49′ 43.17″ N; 88° 33′ 21.57″ E | 1.316 | 1802 | 8.22 |
| 4 | Herobhanga | 21° 59′ 34.32″ N; 88° 41′ 46.52″ E | 1.432 | 1799 | 8.28 |
| 5 | Ajmalmari | 21° 51′ 34.72″ N; 88° 39′ 00.68″E | 1.295 | 1815 | 8.20 |
| 6 | Dhulibasani | 21° 47′ 06.62″ N; 88° 33′ 48.20″E | 1.416 | 1806 | 8.26 |
| 7 | Chulkathi | 21° 41′ 53.62″ N; 88° 34′ 10.31″ E | 1.428 | 1738 | 8.27 |
| 8 | Arbesi | 22° 11′ 43.14″ N; 89° 01′ 09.04″ E | 1.505 | 1775 | 8.33 |
| 9 | Jhilla | 22° 09′ 51.53″ N; 88° 57′ 57.07″ E | 1.493 | 1769 | 8.31 |
| 10 | Pirkhali | 22° 06′ 00.97″ N; 88° 51′ 06.04″ E | 1.489 | 1773 | 8.31 |
| 11 | Panchmukhani | 21° 59′ 41.58″ N; 88° 54′ 14.71″ E | 1.491 | 1778 | 8.32 |
| 12 | Harinbhanga | 21° 57′ 17.85″ N; 88° 59′ 33.24″ E | 1.475 | 1780 | 8.30 |
| 13 | Khatuajhuri | 22° 03′ 06.55″ N; 89° 01′ 05.33″ E | 1.478 | 1800 | 8.30 |
| 14 | Chamta | 21° 53′ 18.56″ N; 88° 57′ 11.40″ E | 1.445 | 1867 | 8.29 |
| 15 | Matla | 21° 53′ 15.30″ N; 88° 44′ 08.74″ E | 1.358 | 1865 | 8.15 |
| 16 | Chandkhali | 21° 51′ 13.59″ N; 89° 00′ 44.68″ E | 1.367 | 1869 | 8.19 |
| 17 | Goashaba | 21° 43′ 50.64″ N; 88° 46′ 41.44″ E | 1.360 | 1795 | 8.18 |
| 18 | Gona | 21° 41′ 15.44″ N; 88° 54′ 31.09″ E | 1.430 | 1799 | 8.28 |
| 19 | Chhotahardi | 21° 44′ 42.24″ N; 88° 44′ 17.79″ E | 1.427 | 1791 | 8.27 |
| 20 | Baghmara | 21° 39′ 04.45″ N; 89° 04′ 40.59″ E | 1.436 | 1782 | 8.29 |

(continued)

**Table 7.15** (continued)

| Stn. No. | Sampling station | Coordinates | Total biomass (kg dry wt. $m^{-2}$) | DIC (μmol/L) | pH value |
|---|---|---|---|---|---|
| 21 | Mayadwip | 21° 35′ 50.23″ N; 88° 47′ 09.95″ E | 1.440 | 1776 | 8.30 |
| 22 | Jambu Island | 21° 35′ 42.03″ N; 88° 10′ 22.76″ E | 0.987 | 2024 | 7.99 |
| 23 | Lothian | 21° 38′ 21.20″ N; 88° 20′ 29.32″ E | 1.512 | 1790 | 8.32 |
| 24 | Sagar Island | 21° 38′ 51.55″ N; 88° 02′ 20.97″ E | 0.879 | 2132 | 7.98 |

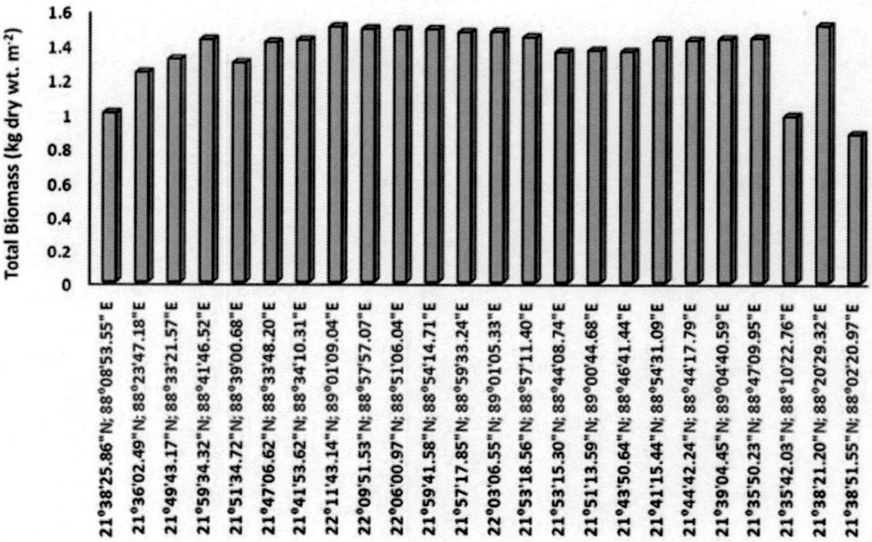


**Fig. 7.15** Total Biomass (kg dry wt. $m^{-2}$) of *Enteromorpha intestinalis* during February 2020

pool of the aquatic phase. Thus, the need of inorganic carbon for seaweed photosynthesis is often met by $HCO_3^-$. This might be the reason for the coincidence of relatively high pH in the estuarine water of Indian Sundarbans with the sampling stations sustaining high biomass of the seaweed species *E. intestinalis*. The bottom line of the entire discussion is the unique potential of seaweeds to transform DIC into organic carbon through the process of photosynthesis, which can decrease the $pCO_2$ in the aquatic phase and retard the pace of acidification.

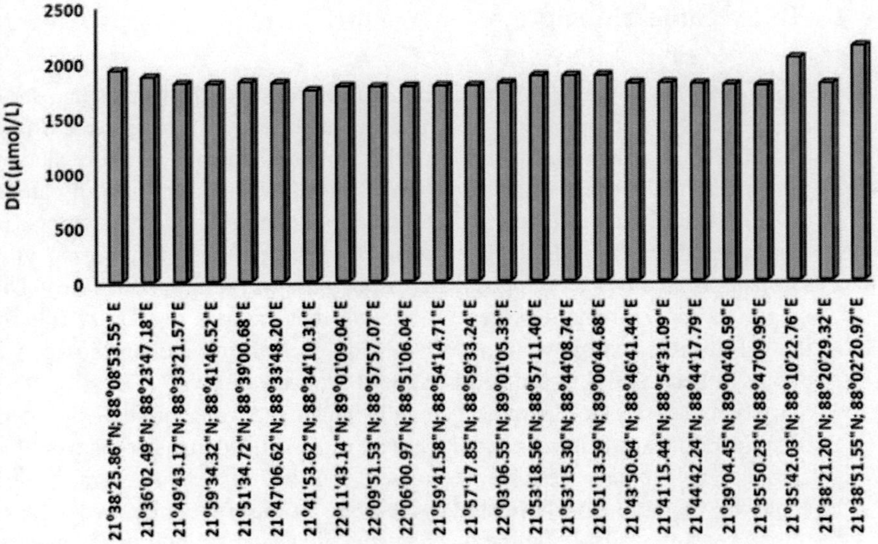


**Fig. 7.16** DIC (μmol/L) of aquatic phase during February 2020

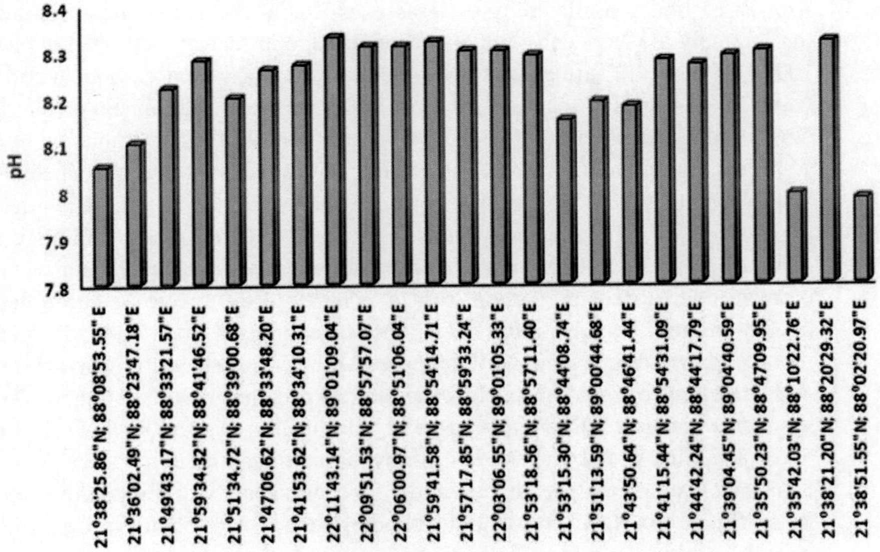


**Fig. 7.17** pH value of ambient aquatic phase during February 2020

## 7.4 Take Home Messages

(a) Mangroves are among the most carbon-dense forests in the tropics and store exceptionally high carbon compared to different forest ecosystems in the world. The ability of mangroves to scrub carbon from the ambient media is the key to regulate the process of acidification. Basically mangroves absorb carbon dioxide from the atmosphere and the water column and store it as carbon in their vegetative and reproductive parts. The principal carbon-fixing enzyme in plants is ribulose 1,5-biphosphate carboxylase/oxygenase (RuBisco). The magnitude of the mangrove vegetation pool thus exerts a regulatory role on the pH of the estuarine water as these halophytes absorb carbon dioxide for photosynthesis and shift the equilibrium towards alkalinity.

(b) Sea grasses are marine plants that generally inhabit the protected shallow waters of temperature and tropical coastal areas and are included in the domain of blue carbon. These plants are not true grasses, and they represent several genera appear to be more closely related to members of the lily family. Very scanty references are available on the mechanisms of carbon fixation in sea grasses and questions concerning the kinetics of the active forms of carbon utilized and possible photosynthetic pathways. Researchers postulated that the exogenous form of carbon assimilated by sea grasses is $HCO_3^-$. This may be attributed to extremely low concentration of carbon dioxide in seawater compared with $HCO_3^-$ (about 200 times less) and the slow rate at which carbon dioxide could be re-supplied to the plant from other forms of carbon. **Salt marsh** communities are found on the shoreward side of mud flats in temperate and subarctic regions of the world. The estuaries of Indian sub-continent sustains an important saltmarsh grass species namely *Porteresia coarctata*, which has a great role in stabilizing the islands. Their long network of roots binds the soil particles very intricately and retards the process of erosion. The growth, biomass and subsequent carbon storage capacity of the species is controlled by the ambient aquatic salinity. The species exhibits a luxuriant growth in relatively low saline regions (salinity range within 5–20 psu) with subsequently higher carbon content, but at salinity around 30 psu, the growth is stunted. The species is noted or retarding the pace of acidification in areas of low salinity in the estuarine stretch, where their growth is significantly high.

(c) Seaweeds are macroalgae and are noted for their capacity to utilize dissolved inorganic carbon (DIC) from the surrounding seawater for photosynthesis and growth, resulting in a decrease in the DIC concentration of seawater and a drop in the partial pressure of carbon dioxide ($pCO_2$) below atmospheric levels. Our ground zero observations show that sites with high biomass of seaweeds coincide with low DIC and high pH level of the estuarine water.

# Annexure 7: Worldwide Common Mangrove and Associate Species

| S. No. | Species | |
|---|---|---|
| 1 | *Acanthus ebracteatus* |  |
| 2 | *Acanthus ilicifolius* |  |
| 3 | *Acanthus volubilis* |  |
| 4 | *Cerbera manghas* |  |
| 5 | *Nypa fruticans* |  |

(continued)

(continued)

| S. No. | Species | |
|---|---|---|
| 6 | *Phoenix paludosa* |  |
| 7 | *Finlaysonia obovata* |  |
| 8 | *Sarcolobus globosus* |  |
| 9 | *Sarcolobus carinatus* |  |
| 10 | *Tylophora tenuis* |  |

(continued)

(continued)

| S. No. | Species | |
|---|---|---|
| 11 | *Avicennia alba* |  |
| 12 | *Avicennia bicolour* |  |
| 13 | *Avicennia germinans* |  |
| 14 | *Avicennia integra* |  |
| 15 | *Avicennia marina* |  |

(continued)

(continued)

| S. No. | Species | |
|---|---|---|
| 16 | *Avicennia var. australasica* |  |
| 17 | *Avicennia var. eucalyptifolia* |  |
| 18 | *Avicennia officinalis* |  |
| 19 | *Avicennia rumphiana* |  |
| 20 | *Avicennia schaueriana* |  |

(continued)

(continued)

| S. No. | Species | |
|---|---|---|
| 21 | *Dolichandrone spathacea* |  |
| 22 | *Camptostemon philippinensis* |  |
| 23 | *Camptostemon schultzii* |  |
| 24 | *Caesalpinia bonduc* |  |
| 25 | *Caesalpinia cristata* |  |

(continued)

(continued)

| S. No. | Species | |
|---|---|---|
| 26 | *Cynometra iripa* |  |
| 27 | *Cynometra ramiflora* |  |
| 28 | *Intsia bijuga* |  |
| 29 | *Conocarpus erectus* |  |
| 30 | *Laguncularia racemosa* |  |

(continued)

(continued)

| S. No. | Species | |
|---|---|---|
| 31 | *Lumnitzera littorea* |  |
| 32 | *Lumnitzera racemosa* |  |
| 33 | *Ipomoea tuba* |  |
| 34 | *Fimbristylis ferruginea* |  |
| 35 | *Scirpus littoralis* |  |

(continued)

(continued)

| S. No. | Species | |
|---|---|---|
| 36 | *Excoecaria agallocha* |  |
| 37 | *Excoecaria indica* |  |
| 38 | *Excoecaria ovate* |  |
| 39 | *Flagellaria indica* |  |
| 40 | *Pemphis acidula* |  |

(continued)

(continued)

| S. No. | Species | |
|---|---|---|
| 41 | *Hibiscus tiliaceous* |  |
| 42 | *Thespesia populneoide* |  |
| 43 | *Thespesia populnea* |  |
| 44 | *Aglaia cucullata* |  |
| 45 | *Xylocarpus granatum* |  |

(continued)

(continued)

| S. No. | Species | |
|---|---|---|
| 46 | *Xylocarpus mekongensis* |  |
| 47 | *Xylocarpus moluccensis* |  |
| 48 | *Aegiceras corniculatum* |  |
| 49 | *Aegiceras floridum* |  |
| 50 | *Osbornia octodonta* |  |

(continued)

(continued)

| S. No. | Species | |
|---|---|---|
| 51 | *Dalbergia spinosa* |  |
| 52 | *Derris scandens* |  |
| 53 | *Derris heterophylla* |  |
| 54 | *Mucuna gigantea* |  |
| 55 | *Pelliciera rhizophorae* |  |

(continued)

(continued)

| S. No. | Species | |
|---|---|---|
| 56 | *Aegialitis rotundifolia* |  |
| 57 | *Aegialitis annulata* |  |
| 58 | *Myriostachya wightiana* |  |
| 59 | *Porteresia coarctata* |  |
| 60 | *Urochondra setulosa* |  |

(continued)

(continued)

| S. No. | Species | |
|---|---|---|
| 61 | *Acrostichum aureum* |  |
| 62 | *Acrostichum danaeifolium* |  |
| 63 | *Acrostichum speciosum* |  |
| 64 | *Bruguiera cylindrica* |  |
| 65 | *Bruguiera exaristata* |  |

(continued)

(continued)

| S. No. | Species | |
|---|---|---|
| 66 | *Bruguiera gymnorhiza* |  |
| 67 | *Bruguiera hainesii* |  |
| 68 | *Bruguiera parviflora* |  |
| 69 | *Bruguiera sexangula* |  |
| 70 | *Ceriops australis* |  |

(continued)

(continued)

| S. No. | Species | |
|---|---|---|
| 71 | *Ceriops decandra* |  |
| 72 | *Ceriops tagal* |  |
| 73 | *Kandelia candel* |  |
| 74 | *Rhizophora apiculata* |  |
| 75 | *Rhizophora harrisonii* |  |

(continued)

(continued)

| S. No. | Species | |
|---|---|---|
| 76 | *Rhizophora mangle* |  |
| 77 | *Rhizophora mucronata* |  |
| 78 | *Rhizophora racemosa* |  |
| 79 | *Rhizophora samoensis* |  |
| 80 | *Rhizophora stylosa* |  |

(continued)

(continued)

| S. No. | Species | |
|---|---|---|
| 81 | *Scyphiphora hydrophyllacea* |  |
| 82 | *Merope angulata* |  |
| 83 | *Sonneratia alba* |  |
| 84 | *Sonneratia apetala* |  |
| 85 | *Sonneratia caseolaris* |  |

(continued)

(continued)

| S. No. | Species | |
|---|---|---|
| 86 | *Sonneratia griffithii* |  |
| 87 | *Sonneratia lanceolata* |  |
| 88 | *Sonneratia ovate* |  |
| 89 | *Sonneratia gulngai* |  |
| 90 | *Sonneratia urama* |  |

(continued)

(continued)

| S. No. | Species | |
|---|---|---|
| 91 | *Heritiera fomes* |  |
| 92 | *Heritiera globosa* |  |
| 93 | *Heritiera kanikensis* |  |
| 94 | *Heritiera littoralis* |  |
| 95 | *Brownlowia tersa* |  |

(continued)

(continued)

| S. No. | Species | |
|---|---|---|
| 96 | *Clerodendrum inerme* |  |

## References

Beer S, Eschel A, Waisel Y (1977) Carbon metabolism in seagrasses. J Exp Bot 28:1180–1189
Beer S, Waisel Y (1979) Some photosynthetic carbon fixation properties of sea grasses. Aquat Bot 7:129–138
Benedict CR, Scott JR (1976) Photosynthetic carbon metabolism of a marine grass. Plant Physio 57:876–880
Bouillon S, Rivera Monroy VH, Twilley RR, Kairo JG (2009) Mangroves. In: Laffoley D, Grimsditch (eds) The management of natural coastal carbon sinks. IUCN
Bowes G, Holaday AS, Van TK, Hallen WT (1978) Photosynthetic and photorespiratory carbon metabolism in aquatic plants. In: Hall DO, Coombs J, Goodwin TW (eds) Photosynthesis. The Biochemical Society, London, pp 289–298
Donato DC, Kauffman JB, Murdiyarso D, Kurnianto S, Stidham M, Kanninen, M (2011) Mangroves among the most carbon-rich forests in the tropics. Nature Geosci 4
Doohan ME, Newcomb EH (1976) Leaf ultrastructure and 8 Is C values of three seagrasses from the great barrier reef. Aust J Plant Physiol 3:9–23
Downton WJS, Bishop DG, Larkum AWD, Osmond CB (1976) Oxygen inhibition of photosynthetic oxygen evolution in marine plants. Aust J Plant Physiol 3:73–79
FAO/Forest Trends (2007) Investing in the future: an assessment of private sector demand for engaging in markets & payments for ecosystem services, by Waage S, with contributions from Mulder I, Ten Kate K, Sherr S, Roberts JP, Hawn A, Hamilton K, Bayon R, Carroll N, Forthcoming in PESAL (Payments for Environmental Services from Agricultural Landscapes) papers series. Rome, FAO and Washington, DC, Forest Trends
Hough RA (1976) Light and dark respiration and release of organic carbon in marine macrophytes of the Great Barrier Reef region. Aust J Plant Physio 3:63–68
Mitra A, Zaman S (2016) Basics of marine and estuarine ecology; Publisher, Springer India, pp 1-481. Book ISBN 978-81-322-2707-6
Parker PL (1964) The biogeochemistry of the stable isotopes of carbon in a marine bay. Geochim Cosmochim Acta 28:1155–1164
Smith BN, Epstein S (1971) Two categories of IsC/~C ratios for higher plants. Plant Physiol 47:380–384
Steemann Nielsen E (1975) Marine photosynthesis. Elsevier, Amsterdam, p 142

## *Internet References*

http://www.bsienvis.nic.in/Database/IndianMangroves_3941.aspx#WORLD. Distribution of Mangroves

http://www.recoftc.org/site/resources/Mangroves-more-Carbon-Rich-and-Important-for-Climate-Change.php

# Chapter 8
# Estuarine Acidification: Management and Mitigation

**Abstract** Acidification is one of the climate change related stressors expected to have multiple impacts on estuarine ecosystem. Considering the magnitude of the impact it is essential to identify its potential impacts to beneficial uses of the ecosystem, cost-effective monitoring strategies and management actions. However, it is important to note that all estuaries are not uniform in terms of morphology, hydrology, and biodiversity status. Hence a generalized management strategy may not yield a fruitful result to combat the process of acidification. Controlling the point sources of industrial discharge by setting Combined Effluent Treatment Plant (CETP) and expanding the dimension of blue carbon along the estuarine stretch might be some measures of first-order scale. Halophyte farming and iron fertilization in closed water bodies coupled with aquaculture are the other options to scrub carbon dioxide from the atmosphere.

## 8.1 Mitigation of Point Sources of Carbon Dioxide Through Clean Energy

Emission of carbon dioxide occurs from industries, urban and transport sectors and change of land use pattern for reasons like expansion of tourism, aquaculture, ports and harbours etc. All these activities cannot be completely stopped and uprooted as there is a growing demand to meet the needs of the rapidly rising population of the world. The scenario of the present day world has changed a lot since the last few centuries. The population has increased exponentially (Table 8.1) while the area (space) remained almost constant resulting in the rise of population density (Table 8.2). At present the population of the World is somewhat around 7.5 billion (https://www.census.gov/newsroom/stories/world-population-day.html) and predictions state that the figure will touch 9.0 billion by the year 2050.

The huge population pressure has posed challenges on leaving space, resources, environmental condition, job opportunity, mobility and the quality of the life. Desperate need for job opportunities/employment has opened the doors for intense industrialization, which has resulted in more emission of GHGs preferably carbon dioxide. As a result, the concentration of dissolved inorganic carbon (DIC) in the



A. Mitra and S. Zaman, *Estuarine Acidification*,
https://doi.org/10.1007/978-3-030-84792-0_8

**Table 8.1** Increase of population in 1 year

| Country | 2017 | 2018 |
|---|---|---|
| Aruba | 105,366 | 105,845 |
| Afghanistan | 36,296,400 | 37,172,386 |
| Angola | 29,816,748 | 30,809,762 |
| Albania | 2,873,457 | 2,866,376 |
| Andorra | 77,001 | 77,006 |
| Arab World | 411,898,965 | 419,790,588 |
| United Arab Emirates | 9,487,203 | 9,630,959 |
| Argentina | 44,044,811 | 44,494,502 |
| Armenia | 2,944,809 | 2,951,776 |
| American Samoa | 55,620 | 55,465 |
| Antigua and Barbuda | 95,426 | 96,286 |
| Australia | 24,601,860 | 24,992,369 |
| Austria | 8,797,566 | 8,847,037 |
| Azerbaijan | 9,854,033 | 9,942,334 |
| Burundi | 10,827,024 | 11,175,378 |
| Belgium | 11,375,158 | 11,422,068 |
| Benin | 11,175,204 | 11,485,048 |
| Burkina Faso | 19,193,284 | 19,751,535 |
| Bangladesh | 159,670,593 | 161,356,039 |
| Bulgaria | 7,075,947 | 7,024,216 |
| Bahrain | 1,494,074 | 1,569,439 |
| Bahamas, The | 381,761 | 385,640 |
| Bosnia and Herzegovina | 3,351,527 | 3,323,929 |
| Belarus | 9,498,264 | 9,485,386 |
| Belize | 375,769 | 383,071 |
| Bermuda | 63,874 | 63,968 |
| Bolivia | 11,192,854 | 11,353,142 |
| Brazil | 207,833,831 | 209,469,333 |
| Barbados | 286,233 | 286,641 |
| Brunei Darussalam | 424,473 | 428,962 |
| Bhutan | 745,568 | 754,394 |
| Botswana | 2,205,128 | 2,254,126 |
| Central African Republic | 4,596,028 | 4,666,377 |
| Canada | 36,540,268 | 37,058,856 |
| Central Europe and the Baltics | 102,738,854 | 102,511,922 |
| Switzerland | 8,451,840 | 8,516,543 |

(continued)

**Table 8.1** (continued)

| Country | 2017 | 2018 |
|---|---|---|
| Channel Islands | 168,665 | 170,499 |
| Chile | 18,470,439 | 18,729,160 |
| China | 1,386,395,000 | 1,392,730,000 |
| Cote d'Ivoire | 24,437,469 | 25,069,229 |
| Cameroon | 24,566,045 | 25,216,237 |
| Congo, Dem. Rep. | 81,398,764 | 84,068,091 |
| Congo, Rep. | 5,110,702 | 5,244,363 |
| Colombia | 48,901,066 | 49,648,685 |
| Comoros | 813,892 | 832,322 |
| Cabo Verde | 537,497 | 543,767 |
| Costa Rica | 4,949,954 | 4,999,441 |
| Caribbean small states | 7,314,990 | 7,358,965 |
| Cuba | 11,339,259 | 11,338,138 |
| Curacao | 160,175 | 159,849 |
| Cayman Islands | 63,382 | 64,174 |
| Cyprus | 1,179,680 | 1,189,265 |
| Czech Republic | 10,594,438 | 10,625,695 |
| Germany | 82,657,002 | 82,927,922 |
| Djibouti | 944,097 | 958,920 |
| Dominica | 71,458 | 71,625 |
| Denmark | 5,764,980 | 5,797,446 |
| Dominican Republic | 10,513,131 | 10,627,165 |
| Algeria | 41,389,198 | 42,228,429 |
| East Asia & Pacific (excluding high income) | 2,068,155,660 | 2,081,651,801 |
| Early-demographic dividend | 3,207,188,541 | 3,249,140,605 |
| East Asia & Pacific | 2,314,202,003 | 2,328,220,870 |
| Europe & Central Asia (excluding high income) | 415,710,935 | 417,797,257 |
| Europe & Central Asia | 915,420,161 | 918,793,590 |
| Ecuador | 16,785,361 | 17,084,357 |
| Egypt, Arab Rep. | 96,442,593 | 98,423,595 |
| Euro area | 341,164,362 | 341,783,171 |
| Eritrea | | |
| Spain | 46,593,236 | 46,723,749 |
| Estonia | 1,317,384 | 1,320,884 |
| Ethiopia | 106,400,024 | 109,224,559 |
| European Union | 512,191,098 | 513,213,363 |

(continued)

**Table 8.1** (continued)

| Country | 2017 | 2018 |
|---|---|---|
| Fragile and conflict affected situations | 504,119,229 | 515,215,936 |
| Finland | 5,508,214 | 5,518,050 |
| Fiji | 877,459 | 883,483 |
| France | 66,865,144 | 66,987,244 |
| Faroe Islands | 48,331 | 48,497 |
| Micronesia, Fed. Sts. | 111,459 | 112,640 |
| Gabon | 2,064,823 | 2,119,275 |
| United Kingdom | 66,058,859 | 66,488,991 |
| Georgia | 3,728,004 | 3,731,000 |
| Ghana | 29,121,471 | 29,767,108 |
| Gibraltar | 33,728 | 33,718 |
| Guinea | 12,067,539 | 12,414,318 |
| Gambia, The | 2,213,894 | 2,280,102 |
| Guinea-Bissau | 1,828,146 | 1,874,309 |
| Equatorial Guinea | 1,262,001 | 1,308,974 |
| Greece | 10,754,679 | 10,727,668 |
| Grenada | 110,874 | 111,454 |
| Greenland | 56,171 | 56,025 |
| Guatemala | 16,914,936 | 17,247,807 |
| Guam | 164,281 | 165,768 |
| Guyana | 775,221 | 779,004 |
| High income | 1,204,429,565 | 1,210,312,147 |
| Hong Kong SAR, China | 7,391,700 | 7,451,000 |
| Honduras | 9,429,013 | 9,587,522 |
| Heavily indebted poor countries (HIPC) | 759,106,221 | 780,234,406 |
| Croatia | 4,124,531 | 4,089,400 |
| Haiti | 10,982,366 | 11,123,176 |
| Hungary | 9,787,966 | 9,768,785 |
| IBRD only | 4,731,120,193 | 4,772,284,113 |
| IDA & IBRD total | 6,335,039,629 | 6,412,522,234 |
| IDA total | 1,603,919,436 | 1,640,238,121 |
| IDA blend | 543,525,897 | 555,830,605 |
| Indonesia | 264,645,886 | 267,663,435 |
| IDA only | 1,060,393,539 | 1,084,407,516 |
| Isle of Man | 83,598 | 84,077 |
| India | 1,338,658,835 | 1,352,617,328 |

(continued)

**Table 8.1** (continued)

| Country | 2017 | 2018 |
|---|---|---|
| Not classified | | |
| Ireland | 4,807,388 | 4,853,506 |
| Iran, Islamic Rep. | 80,673,951 | 81,800,269 |
| Iraq | 37,552,781 | 38,433,600 |
| Iceland | 343,400 | 353,574 |
| Israel | 8,713,300 | 8,883,800 |
| Italy | 60,536,709 | 60,431,283 |
| Jamaica | 2,920,853 | 2,934,855 |
| Jordan | 9,779,173 | 9,956,011 |
| Japan | 126,785,797 | 126,529,100 |
| Kazakhstan | 18,037,776 | 18,276,499 |
| Kenya | 50,221,473 | 51,393,010 |
| Kyrgyz Republic | 6,198,200 | 6,315,800 |
| Cambodia | 16,009,414 | 16,249,798 |
| Kiribati | 114,158 | 115,847 |
| St. Kitts and Nevis | 52,045 | 52,441 |
| Korea, Rep. | 51,466,201 | 51,635,256 |
| Kuwait | 4,056,097 | 4,137,309 |
| Latin America & Caribbean (excluding high income) | 603,254,104 | 609,013,934 |
| Lao PDR | 6,953,035 | 7,061,507 |
| Lebanon | 6,811,873 | 6,848,925 |
| Liberia | 4,702,228 | 4,818,977 |
| Libya | 6,580,724 | 6,678,567 |
| St. Lucia | 180,955 | 181,889 |
| Latin America & Caribbean | 635,372,515 | 641,357,515 |
| Least developed countries: UN classification | 986,365,080 | 1,009,662,578 |
| Low income | 687,449,530 | 705,417,321 |
| Liechtenstein | 37,800 | 37,910 |
| Sri Lanka | 21,444,000 | 21,670,000 |
| Lower middle income | 2,981,420,591 | 3,022,905,169 |
| Low & middle income | 6,306,560,891 | 6,383,958,209 |
| Lesotho | 2,091,412 | 2,108,132 |
| Late-demographic dividend | 2,276,319,334 | 2,288,665,963 |
| Lithuania | 2,828,403 | 2,789,533 |
| Luxembourg | 596,336 | 607,728 |
| Latvia | 1,942,248 | 1,926,542 |

(continued)

**Table 8.1** (continued)

| Country | 2017 | 2018 |
|---|---|---|
| Macao SAR, China | 622,585 | 631,636 |
| St. Martin (French part) | 36,560 | 37,264 |
| Morocco | 35,581,294 | 36,029,138 |
| Monaco | 38,392 | 38,682 |
| Moldova | 3,549,196 | 3,545,883 |
| Madagascar | 25,570,540 | 26,262,368 |
| Maldives | 496,402 | 515,696 |
| Middle East & North Africa | 441,255,234 | 448,912,859 |
| Mexico | 124,777,324 | 126,190,788 |
| Marshall Islands | 58,058 | 58,413 |
| Middle income | 5,619,111,361 | 5,678,540,888 |
| North Macedonia | 2,081,996 | 2,082,958 |
| Mali | 18,512,394 | 19,077,690 |
| Malta | 467,999 | 483,530 |
| Myanmar | 53,382,581 | 53,708,395 |
| Middle East & North Africa (excluding high income) | 376,546,755 | 382,896,715 |
| Montenegro | 622,373 | 622,345 |
| Mongolia | 3,113,779 | 3,170,208 |
| Northern Mariana Islands | 56,562 | 56,882 |
| Mozambique | 28,649,007 | 29,495,962 |
| Mauritania | 4,282,574 | 4,403,319 |
| Mauritius | 1,264,613 | 1,265,303 |
| Malawi | 17,670,260 | 18,143,315 |
| Malaysia | 31,105,028 | 31,528,585 |
| North America | 361,751,263 | 364,290,258 |
| Namibia | 2,402,603 | 2,448,255 |
| New Caledonia | 280,350 | 284,060 |
| Niger | 21,602,472 | 22,442,948 |
| Nigeria | 190,873,311 | 195,874,740 |
| Nicaragua | 6,384,855 | 6,465,513 |
| Netherlands | 17,131,296 | 17,231,017 |
| Norway | 5,276,968 | 5,314,336 |
| Nepal | 27,627,124 | 28,087,871 |
| Nauru | 12,876 | 12,704 |
| New Zealand | 4,793,900 | 4,885,500 |
| OECD members | 1,296,225,760 | 1,303,529,456 |

(continued)

**Table 8.1** (continued)

| Country | 2017 | 2018 |
|---|---|---|
| Oman | 4,665,935 | 4,829,483 |
| Other small states | 30,148,800 | 30,758,989 |
| Pakistan | 207,896,686 | 212,215,030 |
| Panama | 4,106,771 | 4,176,873 |
| Peru | 31,444,297 | 31,989,256 |
| Philippines | 105,173,264 | 106,651,922 |
| Palau | 17,808 | 17,907 |
| Papua New Guinea | 8,438,029 | 8,606,316 |
| Poland | 37,974,826 | 37,978,548 |
| Pre-demographic dividend | 894,512,725 | 919,485,393 |
| Puerto Rico | 3,325,001 | 3,195,153 |
| Korea, Dem. People's Rep. | 25,429,985 | 25,549,819 |
| Portugal | 10,300,300 | 10,281,762 |
| Paraguay | 6,867,062 | 6,956,071 |
| West Bank and Gaza | 4,454,805 | 4,569,087 |
| Pacific island small states | 2,422,086 | 2,457,367 |
| Post-demographic dividend | 1,106,035,186 | 1,109,997,273 |
| French Polynesia | 276,103 | 277,679 |
| Qatar | 2,724,724 | 2,781,677 |
| Romania | 19,587,491 | 19,473,936 |
| Russian Federation | 144,496,740 | 144,478,050 |
| Rwanda | 11,980,937 | 12,301,939 |
| South Asia | 1,792,835,608 | 1,814,388,744 |
| Saudi Arabia | 33,099,147 | 33,699,947 |
| Sudan | 40,813,396 | 41,801,533 |
| Senegal | 15,419,381 | 15,854,360 |
| Singapore | 5,612,253 | 5,638,676 |
| Solomon Islands | 636,038 | 652,858 |
| Sierra Leone | 7,488,431 | 7,650,154 |
| El Salvador | 6,388,122 | 6,420,744 |
| San Marino | 33,671 | 33,785 |
| Somalia | 14,589,119 | 15,008,154 |
| Serbia | 7,020,858 | 6,982,084 |
| Sub-Saharan Africa (excluding high income) | 1,050,057,829 | 1,078,209,758 |
| South Sudan | 10,910,759 | 10,975,920 |
| Sub-Saharan Africa | 1,050,153,672 | 1,078,306,520 |

(continued)

**Table 8.1** (continued)

| Country | 2017 | 2018 |
|---|---|---|
| Small states | 39,885,876 | 40,575,321 |
| Sao Tome and Principe | 207,089 | 211,028 |
| Suriname | 570,496 | 575,991 |
| Slovak Republic | 5,439,232 | 5,447,011 |
| Slovenia | 2,066,388 | 2,067,372 |
| Sweden | 10,057,698 | 10,183,175 |
| Eswatini | 1,124,753 | 1,136,191 |
| Sint Maarten (Dutch part) | 40,574 | 40,654 |
| Seychelles | 95,843 | 96,762 |
| Syrian Arab Republic | 17,068,002 | 16,906,283 |
| Turks and Caicos Islands | 37,115 | 37,665 |
| Chad | 15,016,773 | 15,477,751 |
| East Asia & Pacific (IDA & IBRD countries) | 2,042,687,863 | 2,056,064,424 |
| Europe & Central Asia (IDA & IBRD countries) | 457,810,292 | 459,865,205 |
| Togo | 7,698,475 | 7,889,094 |
| Thailand | 69,209,858 | 69,428,524 |
| Tajikistan | 8,880,268 | 9,100,837 |
| Turkmenistan | 5,757,669 | 5,850,908 |
| Latin America & the Caribbean (IDA & IBRD countries) | 619,460,244 | 625,569,713 |
| Timor-Leste | 1,243,261 | 1,267,972 |
| Middle East & North Africa (IDA & IBRD countries) | 372,091,950 | 378,327,628 |
| Tonga | 101,998 | 103,197 |
| South Asia (IDA & IBRD) | 1,792,835,608 | 1,814,388,744 |
| Sub-Saharan Africa (IDA & IBRD countries) | 1,050,153,672 | 1,078,306,520 |
| Trinidad and Tobago | 1,384,072 | 1,389,858 |
| Tunisia | 11,433,443 | 11,565,204 |
| Turkey | 81,101,892 | 82,319,724 |
| Tuvalu | 11,370 | 11,508 |
| Tanzania | 54,663,906 | 56,318,348 |
| Uganda | 41,162,465 | 42,723,139 |
| Ukraine | 44,831,135 | 44,622,516 |
| Upper middle income | 2,637,690,770 | 2,655,635,719 |
| Uruguay | 3,436,646 | 3,449,299 |
| United States | 325,147,121 | 327,167,434 |
| Uzbekistan | 32,388,600 | 32,955,400 |
| St. Vincent and the Grenadines | 109,827 | 110,210 |

(continued)

**Table 8.1** (continued)

| Country | 2017 | 2018 |
|---|---|---|
| Venezuela, RB | 29,390,409 | 28,870,195 |
| British Virgin Islands | 29,577 | 29,802 |
| Virgin Islands (U.S.) | 107,268 | 106,977 |
| Vietnam | 94,596,642 | 95,540,395 |
| Vanuatu | 285,510 | 292,680 |
| World | 7,510,990,456 | 7,594,270,356 |
| Samoa | 195,352 | 196,130 |
| Kosovo | 1,830,700 | 1,845,300 |
| Yemen, Rep. | 27,834,821 | 28,498,687 |
| South Africa | 57,000,451 | 57,779,622 |
| Zambia | 16,853,688 | 17,351,822 |
| Zimbabwe | 14,236,745 | 14,439,018 |

near surface ocean increases, which drives a decrease in pH in order to maintain a chemical equilibrium. The various sources of GHG's that triggers the process of Estuarine Acidification (EA) are discussed here in points with particular reference to estuaries of lower Gangetic delta.

1. **Emission of Fluorine Compounds**

   (i) Various fluorine compounds (HF, $F_2$, $SiF_4$, $H_2SiF_6$ etc.) are emitted from phosphate fertilizer manufacturing plants (since large amounts of fluorine are present in phosphate rock).

   (ii) During aluminium extraction (where cryolite i.e., $Na_3AlF_6$ is used to lower the melting point of the mixture of $CaF_2$, $Al_2O_3$ and $Na_3AlF_6$) compounds of fluorine are released in to the atmosphere.

2. **Emission of Oxides of Carbon**

   (i) Cement manufacturing unit is one of the largest sources of industrial carbon dioxide emissions. The gas is produced when calcium carbonate ($CaCO_3$) is heated to produce the calcium oxide (CaO, also called *quick-lime*), which is the main ingredient for cement ($CaSiO_3$) production. The reaction steps in this manufacture are:

$$CaCO_3 \xrightarrow{\Delta} CaO + CO_2$$

$$CaO + SiO_2 \xrightarrow{\Delta} CaSiO_3$$

   $CO_2$ produced in the first step has considerable contribution to global warming. While fossil fuel combustion and deforestation each produce

**Table 8.2** Country-wise area in sq. mi with population density (PD)

| Country | Area (sq. mi) | PD (per sq. mi) |
|---|---|---|
| Afghanistan | 647,500 | 48.0 |
| Albania | 28,748 | 124.6 |
| Algeria | 2,381,740 | 13.8 |
| American Samoa | 199 | 290.4 |
| Andorra | 468 | 152.1 |
| Angola | 1,246,700 | 9.7 |
| Anguilla | 102 | 132.1 |
| Antigua & Barbuda | 443 | 156.0 |
| Argentina | 2,766,890 | 14.4 |
| Armenia | 29,800 | 99.9 |
| Aruba | 193 | 372.5 |
| Australia | 7,686,850 | 2.6 |
| Austria | 83,870 | 97.7 |
| Azerbaijan | 86,600 | 91.9 |
| The Bahamas | 13,940 | 21.8 |
| Bahrain | 665 | 1050.5 |
| Bangladesh | 144,000 | 1023.4 |
| Barbados | 431 | 649.5 |
| Belarus | 207,600 | 49.6 |
| Belgium | 30,528 | 340.0 |
| Belize | 22,966 | 12.5 |
| Benin | 112,620 | 69.8 |
| Bermuda | 53 | 1241.0 |
| Bhutan | 47,000 | 48.5 |
| Bolivia | 1,098,580 | 8.2 |
| Bosnia & Herzegovina | 51,129 | 88.0 |
| Botswana | 600,370 | 2.7 |
| Brazil | 8,511,965 | 22.1 |
| British Virgin is | 153 | 151.0 |
| Brunei | 5770 | 65.8 |
| Bulgaria | 110,910 | 66.6 |
| Burkina Faso | 274,200 | 50.7 |
| Burma | 678,500 | 69.8 |
| Burundi | 27,830 | 290.7 |
| Cambodia | 181,040 | 76.7 |
| Cameroon | 475,440 | 36.5 |
| Canada | 9,984,670 | 3.3 |

(continued)

**Table 8.2** (continued)

| Country | Area (sq. mi) | PD (per sq. mi) |
|---|---|---|
| Cape Verde | 4033 | 104.4 |
| Cayman Islands | 262 | 173.4 |
| Central African Rep. | 622,984 | 6.9 |
| Chad | 1,284,000 | 7.7 |
| Chile | 756,950 | 21.3 |
| China | 9,596,960 | 136.9 |
| Colombia | 1,138,910 | 38.3 |
| Comoros | 2170 | 318.4 |
| Congo, Dem. Rep. | 2,345,410 | 26.7 |
| Congo, Repub. of the | 342,000 | 10.8 |
| Cook Islands | 240 | 89.1 |
| Costa Rica | 51,100 | 79.8 |
| Cote d'Ivoire | 322,460 | 54.8 |
| Croatia | 56,542 | 79.5 |
| Cuba | 110,860 | 102.7 |
| Cyprus | 9250 | 84.8 |
| Czech Republic | 78,866 | 129.8 |
| Denmark | 43,094 | 126.5 |
| Djibouti | 23,000 | 21.2 |
| Dominica | 754 | 91.4 |
| Dominican Republic | 48,730 | 188.5 |
| East Timor | 15,007 | 70.8 |
| Ecuador | 283,560 | 47.8 |
| Egypt | 1,001,450 | 78.8 |
| El Salvador | 21,040 | 324.3 |
| Equatorial Guinea | 28,051 | 19.3 |
| Eritrea | 121,320 | 39.5 |
| Estonia | 45,226 | 29.3 |
| Ethiopia | 1,127,127 | 66.3 |
| Faroe Islands | 1399 | 33.8 |
| Fiji | 18,270 | 49.6 |
| Finland | 338,145 | 15.5 |
| France | 547,030 | 111.3 |
| French Guiana | 91,000 | 2.2 |
| French Polynesia | 4167 | 65.9 |
| Gabon | 267,667 | 5.3 |
| The Gambia | 11,300 | 145.3 |

(continued)

**Table 8.2** (continued)

| Country | Area (sq. mi) | PD (per sq. mi) |
|---|---|---|
| Gaza Strip | 360 | 3968.8 |
| Georgia | 69,700 | 66.9 |
| Germany | 357,021 | 230.9 |
| Ghana | 239,460 | 93.6 |
| Gibraltar | 7 | 3989.7 |
| Greece | 131,940 | 81.0 |
| Greenland | 2,166,086 | 0.0 |
| Grenada | 344 | 260.8 |
| Guadeloupe | 1780 | 254.4 |
| Guam | 541 | 316.1 |
| Guatemala | 108,890 | 112.9 |
| Guernsey | 78 | 838.6 |
| Guinea | 245,857 | 39.4 |
| Guinea-Bissau | 36,120 | 39.9 |
| Guyana | 214,970 | 3.6 |
| Haiti | 27,750 | 299.4 |
| Honduras | 112,090 | 65.4 |
| Hong Kong | 1092 | 6355.7 |
| Hungary | 93,030 | 107.3 |
| Iceland | 103,000 | 2.9 |
| India | 3,287,590 | 333.2 |
| Indonesia | 1,919,440 | 127.9 |
| Iran | 1,648,000 | 41.7 |
| Iraq | 437,072 | 61.3 |
| Ireland | 70,280 | 57.8 |
| Isle of Man | 572 | 131.9 |
| Israel | 20,770 | 305.8 |
| Italy | 301,230 | 193.0 |
| Jamaica | 10,991 | 250.9 |
| Japan | 377,835 | 337.4 |
| Jersey | 116 | 785.2 |
| Jordan | 92,300 | 64.0 |
| Kazakhstan | 2,717,300 | 5.6 |
| Kenya | 582,650 | 59.6 |
| Kiribati | 811 | 130.0 |
| Korea, North | 120,540 | 191.8 |
| Korea, South | 98,480 | 496.0 |

(continued)

**Table 8.2** (continued)

| Country | Area (sq. mi) | PD (per sq. mi) |
|---|---|---|
| Kuwait | 17,820 | 135.7 |
| Kyrgyzstan | 198,500 | 26.3 |
| Laos | 236,800 | 26.9 |
| Latvia | 64,589 | 35.2 |
| Lebanon | 10,400 | 372.5 |
| Lesotho | 30,355 | 66.6 |
| Liberia | 111,370 | 27.3 |
| Libya | 1,759,540 | 3.4 |
| Liechtenstein | 160 | 212.4 |
| Lithuania | 65,200 | 55.0 |
| Luxembourg | 2586 | 183.5 |
| Macau | 28 | 16,183.0 |
| Macedonia | 25,333 | 80.9 |
| Madagascar | 587,040 | 31.7 |
| Malawi | 118,480 | 109.8 |
| Malaysia | 329,750 | 74.0 |
| Maldives | 300 | 1196.7 |
| Mali | 1,240,000 | 9.5 |
| Malta | 316 | 1266.5 |
| Marshall Islands | 11,854 | 5.1 |
| Martinique | 1100 | 396.5 |
| Mauritania | 1,030,700 | 3.1 |
| Mauritius | 2040 | 608.3 |
| Mayotte | 374 | 538.1 |
| Mexico | 1,972,550 | 54.5 |
| Micronesia, Fed. St. | 702 | 153.9 |
| Moldova | 33,843 | 132.0 |
| Monaco | 2 | 16,271.5 |
| Mongolia | 1,564,116 | 1.8 |
| Montserrat | 102 | 92.5 |
| Morocco | 446,550 | 74.4 |
| Mozambique | 801,590 | 24.6 |
| Namibia | 825,418 | 2.5 |
| Nauru | 21 | 632.7 |
| Nepal | 147,181 | 192.2 |
| Netherlands | 41,526 | 397.1 |
| Netherlands Antilles | 960 | 231.0 |

(continued)

**Table 8.2** (continued)

| Country | Area (sq. mi) | PD (per sq. mi) |
|---|---|---|
| New Caledonia | 19,060 | 11.5 |
| New Zealand | 268,680 | 15.2 |
| Nicaragua | 129,494 | 43.0 |
| Niger | 1,267,000 | 9.9 |
| Nigeria | 923,768 | 142.7 |
| N. Mariana Islands | 477 | 172.9 |
| Norway | 323,802 | 14.2 |
| Oman | 212,460 | 14.6 |
| Pakistan | 803,940 | 206.2 |
| Palau | 458 | 44.9 |
| Panama | 78,200 | 40.8 |
| Papua New Guinea | 462,840 | 12.3 |
| Paraguay | 406,750 | 16.0 |
| Peru | 1,285,220 | 22.0 |
| Philippines | 300,000 | 298.2 |
| Poland | 312,685 | 123.3 |
| Portugal | 92,391 | 114.8 |
| Puerto Rico | 13,790 | 284.8 |
| Qatar | 11,437 | 77.4 |
| Reunion | 2517 | 312.9 |
| Romania | 237,500 | 93.9 |
| Russia | 17,075,200 | 8.4 |
| Rwanda | 26,338 | 328.4 |
| Saint Helena | 413 | 18.2 |
| Saint Kitts & Nevis | 261 | 149.9 |
| Saint Lucia | 616 | 273.5 |
| St Pierre & Miquelon | 242 | 29.0 |
| Saint Vincent and the Grenadines | 389 | 303.0 |
| Samoa | 2944 | 60.1 |
| San Marino | 61 | 479.5 |
| Sao Tome & Principe | 1001 | 193.2 |
| Saudi Arabia | 1,960,582 | 13.8 |
| Senegal | 196,190 | 61.1 |
| Serbia | 88,361 | 106.3 |
| Seychelles | 455 | 179.2 |
| Sierra Leone | 71,740 | 83.7 |

(continued)

**Table 8.2** (continued)

| Country | Area (sq. mi) | PD (per sq. mi) |
|---|---|---|
| Singapore | 693 | 6482.2 |
| Slovakia | 48,845 | 111.4 |
| Slovenia | 20,273 | 99.2 |
| Solomon Islands | 28,450 | 19.4 |
| Somalia | 637,657 | 13.9 |
| South Africa | 1,219,912 | 36.2 |
| Spain | 504,782 | 80.0 |
| Sri Lanka | 65,610 | 308.2 |
| Sudan | 2,505,810 | 16.5 |
| Suriname | 163,270 | 2.7 |
| Swaziland | 17,363 | 65.5 |
| Sweden | 449,964 | 20.0 |
| Switzerland | 41,290 | 182.2 |
| Syria | 185,180 | 102.0 |
| Taiwan | 35,980 | 640.3 |
| Tajikistan | 143,100 | 51.2 |
| Tanzania | 945,087 | 39.6 |
| Thailand | 514,000 | 125.7 |
| Togo | 56,785 | 97.7 |
| Tonga | 748 | 153.3 |
| Trinidad & Tobago | 5128 | 207.9 |
| Tunisia | 163,610 | 62.2 |
| Turkey | 780,580 | 90.2 |
| Turkmenistan | 488,100 | 10.3 |
| Turks & Caicos Is | 430 | 49.2 |
| Tuvalu | 26 | 454.2 |
| Uganda | 236,040 | 119.5 |
| Ukraine | 603,700 | 77.4 |
| United Arab Emirates | 82,880 | 31.4 |
| United Kingdom | 244,820 | 247.6 |
| United States | 9,631,420 | 31.0 |
| Uruguay | 176,220 | 19.5 |
| Uzbekistan | 447,400 | 61.0 |
| Vanuatu | 12,200 | 17.1 |
| Venezuela | 912,050 | 28.2 |
| Vietnam | 329,560 | 256.1 |
| Virgin Islands | 1910 | 56.9 |

(continued)

**Table 8.2** (continued)

| Country | Area (sq. mi) | PD (per sq. mi) |
|---|---|---|
| Wallis and Futuna | 274 | 58.5 |
| West Bank | 5860 | 419.9 |
| Western Sahara | 266,000 | 1.0 |
| Yemen | 527,970 | 40.6 |
| Zambia | 752,614 | 15.3 |
| Zimbabwe | 390,580 | 31.3 |

significantly more carbon dioxide in the Earth's atmosphere, cement production alone is responsible for approximately 2.5% of total worldwide emissions from industrial sources.

(iii) Automobiles like trucks, cars, scooters, motor cycles etc. produce nearly two-thirds of the carbon monoxide and one half of the hydrocarbons and nitrous oxides.

(iv) Fossil fuels are the sources of energy for cooking, heating, lightening our houses, washing clothes through washing machine, or for running TV etc. Considerable amount of $CO_2$ are released during coal burning. In majority of the Indian cities, towns and villages cooking is done by burning coals, which is an important cause behind the increment of carbon dioxide concentration in the atmosphere.

(v) Aircrafts are responsible for about 2.5% of the CO emissions and about 1% of the hydrocarbon emissions, along with negligible amounts of the other major air pollutants. The smoke emitted from jet aircraft is composed largely of fine particles approximately 0.5 μm in diameter, which are completely burnt. As these particles scatter light quiet well, they often reduce visibility.

(vi) Burning of forest areas, grasslands etc. for pastures and croplands produces about 60 to 65% of $CO_2$ in the atmosphere.

(vii) In places like Sundarbans where there are not much industrial activities, the situation has been aggravated from the emissions of fishing vessels, trawlers, passenger vessels, brick kilns, removal of antifouling paints from boats and trawlers for conditioning purpose by burning the old paints etc. (Fig. 8.1, 8.2, 8.3, 8.4 and 8.5).

3. **Emission of Oxides of Sulphur**

(i) Electrical power plants, burning fossil fuels, particularly coal and sometimes petrol or diesel, produce two-thirds of the $SO_2$.

(ii) $SO_2$ is also released from various industrial processes. These encompass activities like extracting metal from ore and the burning of fuels with high sulphur content by locomotives, large ships and non-road equipment.

**Fig. 8.1** Antifouling paints (usually coal tar) are applied for conditioning fishing vessels and trawlers. Burning this coal tar releases considerable amount of carbon dioxide in the atmosphere of Sundarbans

**Fig. 8.2** Tourist vessels stranded along the bank are the point sources of carbon dioxide

4. **Emission of Methane**

   (i) Methane is emitted during the production and transport of coal, natural gas, and oil.

**Fig. 8.3** Passenger boats with crowds is a common scene in all the inhabitable islands of Sundarbans, which are the sources of $CO_2$ emission

**Fig. 8.4** Black smoke with carbon dioxide is emitted from tourist vessels

**Fig. 8.5** Brick kilns are the point sources of carbon dioxide

(ii) About 40% of methane is produced from paddy fields, guts of livestock and also from burning of biomass.

The mangrove vegetation of Sundarbans is a unique sink of carbon dioxide, but massive destruction of the mangrove reservoir has squeezed the potential of scrubbing atmospheric carbon dioxide, eventually resulting in the rise of carbon dioxide. The main reasons behind these deforestation activities are the setting of shrimp farms, fish landing units, tourism units etc., in which the EIAs are conducted as per the norms and acts, but EMP is hardly carried out. In many EIA reports uses of technologies to minimize and recycling of wastes are stated along with specific floral species plantation, but in practice they are not followed to that extent.

We carried out extensive studies for three decades on the near-surface carbon dioxide level in the atmosphere of Indian Sundarbans (1984–2016) and observed that our results have touched/exceeded 400 ppm (Fig. 8.6) although there are no industries or power plants in this deltaic complex. However, the Haldia—port-cum industrial complex just adjacent to Indian Sundarbans is an industrial hub (Table 8.3), from where the emissions diffuse in the atmosphere of Sundarbans.

To combat the rise of carbon dioxide it is extremely important to shift from traditional non-renewable energy (primarily coal-based) sources to renewable clean energy phase. The resources which are being continuously consumed by man but are renewed by nature constantly are called as renewable resources. These resources

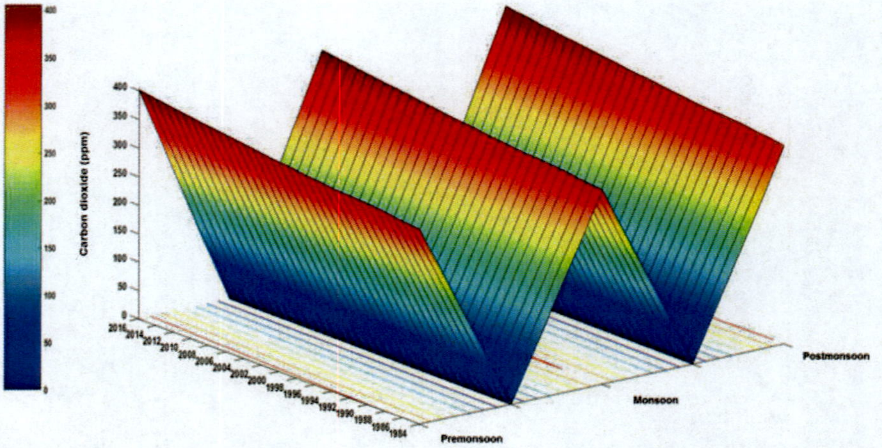


**Fig. 8.6** Near-surface carbon dioxide level in the atmosphere of Indian Sundarbans (1984–2016)

are inexhaustible because they are not exhausted permanently. Renewable resources are also called as 'Non-Conventional' sources of energy. Examples of renewable resources include solar energy, wind energy, tidal energy, hydro power, geothermal energy and biofuels. These sources are highly eco-friendly and do not emit GHGs. In the southern states of Indian sub-continent, wind based energy is widely used to run the wheels of civilization (Fig. 8.7). The emission from renewable sources is practically zero/negligible.

The non-renewable resources do not replenish and cannot be renewed. It takes thousands of years of time to form the non-renewable resources which exist inside the Earth in the form of coal, fossil fuels, etc. Examples of non-renewable resources are coal, mineral ores, metal ores, crude oil, nuclear energy etc. These resources, on burning, emit GHGs, which are the main drivers of acidification of oceans, seas, bays and estuaries. A comparative account of renewable and non-renewable resources is presented in Table 8.4.

For islands and mangrove forest dominated ecosystems like Sundarbans, tidal energy is one of the best options for generation of electric current. The mechanism of working in tide-based energy conversion is not very complicated. During ingression of the water, the tidal energy works via a turbine in which the blade rotates 12–20 times per minute. The rotation of the blade is, however, a function of the tidal strength. The turbine, which is connected to a gear box, turns the generator that finally produces electric current. In this context, we would like to state that the high upfront cost is one of the greatest hindrances to implement tidal-based electric generation. Because of the high density of the water, the turbine need to bear high strength compared to wind turbines. For Sundarban deltaic ecosystem, siltation is a major problem to run the turbine as suspended particulate matter may clog the system.

The maritime state of West Bengal has a gross wind power potential of 450 MW, but the installed capacity, as of March 2013 is 2.5 MW. The Government of West

**Table 8.3** Industries in Haldia port-cum-industrial complex adjacent to Indian Sundarbans

| Sl. No. | Name of industry | Product |
|---|---|---|
| 1 | Indian oil corporation Ltd., Haldia | L.P.G., motor gasoline, naptha, ATF, MTO, HSD, JBO, kerosene, furnace oil, lubes, bitumen |
| 2 | KoPT/Haldia dock complex | Port services |
| 3 | Tata chemicals Ltd., Haldia | Industrial phosphate & acids |
| 4 | Exide industries Ltd., Haldia | Automotive batteries, heavy duty batteries, containers, special types of separators, etc. |
| 5 | Swal corporation Ltd., Haldia | Dimethanate fenithrothion, ethion, malathion |
| 6 | MCC PTA India corpn. pvt. Ltd., Haldia | P.T.A |
| 7 | Haldia petrochemicals Ltd., Haldia | LLDPE, HDPE, naptha cracker etc. |
| 8 | IOCL, Paradip-Haldia oil pipeline | Petroleum storage & transportation |
| 9 | IOC Petronas Ltd., Haldia | L.P.G |
| 10 | Shamon ispat Ltd. | Steel rolling |
| 11 | Dhunseri petrochem & tea Ltd. | PET resin |
| 12 | Greenways shipping agencies Pvt. Ltd. | Containers freight station (CFS) |
| 13 | IOC Ltd., Haldia | Petroleum storage |
| 14 | Hindustan petroleum corporation Ltd. | Petroleum and allied products |
| 15 | Bharat petroleum corporation Ltd., Haldia | Petroleum and allied products |
| 16 | Hindustan unilever limited | Detergents |
| 17 | Marcus oils & chemical Pvt. Ltd. | Polyethylene waxes |
| 18 | IOC Ltd., Haldia | Petroleum storage |
| 19 | Ruchi soya industries Ltd. | Edible oil |
| 20 | Manaksia Ltd. | Aluminum and Steel |
| 21 | Sanjana cryogenic storages Ltd. | Ammonia storage and handling terminal |
| 22 | R. D. B. Rasayans Ltd. | PP jumbo bag and small bag |
| 23 | Reliance industries Limited | Storage & handling petroleum product |
| 24 | Adani wilmar Ltd. | Edible oil refinery |
| 25 | Electrosteel castings Ltd. | Coke oven plant, sponge iron plant, power plant |
| 26 | URAL India Ltd. | Automobile |
| 27 | K.S. oils Ltd. | Edible oil refinery |
| 28 | DPM Net Pvt. Ltd. | Fishing net |
| 29 | Hooghly met coke & power co. Ltd. | Coke oven plant |
| 30 | Ruchi infrastructure Pvt. Ltd. | 3RD Party liquid storage tank terminal |
| 31 | Shree Renuka sugars Ltd. | Sugar refinery and food complex |
| 32 | Gokul refoils & solvent Ltd. | Edible oil refinery |

(continued)

**Table 8.3** (continued)

| Sl. No. | Name of industry | Product |
|---|---|---|
| 33 | Emami biotech Ltd. | Bio-diesel plant |
| 34 | Ennore coke private Ltd. | Coke oven plant |
| 35 | West Bengal waste management Ltd. | Industrial waste/municipal waste management complex |
| 36 | Lalbaba seamless tubes Pvt. Ltd. | Seamless tube |
| 37 | Modern India con-cast Ltd. | Ferro alloy plant |
| 38 | Rohit ferro tech Ltd. | Ferro alloy |

Bengal is planning to set up a wind power station at Dadan Patra Bar near Mandarmoni in East Midnapore district of the state. The West Bengal Government has decided to set up 100 MW project with an estimated investment of INR 500 crore. The project will be developed in phases. The state government has already earmarked 1000 acre of land for the power station and in the first phase of the project a total of 700 acre land will be used. The Dadan Patra Bar area has a wind speed of 19 km per hour and the plant is expected to generate 50 MW power in its phase I (https://www.projectstoday.com/news/west-bengal-to-set-up-wind-power-station).

To mitigate the emission of carbon dioxide, West Bengal Renewable Energy Development Agency (WBREDA) has installed four aero generators each of capacity 5 kW at two existing Solar PV Power Plants at Gayen Bazar 25 kW Solar PV Power Plant and at Moushuni 55 kW Solar PV Power Plant of Sundarbans. WBREDA is facing problem in maintaining the Aero Generators. Three Aero Generators have been badly damaged due to Aila, a super cyclone that hit the lower Gangetic delta on 25th May, 2009. WBREDA already informed the matter to Ministry of New and Renewable Energy (MNRE), Government of India for providing fund in restoring the systems. Performances of Aero Generators are found well particularly for the period from March to September. It is to be noted in this context that Sundarban is a cyclone prone zone and this can pose considerable damage to installed structures for generating non-conventional energy.

To conclude, it can be stated that reduction of $CO_2$ emission through non-conventional energy can see the light of success if few actions as highlighted here are adopted on fast track basis.

1. **Universal access to energy**

While most part of the Indian sub-continent is on track to reach near universal access to energy, there are several islands in Indian Sundarban region with extremely low access rates. These islands require particular attention and institutional support from the side of their respective Governments, as well as from international development partners. Issues of quality and reliability of energy access, as well as availability of reliable data for monitoring of this indicator need to be controlled by the policy makers.

**Fig. 8.7** Wide use of wind energy in the southern parts of Indian sub-continent. *Source* Mitra and Zaman (2020)

2. **Clean energy**

National targets for clean energy and technologies need to be established on priority basis and clean energy should be integrated into energy policy frameworks of the country.

3. **Increasing the share of renewables in the energy mix**

Considering the quasi stagnation of the share of renewable energy in the energy matrix of the region, efforts at promoting renewable sources of energy should be the

**Table 8.4** Comparative account of renewable and non-renewable resources

| Point of comparison | Renewable resource | Non-renewable resource |
|---|---|---|
| Presence | The renewable resources are mostly present in the atmosphere of the Earth | The non-renewable resources are typically found in the underground layers of the Earth |
| Replaceable | The renewable resources are replaced by nature itself in a very short period | The non-renewable resources cannot be replaced by nature during the time of human life span |
| Availability | The renewable energy resources are plentiful and abundant in nature | The non-renewable resources are scarce resources and not available in an abundant manner in nature |
| Cost | The renewable resources are obtained free of cost or at very less cost in nature | The non-renewable resources are very costly and not easily available |
| Impact on environment | The renewable resources do not affect the environment of the Earth and do not cause any climate changes in the atmosphere | The non-renewable resources seriously affect the environment and cause climate changes in the environment like burning of fossil fuels cause Green House Gas emission |
| Pollution | The renewable resources do not cause pollution in the environment and do not release any pollutants into the environment | The non-renewable resources pollute the Earth by releasing various types of pollutants into the air, water, soil, etc. when fossil fuels are burned |
| Impact on atmosphere | The renewable resources are called as 'Clean and Green' energy sources because they do not produce any adverse impact to the environment | The non-renewable resources release Green House Gases into the atmosphere which leads to global warming |
| Impact on health | The renewable resources do not cause any health problems to the living beings of the Earth | The non-renewable resources adversely affect the health of the living beings by releasing smoke, radiations, carcinogenic or cancer causing elements into the environment |
| Impact on nature | The use of renewable resources promotes the balance in the nature and natural habitat of the Earth | The use of non-renewable resources disrupts the balance in nature which is due to digging the earth to take out coal, minerals, fuels, etc. |

After Mitra and Zaman (2020)

need of the hour. This is primarily because of the rapid rate of industrialization and urbanization, which are presently fully fuel dependent in India. Moreover the effort can be a practical step to curb emission of GHGs.

4. **Energy efficiency**

Energy efficiency measures should be considered with high priority in the transport, industry, tourism, aquaculture and building sectors.

5. **Intergovernmental and Trans-boundary cooperation**

Member states with the support of Economic and Social Commission for Asia and the Pacific (ESCAP) and other relevant regional institutions need to develop an intergovernmental mechanism that would promote trans-boundary electricity power trade and connectivity as one of the building blocks to attain sustainable development. This is especially applicable for single large mangrove chunk like Sundarbans, in which 62% lies in Bangladesh and the remaining 38% in India.

6. **Removing policy barriers and promoting funds to innovative technology based research**

To facilitate renewable energy and energy efficiency expansion, member States need to address existing policy barriers such as restrictive permit procedures, monopolistic utility positions and restrictions in energy prices. After successful implementation of these measures, the introduction of financial incentives for clean energy projects will bring positive results. While a number of member States have made strong efforts to incentivize clean energy and have committed to phasing out fossil fuel subsidies, more improvements need to be made to reach the respective targets. Despite significant technological innovation in many industries, international energy cooperation efforts to deploy innovative technologies are limited due to lack of a coherent approach among the member states. Cutting edge innovation in the sphere of solar energy, wind energy, biomass energy, geo-thermal energy, tidal energy should be encouraged through adequate funding after a comprehensive SWOT analysis.

## 8.2 Expansion of Blue Carbon

Blue carbon refers to the coastal vegetation which includes mangroves, sea grass, salt marsh grass etc. In our previous publication we have also included phytoplankton and seaweeds under the domain of blue carbon (Mitra and Zaman 2016). Blue carbon has specific status in combating the process of acidification as they scrub dissolved form of carbon from the ambient aquatic phase. It is no exaggeration to state that increased level of carbon dioxide has shifted the pH of ocean, seas, bays and estuaries from the normal value ranging between 8.1–8.3 and 8.0–7.8. This shifting has posed several adverse impacts on marine and estuarine systems especially on the molluscan community with calcareous structures. It is therefore the need of the hour to expand the dimension of blue carbon and other mangrove associate species. Presently more than 90 varieties of mangrove and associate species are thriving luxuriantly in different coastal zones of the planet Earth. However, all the species do not prefer similar type of salinity. Many species like *Heritiera fomes, Sonneretia apetala, Nypa fruticans* etc. thrive in low saline zone of the estuary/coastal zone water, while some other species like *Phoenix paludosa, Rhizophora* spp. Prefer high saline water. Some species like *Exoecaria agallocha* survive and grow luxuriantly in all categories of salinity. A case study conducted by us during 2014–2016 in Indian Sundarbans exhibited the growth (in terms of biomass and carbon sequestration potential) of dominant species of mangroves in different gradients of salinity leading us to conclude that survival and growth of mangroves is significantly salinity -specific (*Vide* Annexure 8 for details).

Basically the producers of the ecosystem drive the global carbon cycle by sequestering (storing) carbon dioxide through photosynthesis and releasing it through respiration. When the uptake of carbon dioxide (photosynthesis) exceeds losses via respiration, harvest and management, then forests store carbon (C sinks). In an undisturbed forest ~74% of the carbon dioxide is stored in live stems and branches, 16% is stored in roots and 10% in soils. The global sink in forest vegetation and soils is estimated to be 1200 Gt of carbon (1 Gt $= 10^9$ tonnes). This increases at a rate of 1–3 Gt annually. Several researchers have estimated the carbon stored in forest vegetation in the pre-industrial and present times (Tables 8.5 and 8.6).

The tropical forest zone encompasses 1.76 billion ha or 17,600,000 Km$^2$, and is divided into six ecofloristic zones: the tropical rain forests, the moist deciduous forests, the dry zone, the very dry zone, the desert zone and the hill and mountain forests. Tropical rain forests are found in areas with more than 2500 mm of annual

**Table 8.5** Previous global carbon storage estimates for vegetation

| Storage (Gt C) | Reservoir type | Author(s) |
|---|---|---|
| 827 Gt (1) | Present actual land vegetation | Whittaker and Likens (1975)(s) |
| 560 Gt (2) | Present actual land vegetation | Olson et al. (1983) |
| 550 Gt (3) | Present actual (1980s) land vegetation | IPCC (1990)(s) |
| 610 Gt (3) | Pre-industrial (pre-1700) vegetation | IPCC (1990)(s) |
| 1080 Gt (4) | Land vegetation, 'prehistoric' times | Bazilevich et al. (1971) |
| 924 Gt | Present potential ('prehistoric') vegetation | Adams et al. (1990) |
| 343 Gt (5) | Last glacial maximum vegetation | Adams et al. (1990) |
| 350 Gt | Coarse woody debris (present potential) | Harmon et al. (pers. Comm., 1990) |
| 591 Gt (8) | Present-actual land vegetation | Ajtay et al. (1979)(s) |

**Table 8.6** Carbon storage totals for global soils

| Storage (Gt C) | Reservoir type | Author(s) |
|---|---|---|
| 1115 Gt | Soils, present potential ('prehistoric') | Adams et al. (1990) |
| 1395 Gt (1) | Peats + soils, present potential | Adams et al. (1990) |
| 1405 Gt (3) | Soils, present-day | Bazilevich (1974)(s) |
| 3000 Gt (4) | Soils (+ peats ?), present-day | Bohn (1978)(s) |
| 1672 Gt (3) | Soils, present-day | Bolin et al. (1979)(s) |
| 1477 Gt (5) | Soils, present-day | Buringh (1983)(s) |
| 1515 Gt (6) | Soils (+peat lands?) present-day | Schlesinger (1984) |
| 787 Gt | Forestsoils only (+fine debris) | Dixon and Krankina (1993) |
| 1500 Gt (7) | Soils, in 1989 | IPCC (1990)(s) |
| 1560 Gt (7) | Soils, in 'pre-industrial' era | IPCC (1990)(s) |
| 860 Gt (8) | Peats, present-day | Bohn (1976)(s) |
| 300 Gt (8) | Peats | Sjors (1980)(s) |
| 202 Gt (8) | Peats | Post et al. (1982) |
| 377 Gt (8) | Peats | Bohn (1976, 82) |
| 180–227 Gt (8) | Peats | Gorham (1990)(s) |
| 461 Gt (9) | Subarctic and boreal peat | Gorham and Janssens (1992) |
| 1576 Gt (10) | Global soils (present-day) | Eswaran et al. (1993) |
| 500 Gt (11) | Global peats | Markov et al. (1988)(s) |

rainfall. They are evergreen, luxuriant and rich in animal and plant species. More than half the world's 718.3 million ha of rain forests are located in two countries: Brazil (41%) and Indonesia (13%). Rain forest composition and structure vary with distance from the ocean, distance from rivers, altitude and geographic position. Moist deciduous forests occur in areas with an annual rainfall of 1000 to 2000 mm. Forest structure varies depending on the amount and distribution of rain, the type of soil and the length of the dry season. Some dominant tree species may lose their leaves towards the end of the dry season. This forest type is generally less diverse than rain forest. Dry zone forests are found in tropical areas receiving between 500 and 1000 mm of rainfall per year. They are relatively open and include thorn land, shrub land, savannah and other short and sparse woody vegetation. Dry zone forests tend to be fragile and are easily degraded. More than half are in Africa. Dry forest types include oak, mesquite, piñon-juniper, maquis and acacia. Tropical upland forests are forests above 800 m and include cloud forests (montane rain forests), which are shorter, floristically simpler and more heavily laden with mosses and lichens than lowland rain forests. Tropical upland species are similar to temperate forest species. The upland zone covers the Himalayas, parts of Myanmar, Thailand and Vietnam, the highlands of Mexico, the Andes and the highlands of Ethiopia and mountains around Lake Victoria.

Acidification of coastal waters, bays, estuaries and seas are mostly regulated and controlled by blue carbon, whose expansion is needed as a part of managing the heath of the aquatic ecosystem. We monitored the impact of massive mangrove plantation in and around the Nayachar island (latitudes 21°54′41″ N–22°01′28″ N and longitudes 88°03′02″ E–88°08′43″ E) in the western sector of lower Gangetic delta along the Hooghly estuary considering relevant parameters like DIC and surface water pH for more than three decades (1984–2016). The island is located about 56 nautical miles south of the mega city of Kolkata, directly facing the port-cum-industrial complex of Haldia and has an area of about 47 $km^2$ and a maximum elevation of 6.5 m above mean sea level (MSL). Length of the Island centrally measured from North to South is 15,861 m and from East to West is 4340 m (Fig. 8.8). The land topography is generally flat, with elevation ranging from a low of 1.08 m MSL to as high as 3.68 m MSL along the coastal region of the island. Average ground level of the island is 1.366 m MSL. The island is convex upward with gently sloping eastern and western flanks. The eastern flank is undergoing erosion at present while the western edge is accreting. According to recent estimation, the present accretion rate is about 1 $km^2$/yr on the Haldia port area and the erosion rate is about 0.05 $km^2$/yr on the eastern part of Nayachar. The island is exposed to strong scouring action of the running Hooghly River and hence a programme was undertaken by Calcutta Port Trust in eighties to stabilize the island and increase the draft of the adjacent navigational channel. 11 true mangrove species were planted inside and along the border of the island on geo-jute matrix and within a span of two years excellent growth of the species (particularly *Sonneratia apetala* and *Bruguiera gymnorhiza*) stabilized the boulders with considerable accretion of silt particles (Mitra 2013) (Fig. 8.9).

The biomass of the mangrove vegetation ($tha^{-1}$) was estimated simultaneously along with DIC and aquatic pH to evaluate the role mangrove vegetation in the domain of Estuarine Acidification (EA).

We observed that with the passage of time as the biomass of the mangroves increased in the island, the DIC values exhibited a gradual decrease (Fig. 8.10) with subsequent increase of pH values in the ambient aquatic phase.

Expansion of blue carbon can be achieved by several technologies like (i) massive mangrove afforestation programme considering the salinity of the region (salinity-specific mangrove afforestation programme), (ii) halophyte farming and (iii) triggering phytoplankton growth through artificial method like iron fertilization. Each of these methods is briefly discussed here.

### (i) Salinity-specific mangrove afforestation programme

The survival, growth and metabolic activities of mangroves are extremely salinity-specific. In Indian Sundarbans we have recorded significant growth of *Sonneratia apetala*, *Sonneratia caseolaris*, *Heritiera fomes* and *Nypa fruticans* in low saline belt (aquatic salinity = ~5 psu). These true mangrove floral species are best suited in hypo-saline environment (Figs. 8.11, 8.12, 8.13 and 8.14).

Species like *Avicennia marina*, *Avicennia officinalis*, *Avicennia alba*, *Phoenix paludosa* thrive best in hyper-saline environment (Figs. 8.15, 8.16, 8.17 and 8.18).

**Fig. 8.8** Location of Nayachar island along the Hooghly estuary (*Source* Mitra 2013)

**Fig. 8.9** Stand of *Bruguiera gymnorhiza* around Nayachar island along the Hooghly estuary

We have observed that *Excoecaria agallocha* (Fig. 8.19) grows luxuriantly in different salinity regimes.

This wide range of tolerance to salinity by the species must be considered with due weightage while transplanting mangroves from the nursery (Fig. 8.20) to the field.

The survival and growth of mangroves in the field condition can only shift the pH value of the ambient aquatic phase to its equilibrium by retarding the process of acidification.

Basically mangroves absorb carbon dioxide from the atmosphere and the water column and store it as carbon in their vegetative and reproductive parts. The principal carbon-fixing enzyme in plants is ribulose 1, 5- biphosphate carboxylase/oxygenase (RuBisco). The magnitude of the mangrove vegetation pool thus exerts a regulatory role on the pH of the estuarine water as these halophytes absorb carbon dioxide for photosynthesis and shifts the equilibrium towards alkalinity.

The fish bone model of distributing water from the source (estuaries, bays or coastal regions) can successfully result in the growth of mangroves. In this method brackish/saline water from the source is diverted to the targeted site through the feeder canal and field channels and subsequently help transforming the barren land with high salt content into the low saline zone that supports the survival of mangrove plant species. The shape of model is just like the bone of the fish with side bones which allows the brackish water to reach the region where the mangroves need to be

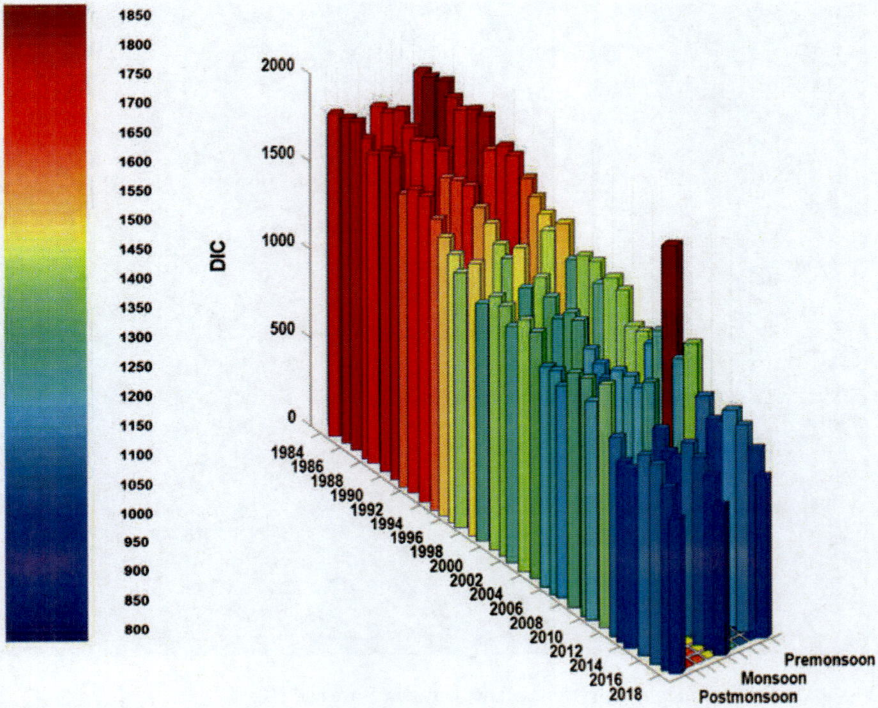


**Fig. 8.10** Gradual decrease of DIC (μmol/L) in the aquatic phase around Nayachar island with the increase of mangrove vegetation (in terms of biomass)

planted. This technology has helped to eco-restore the mangrove cover in the Krishna Wild Life Sanctuary (KWS) covering about 6,000 ha since 2006. The model has been successfully standardized by M.S Swaminathan Research foundation (MSSRF).

In Indian Sundarbans brackish water channels have been excavated to transfer the source water to the plantation site and survival rate was monitored after 2 years of plantation in 2017. We conclude that the survival rates are significantly governed by salinity and hence site-specific. We observed that survival rates of *Sonneretia apetala, Aegiceros corniculatum, Bruguiera gymnorrhiza, Xylocarpus granatum, Nypa fruticans, Heritiera fomes, Derreis trifoliate* are significantly high in the low saline belt of western Indian Sundarbans (marked in the graph as Stns. 1–5). The survival rates of these hyposalinophilic species are highlighted in Figs. 8.21, 8.22, 8.23, 8.24, 8.25, 8.26 and 8.27.

In hypersaline environment of central Indian Sundarbans the survival rates of mangrove species like Avicennia alba, Avicennia marina, Avicennia officinalis, Aegiceros ilicifolius, Excoecaria agallocha, Phoenix paludosa, Ceriops decandra, Rhizophora mucronata and Aegialitis rotundifolia are relatively high (Figs. 8.28, 8.29, 8.30, 8.31, 8.32, 8.33, 8.34, 8.35 and 8.36).

**Fig. 8.11** *Sonneratia apetala* prefers salinity between 5 psu and 10 psu

These results point out the fact that expansion of mangroves and other coastal vegetation need to be done considering the salinity as the primary driver.

## 8.3 Acceleration of Resilience of Affected Components

Estuarine acidification (EA), being a major stressor poses adverse impacts on the biotic community as well as on the ambient environmental quality. Thus both biotic and abiotic components are exposed to negative impacts of acidification. There are methods by which stressors can be reduced by employing resilience management to assist the ecological systems and dependent stakeholders, industrials and communities in resisting and recovery from EA. Very scanty works have been conducted on the specific biological attributes that can offer resilience against EA in the lower Gangetic delta region. Till date, there is limited understanding of the major drivers that impart estuarine ecosystem and its services more resilience to estuarine acidification. It is, however, confirmed that EA poses an adverse impact on the socio-economic profile of the region as the natural resources drive the economic cycle of the coastal population. In this context it is extremely important to develop a comprehensive strategy involving the major stakeholders and their respective action plans as reflected in Table 8.7.

**Fig. 8.12** *Sonneratia caseolaris* prefers salinity between 1.0 psu and 8.0 psu

EA can be minimized to a great extent by following steps like imposing restrictions on recreational activities, shrimp farming at the expense of blue carbon, industrial discharges, land based activities and emissions from the surroundings. Waste discharges from fishing vessels/trawlers and fish landing stations need to be minimized or properly treated before releasing them to ambient aquatic phase. Regular monitoring needs to be undertaken to manage risks from invasive species, ballast discharge etc. Temporary closure may be done to safeguard the vulnerable spots. A schematic diagram to manage the EA in context to lower Gangetic delta estuaries is highlighted in Fig. 8.37.

To sum up this section, we feel that in Indian Sundarbans out of 10 stations we surveyed since more than three decades (Table 8.8), special restrictions should be imposed on canning in the central sector as the magnitude of stressors is highest in this station owing to presence of busy markets, fish landing sites, passenger jetties, fishing trawler and boat repairing units etc. Moreover this station is also the gateway of Indian Sundarbans forest area, which attracts large number of tourists. All these stressors have high probability to trigger EA in this station as represented in Fig. 8.38.

On contrary, due to presence of luxuriant mangrove vegetation and abundance of mangrove associate species, the environment is quiet safe in terms of EA as represented in Fig. 8.39.

The water quality in the lower Gangetic delta can be improved in context to EA through proper conservation of blue carbon. Studies throughout the world have

**Fig. 8.13** *Hertiera fomes* prefers salinity between 1.0 psu and 5.0 psu

**Fig. 8.14** *Nypa fruticans* prefers salinity between 4.0 psu and 12.0 psu

**Fig. 8.15** *Avicennia marina* prefers salinity between 15.0 psu and 30.0 psu

shown that wetlands chemically, physically and biologically remove pollutants, sediments and nutrients from water flowing through them (Zhang 1995; Day et al. 2004; Alexander and Dunton 2006; Conkle et al. 2008; Kadlec and Wallace 2009; Vymazal 2010; Shaffer et al. 2015).

In context to the present scales represented in Figs. 8.38 and 8.39, it is therefore utmost important to carry out mass scale mangrove plantation in central sector preferably in Canning (along the Matla estuary). The barrier of high salinity can be overcome through interlinking of Matla and Hooghly estuary after a thorough EIA. Rain water harvesting in an around the Canning must be encouraged from where the stored rain water should be channelized to mangrove plantation site.

(ii) **Halophyte farming**

Halophtye farming has dual benefits in the sense that the farmed species have edible value and may serve as source material for alternative livelihood and at the same time may act as carbon sequestering mass leading to inhibition of EA. Literature survey indicates that *Suaeda maritima* and *Salicornia brachiata* when cultivated in high saline environment yield good results, which indicates that there is scope to grow these halophytes in high saline soil or sea water inundated region. The soil usually has high sodium and potassium level. Large scale farming of these mangrove associate species can help retard EA. The field level technology to grow these mangrove associate species is highlighted here in steps.

**Fig. 8.16** *Avicennia officinalis* prefers salinity between 12.0 psu and 28.0 psu

### 1st Step

In the 1st step *Suaeda maritima* and *Salicornia brachiata* are raised in the nursery for 45 days in 3″ × 5″ polythene bags filled with garden soil. Shade net (50%) are used to reduce the light and temperature. Fresh water is used to raise the saplings in the nursery.

### 2nd Step

In this step the 45-day-old seedlings are transplanted and each species are grown in the saline land preferably in the supra-littoral zone. Initially fresh water is used for irrigation (2 wettings) and then brackish water from nearby creeks is used.

***Remarks*** Since these mangrove associate plants are highly saline- and- drought-tolerant, two wettings with brackish water are given. The salinity of the brackish water used for the first wetting is around 28–30 psu whereas salinity of the brackish water given for the second wetting is around 20–25 psu.

### 3rd Step

In this step the final harvesting is done before the onset of the monsoon since these halophytes will die thereafter. It is to be noted in this context that these mangrove associate species thrive and grow luxuriantly in high saline condition and die when

**Fig. 8.17** *Avicennia allba* prefers salinity between 10.0 psu and 30.0 psu

the dilution factor of the estuarine and coastal water increases during the monsoon season.

Halophyte farming has several benefits in terms of livelihood, but in context to $CO_2$ mitigation, the magnitude of stored carbon in *Suaeda maritima* and *Salicornia brachiata* (Table 8.9) speaks in favor of scrubbing the GHG at the local scale.

**(iii) Triggering phytoplankton growth through artificial method like iron fertilization**

Iron fertilization in closed water body coupled with pisciculture/aquaculture can be a food for thought to regulate EA. The recent researches on iron fertilization put forward the concept that phytoplankton growth can be stimulated by adding iron to **HNLC** waters (High Nitrate Low Chlorophyll). Many scientists also forwarded the view that the efficiency of this biological pump can be enhanced by fertilizing the oceans with iron. Increased efficiency means sucking up more carbon dioxide from the atmosphere into the oceans. This can be an effective roadmap to mitigate climate change due to rise of carbon dioxide in the atmosphere. However, such researches are based on an incomplete understanding and highly simplified interpretation of current scientific knowledge. The major drawbacks of the concept of reducing carbon dioxide level in the atmosphere by triggering the growth of phytoplankton of the ocean through iron fertilization are listed here in points:

**Fig. 8.18** *Phoenix paludosa* prefers salinity between 18.0 psu and 32.0 psu

1. Ecological perturbations in the marine and estuarine environments due to iron fertilization have not been given proper importance.
2. The geophysical changes of the ecosystem that may result due to large scale iron fertilization have not been accounted in this research programme.
3. In order to remove significant amount of atmospheric carbon dioxide, a huge area of the ocean needs to be fertilized which may not be particularly feasible.
4. Iron fertilization if done, on commercial scale, may lead to harmful algal bloom (HAB), which has adverse effect on other marine lives.
5. Iron addition and subsequent phytoplankton bloom associated with increased particulate organic export and remineralization, could reduce dissolved oxygen (DO) level in sub-surface water. This has high probability to pose a negative impact on the marine organisms.
6. As the bloom will progress due to iron fertilization, the nutrient load of the ambient water will sharply decrease. This could result in a reduction of these nutrients down current from an iron fertilized area. Finally, lack of nutrients may cause a detrimental effect on phytoplankton resulting in a reduction in overall biological productivity. Predictive mathematical models have suggested that commercial scale iron fertilization of the oceans could have a significant adverse impact on marine fisheries.
7. There is also the concern that the changes in water chemistry caused by iron fertilization could lead to higher emission of nitrous oxide due to decomposition

**Fig. 8.19** *Excoecaria agallocha* prefers salinity between 5.0 psu and 28.0 psu

of organic matter in deep water. Some researchers have contradicted the concern with the idea of fertilizing low-nutrient regions of the oceans instead of high—nutrient ones. This scientific school believes that excess phytoplankton density (bloom condition) could create a larger aerosol effect leading to block incoming solar radiation that may cause global cooling.

8. It is difficult to track phytoplankton blooms in the vast ocean with dynamic circulation pattern. The added iron rapidly dilutes, sinks, and reacts with seawater, becoming virtually undetectable after a few days. Researchers therefore have to add minute amounts of an inert tracer like sulphur hexafluoride ($SF_6$), which

**Fig. 8.20** Nursery of mangroves prior to translocation on the mudflats

itself is a potent greenhouse gas. One kilogram of $SF_6$ added in an experiment is equivalent to releasing 7 tonnes of carbon dioxide.

Considering the drawbacks of iron fertilization in Open Ocean, we suggest similar approach in closed water body stocked with estuarine/coastal water. The merits of the programme are as follows:

1. The question of huge input of iron does not arise because the water body has a limited surface area or volume.
2. Standardization of iron fertilization can be easily done because of the closed nature of the system due to which the wave action, tidal phenomenon and other parameters can be controlled.
3. Ecological perturbations will be negligible in aquaculture ponds because unlike open oceans the species richness is not too high in this closed system. Only the target species is cultured in the pond and essentially the food chain consists of phytoplankton, zooplankton and target (cultured) species.
4. The aquaculture ponds are closed systems with depth ranging from 1 to 5 m. The substratum is basically soil. The question of massive geophysical changes does not arise when iron fertilization is done.
5. In aquaculture pond the algal bloom can be converted into animal biomass as the cultured species feed and phytoplankton and the question of dissolved oxygen (DO) fall does not arise.

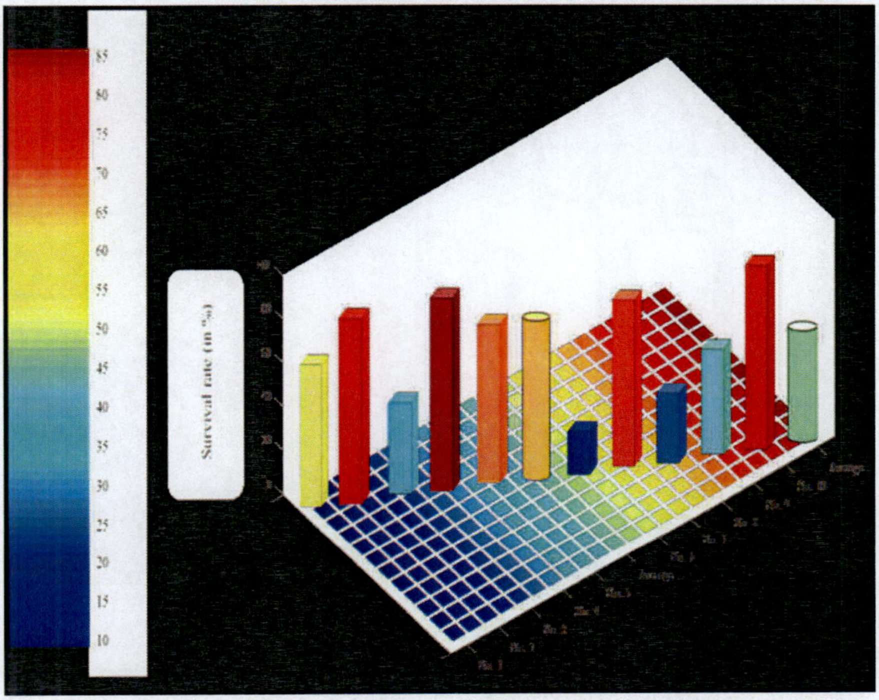


**Fig. 8.21** Survival rate of *Sonneretia apetala*

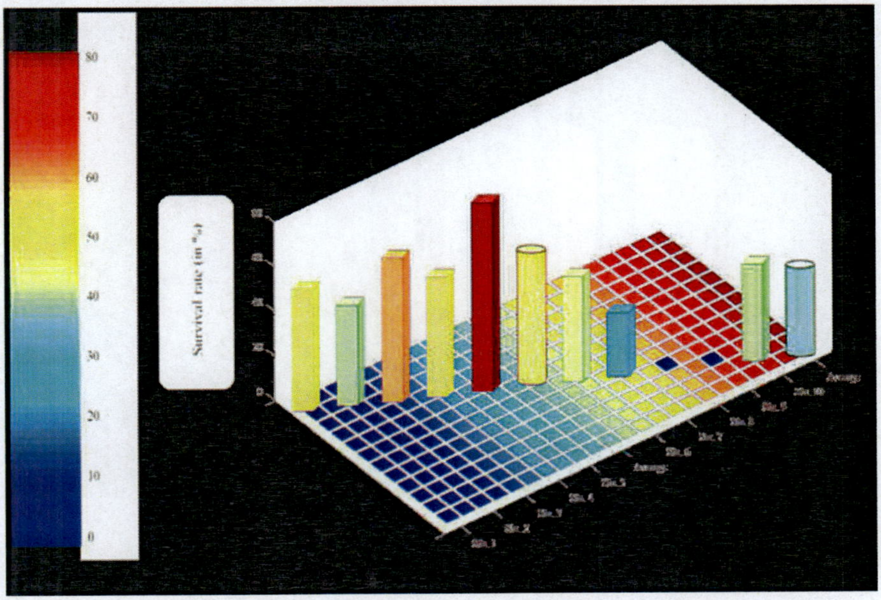


**Fig. 8.22** Survival rate of *Aegiceros corniculatum*

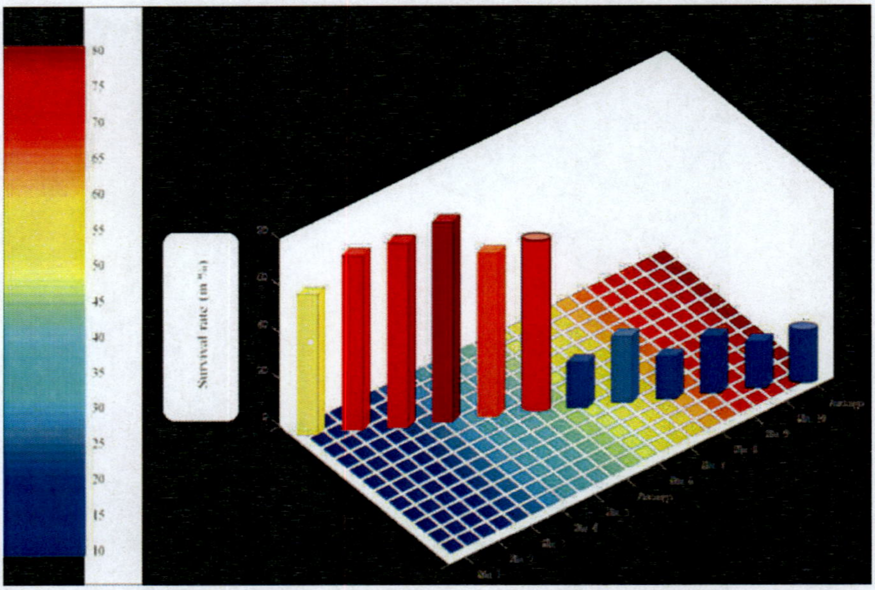


**Fig. 8.23** Survival rate of *Bruguiera gymnorrhiza*

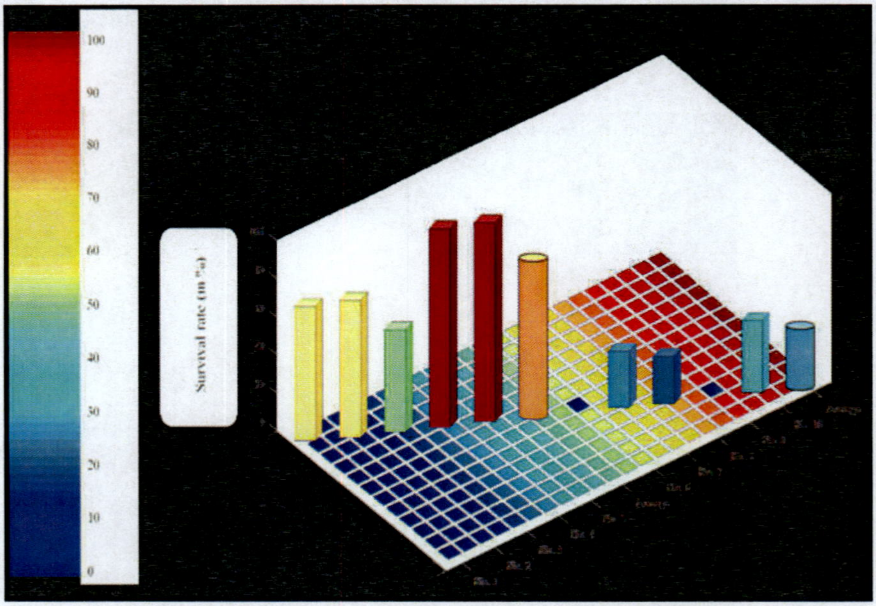


**Fig. 8.24** Survival rate of *Xylocarpus granatum*

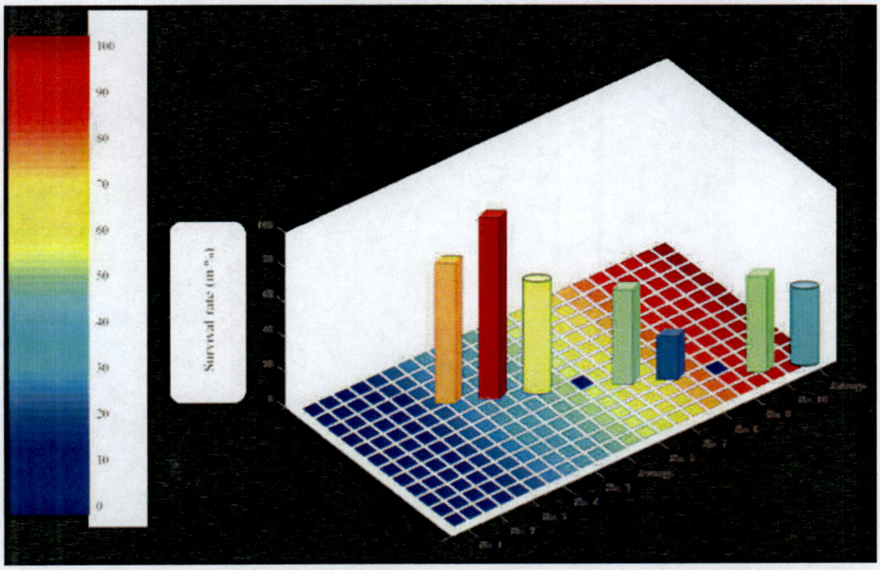


Fig. 8.25 Survival rate of *Nypa fruticans*

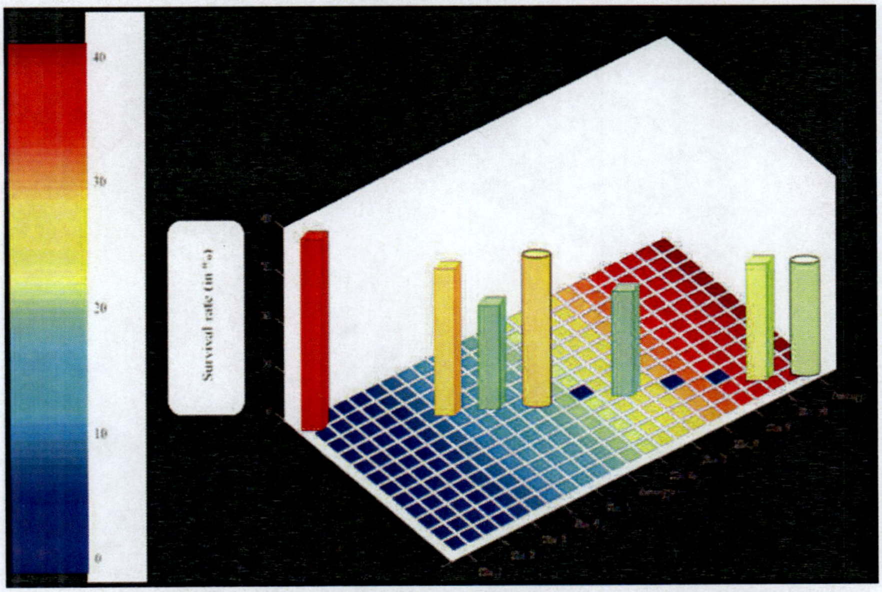


Fig. 8.26 Survival rate of *Heritiera fomes*

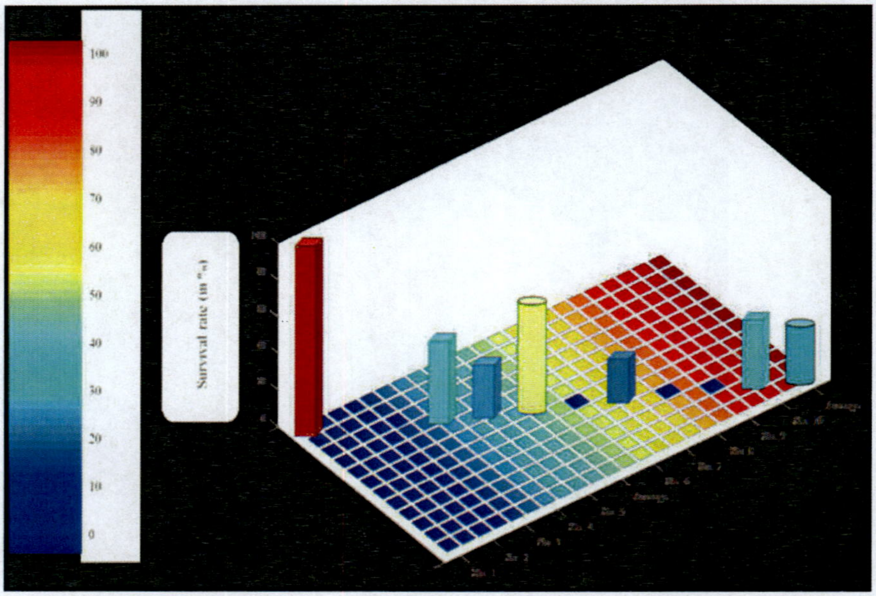


**Fig. 8.27** Survival rate of *Derris trifoliate*

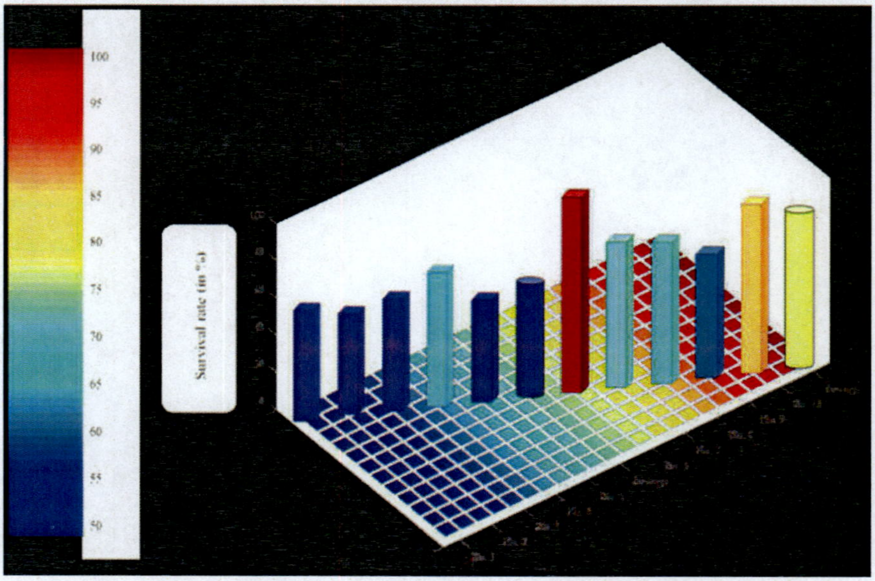


**Fig. 8.28** Survival rate of *Avicennia alba*

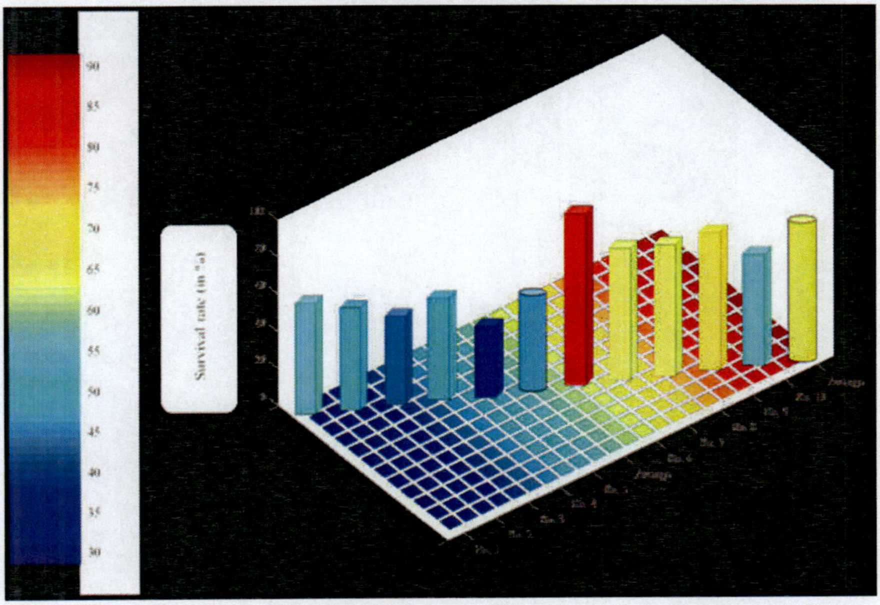


**Fig. 8.29** Survival rate of *Avicennia marina*

**Fig. 8.30** Survival rate of *Avicennia officinalis*

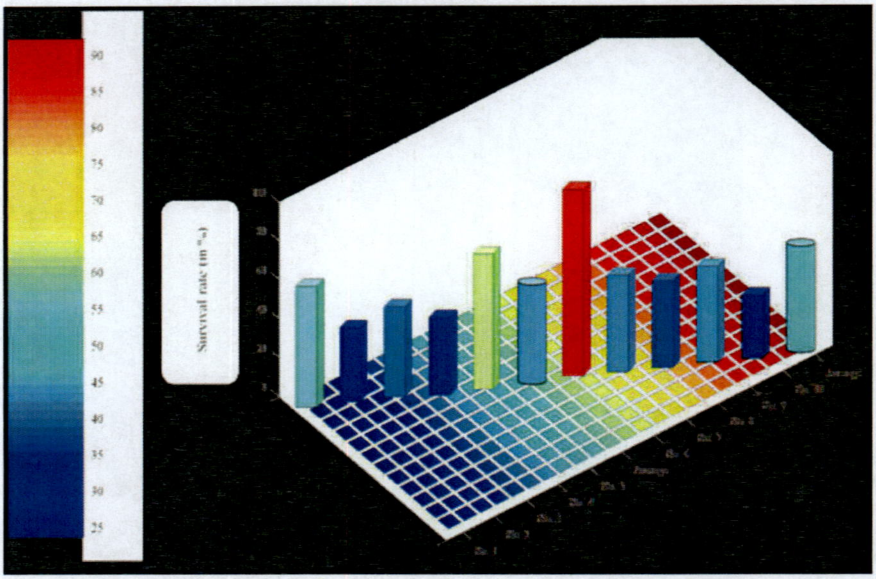

**Fig. 8.31** Survival rate *Aegiceros ilicifolius*

**Fig. 8.32** Survival rate of *Excoecaria agallocha*

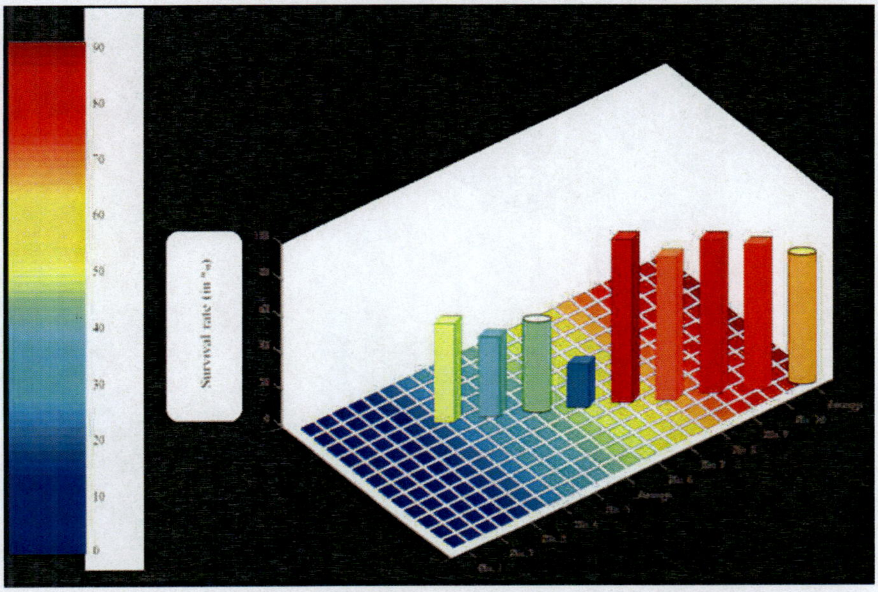


**Fig. 8.33** Survival rate of *Phoenix paludosa*

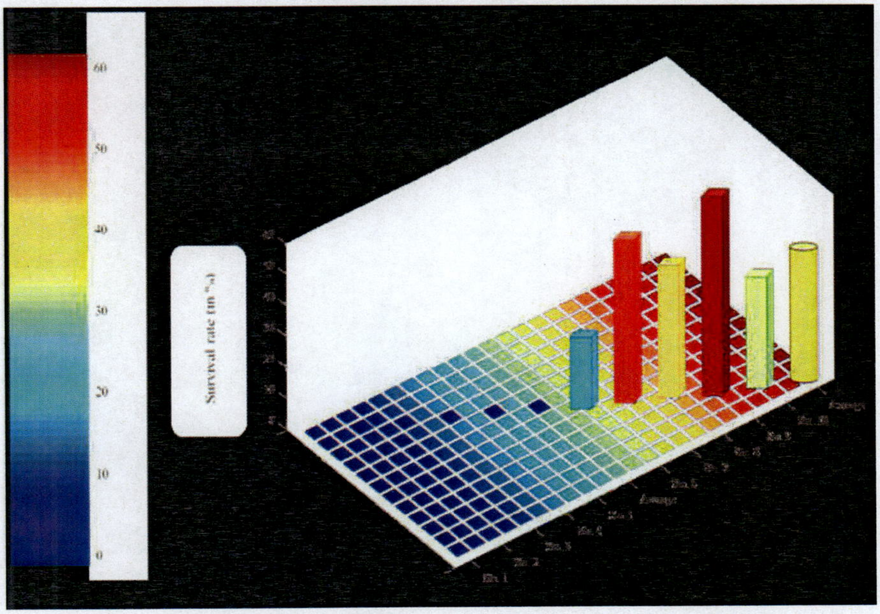


**Fig. 8.34** Survival rate of *Ceriops decandra*

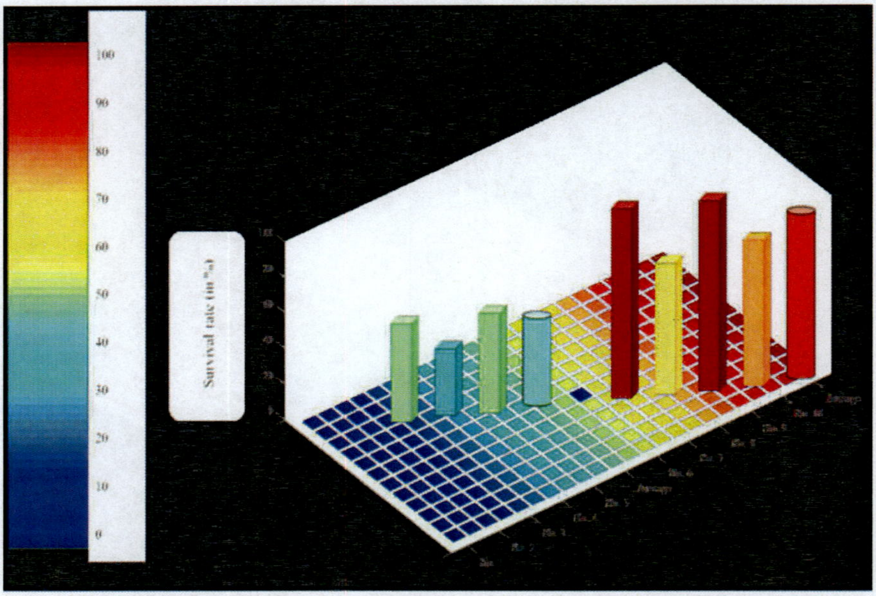


**Fig. 8.35** Survival rate of *Rhizophora mucronata*

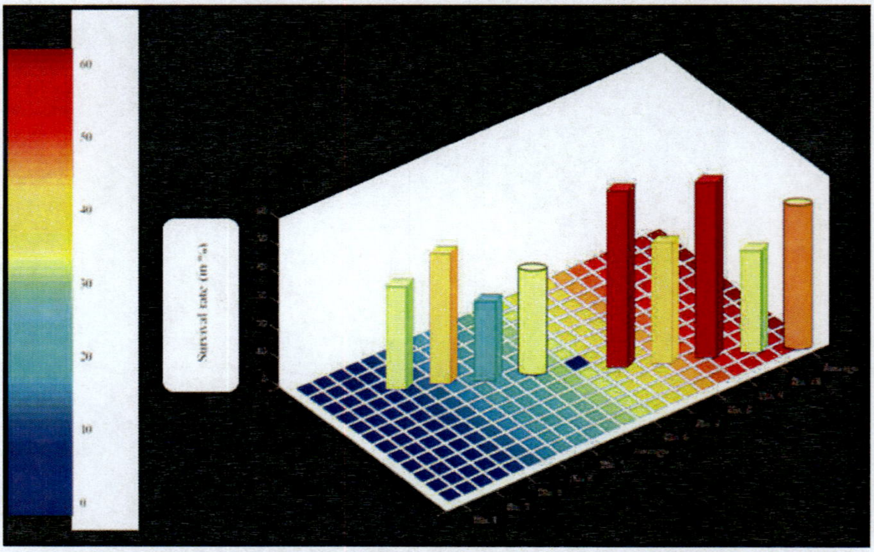


**Fig. 8.36** Survival rate of *Aegialitis roduntifolia*

**Table 8.7** Comprehensive strategies to manage EA with pinpointed questions and indicators in context to estuaries of lower Gangetic region

| Impact | Questions | Indicators/approaches |
|---|---|---|
| Local anthropogenic factors like industrial discharges, municipal wastes pose considerable impact on EA. This local effect may be magnified by phenomenon like eutrophication—enhanced acidification in some locations in the lower stretch of the estuary | What are the sources, of altered water quality preferably pH?<br>What is the relative importance of low-pH water compared to other stressors?<br>What is the contribution of local anthropogenic sources of $CO_2$ like brick kiln or industrial emissions to low pH water?<br>How massive deforestation due to shrimp culture activities or mushrooming of tourism units contribute to the process of EA? | • Conduct chemical and physical monitoring via cruises/surveys and moored sensors<br>• Develop and test calibrated carbon models, conceptual models, and mass balance calculations<br>• Obtain information on point sources of $CO_2$ at multiple locations<br>• Obtain time series information on replication/proliferation of shrimp farms and tourism units at the cost of mangroves |
| Changes in phytoplankton community and food quality Adverse impact on filter feeding molluscan community | To what extent can exposure to lower-pH water change the phytoplankton community or food quality?<br>How the molluscs preferably *Saccostrea* spp. are impacted due to compositional variation or abundance of phytoplankton? | • Evaluate changes in phytoplankton distribution and diversity along with chlorophyll content of the aquatic phase |
| Increased toxins from harmful algal blooms and effects on higher trophic levels | To what extent can exposure to lower-pH water increase the production of (harmful algal bloom) HAB toxins? | • Assess HABs and toxins in the relation to EA and evaluate inter-relationship preferably through regression model |
| Impact on fisheries and shellfish community | To what extent does exposure to lower-pH water decrease shellfish populations and fisheries?<br>Is there any observation on oyster shell thinning? | • Assess larval quality, recruitment, and juvenile growth of bivalves, crabs, etc.<br>• Use EA as a lens through which habitat is valued for upper trophic levels<br>• Assess finfish otoliths<br>• Assess oyster shell weight and thickness |
| Local mitigation of EA impacts by submerged aquatic vegetation (SAV) | To what extent can existing or restored SAV buffer for lower pH exposure?<br>What is the carbon sequestration potential for different natural and restored vegetative habitats like mangrove, sea grass, saltmarsh grass etc.? | • Assess SAV health and buffer capacity, and quantify habitat suitability through evaluation of ecological indices<br>• Overlay EA "hotspots" over the habitat- suitability maps for SAV to help prioritize restoration or other projects |

(continued)

**Table 8.7** (continued)

| Impact | Questions | Indicators/approaches |
|---|---|---|
| Other considerations | Have any indicators/proxy in relation to EA in the lower Gangetic delta been standardized? | • Couple biological, chemical, and physical monitoring in association with living shoreline projects<br>• Assess nested indicators in different trophic levels<br>• Assess microbial community and function<br>• Assess the buffering capacity of molluscan shells (preferably oyster) as they dissolve in low pH condition<br>• Evaluate benthic foraminifera characteristics |

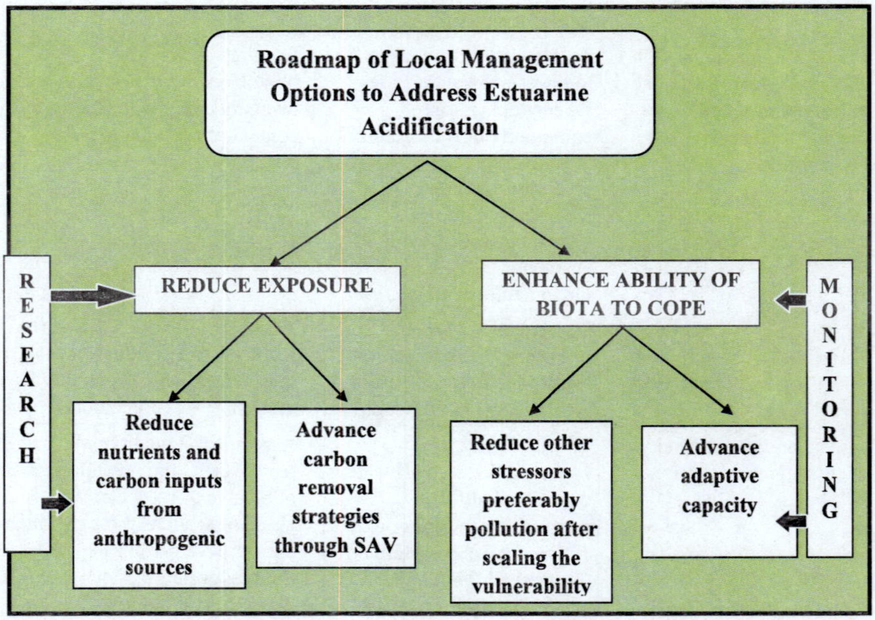


**Fig. 8.37** Local management options to address estuarine acidification in the lower Gangetic estuaries

6. The HAB species can be detected and managed as aquaculturists monitor the water quality and check the pond health almost on hourly basis.
7. Hydrogen sulphide gas produced by microbial activity in the aquaculture ponds can be precipitated through application of ferrous sulphate, which may otherwise be toxic for the cultured species.

**Table 8.8** Sampling stations with salient features surveyed in context to EA

| Study site | Latitude and longitude | Site description |
|---|---|---|
| Harinbari (Stn. 1) | 21°46′53.07″ N<br>88°04′22.88″ E | Situated in the western region of Indian Sundarbans almost in the middle of the Sagar Island; receives the water of the Hugli River |
| Chemaguri (Stn. 2) | 21°39′42.88″ N<br>88°08′49.01″ E | Situated on the south-eastern side of Sagar Island and receives the water of the Mooriganga River |
| Sagar South (Stn. 3) | 21°37′49.90″ N<br>88°04′0.51″ E | Situated on the south-western part of the Sagar Island at the confluence of the River Hugli and the Bay of Bengal. Anthropogenically stressed zone due to presence of passenger jetties, fishing activities and pilgrimage |
| Lothian island (Stn. 4) | 21°39′08.04″ N<br>88°19′8.47″ E | Situated east of Bakkhali island; a Wildlife sanctuary; faces the River Saptamukhi |
| Prentice island (Stn. 5) | 21°42′43.31″ N<br>88°17′3.62″ E | Situated north of Lothian island; receives the water of the Saptamukhi River |
| Canning (Stn. 6) | 22°19′03.20″ N<br>88°41′04.43″ E | Situated in the central part of the Indian Sundarbans and faces the mighty River Matla, a tide-fed river. Due to presence of fish landing stations, passenger jetties and busy market, the area is anthropogenically stressed |
| Sajnekhali (Stn. 7) | 22°06′34.19″ N<br>88°48′15.78″ E | A Wildlife Sanctuary and a part of Sundarban Tiger Reserve; adjacent to River Bidhya and Gomor. Tourism pressure is extremely high in this station particularly during postmonsoon |
| Chotomollakhali (Stn. 8) | 22°10′40.00″ N<br>88°54′26.71″ E | Situated in the upper portion of Central Indian Sundarban adjacent to Jhila forest; receives the water of Rangabelia and Korankhali Rivers |
| Satjelia (Stn. 9) | 22°05′17.86″ N<br>88°52′49.51″ E | Situated adjacent to River Duttar in the upper region of Central Indian Sundarban facing western part of the Jhilla forest |
| Pakhiralaya (Stn. 10) | 22°07′07.23″ N<br>88°48′29.00″ E | Situated adjacent to River Gomor; opposite to Sajnekhali Wild Life Sanctuary |

Ocean acidification has presently taken a faster pace, approximately 30 to100 times faster than at any time during the last several million years driven by the rapid growth rate atmospheric $CO_2$ that is almost unprecedented over geologic history. According to the Intergovernmental Panel on Climate Change (IPCC), economic and population scenarios predict that atmospheric $CO_2$ levels could reach 500 ppm by 2050 and 800 ppm or more by the end of the century. This will not only lead to significant temperature increases in the atmosphere and ocean, but will further acidify

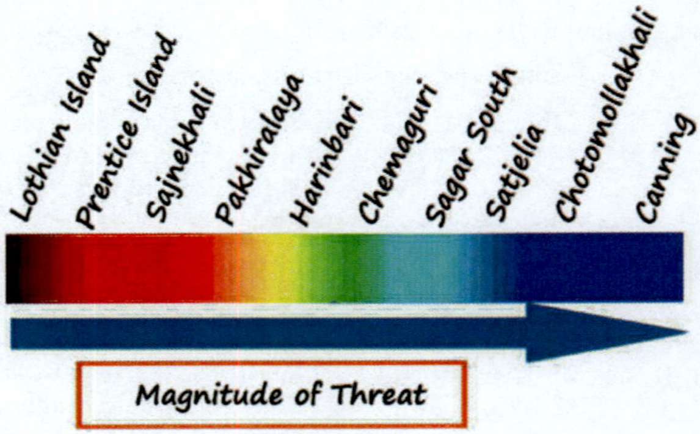


**Fig. 8.38** Scaling the spatial threat on mangroves; note the order of threat is Canning > Pakhiralaya > Chotomollakhali > Satjelia > Chemaguri > Harinbari > Sagar South > Prentice island > Sajnekhali > Lothian island

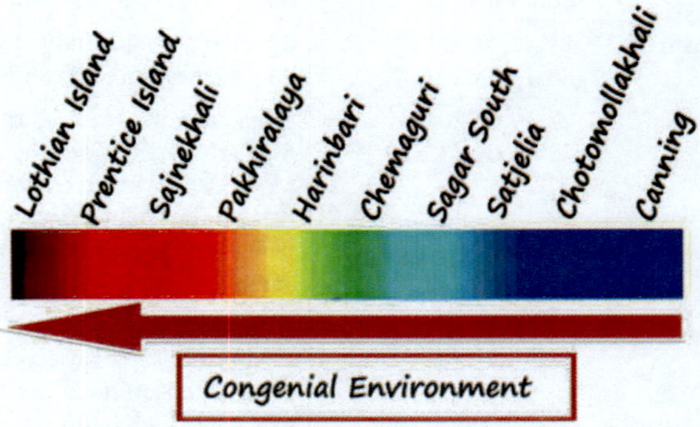


**Fig. 8.39** Scaling the health and water quality in Indian Sundarban mangroves; note the order is Lothian island > Prentice island > Sajnekhali > Pakhiralaya > Harinbari > Chemaguri > Sagar South > Satjelia > Chotomollakali > Canning

ocean water, reducing the pH an estimated 0.3 to 0.4 units by 2100, a 150% increase in acidity over preindustrial times (https://www.whoi.edu/know-your-ocean/ocean-topics/ocean-chemistry/ocean-acidification/). It's a fact that phytoplankton community is a biological pump and will drag more $CO_2$ in the aquatic system due to iron fertilization, but if it can be scientifically carried out in closed aquacultural ponds, then these tiny floral particles can be the primary source of food for cultured species (preferably filter feeders like oysters) and will serve dual benefits (i) local scale reduction of atmospheric $CO_2$ thereby reducing the dissolving potential of atmospheric $CO_2$ and (ii) production of edible finfish and shell fish for human consumption.

In this context it is interesting to note that an experiment was performed during 2011 in Indian Sundarbans in closed brackish water body where different doses of iron were mixed to trigger up the bloom of phytoplankton and prawn species *Macrobrachium rosenbergii* was introduced simultaneously (Fig. 8.40).

**Table 8.9** Biomass and Stored Carbon in *Suaeda maritima* and *Salicornia brachiata* sampled from Jharkhali region of Central Indian Sundarbans during 2018

| Species | Total Biomass ($kgm^{-2}$) | Total Carbon ($kgm^{-2}$) |
|---|---|---|
|  *Suaeda maritima* | 4.86 | 1.86 (38.2%)* |
|  *Salicornia brachiata* | 3.52 | 1.25 (35.6%)* |

**Note* The values within bracket represent stored carbon percentage of the biomass of the species

**Fig. 8.40** Phytoplankton bloom in the experimental pond after adding ferrous sulphate

**Fig. 8.41** Growth of *Macrobrachium rosenbergii* in the iron fertilized brackish water pond

Best growth of prawn (in terms of biomass) was observed after a culture period of 9 months in the culture pond when 1.5 ppm ferrous sulphate was applied in the aquatic phase (Fig. 8.41). In this experiment salinity was uniformly maintained (3.5–4.2 psu) during the experimental period.

In conclusion we strongly believe that retardation of the pace of EA requires micro-level strategies at local level as many of the estuaries flowing through different districts/states are inter-connected. These strategies need to be developed through an inter-departmental approach by involving all the relevant actors as stated in Table 8.10.

## 8.4 Take Home Messages

A. To retard the pace of rise of carbon dioxide it is extremely important to shift from traditional non-renewable energy (primarily coal-based) sources to renewable clean energy phase. The resources which are being continuously consumed by man but are renewed by nature constantly are called as renewable resources. These resources are inexhaustible because they cannot be exhausted permanently. Renewable resources are also called as 'Non-Conventional' sources of energy. Examples of renewable resources include solar energy, wind energy,

**Table 8.10** Ten micro-level approaches to control EA

| S. No. | Estuarine acidification (EA) management strategy | Institutional capacity at present | Financial investment required (INR) | Actors |
|---|---|---|---|---|
| 1 | Set up initiatives in each of the Indian coastal districts/states to assess the threat of EA to ecosystem health and services along with livelihoods of coastal people and island dwellers and develop strategies to mitigate local drivers | Partial | 20 Crores | State agencies, science, policy and capacity building NGOs, academic and research institutes, port authorities, coast guard, state environment department, state fisheries department and district administration |
| 2 | For regional EA issues, leverage existing inter-district/state coalitions or form new inter-state/district task forces to facilitate multi stakeholder solutions | Absent | 20 Crores | State agencies, science and policy NGOs, academic and research institutes, port authorities, coast guard, district administration |
| 3 | Continue developing a coordinated regional network of monitoring stations to map the vulnerability of coastal areas to EA, extend monitoring to near-shore systems relevant to management jurisdictions, and establish biological indicators as proxies for ecosystem health | Adequate | 10 Crores | State agencies, science and policy NGOs, academic and research institutes, remote sensing agencies |

(continued)

**Table 8.10** (continued)

| S. No. | Estuarine acidification (EA) management strategy | Institutional capacity at present | Financial investment required (INR) | Actors |
|---|---|---|---|---|
| 4 | Launch public education programs on EA and its causes to develop public awareness and stakeholder involvement in the process | Adequate | 2 Crores | State agencies, NGOs meant for awareness building and science education, resources persons from academic and research institutes |
| 5 | Coordinate domestic research and communication efforts with international programs to focus attention to EA, increase international collaboration, and standardize Inter-State monitoring Protocols | Partial | 5 Crores | State agencies, science and policy NGOs, academic and research institutes, port authorities, foreign high commissions |
| 6 | Integrate the threat of EA into existing and new climate change programs and state-level coastal programs under the ICZM umbrella | Mostly absent | – | State agencies, academic and research institutes, port authorities, coast guard, central and sate pollution control boards, ministry of environment & forests, ministry of earth science, ICZM authorities |
| 7 | Enforce existing regulatory measures to regulate water quality that affects enhanced EA | Absent | – | State agencies, juridical universities, advocacy or policy NGOs, central and sate pollution control boards, state police department |

(continued)

**Table 8.10** (continued)

| S. No. | Estuarine acidification (EA) management strategy | Institutional capacity at present | Financial investment required (INR) | Actors |
|---|---|---|---|---|
| 8 | Include the impact of permitting actions on EA when conducting environmental impact assessments under the National Environmental Policy Act or under state-level environmental policy acts. EMP should be strong enough to mitigate EA | Absent | – | State agencies, academic and research institutes, port authorities, coast guard, central and sate pollution control boards, ministry of environment & forests, ministry of earth science |
| 9 | Consider and incorporate information about coastal gene pool within EA hotspots into marine protected area planning processes | Partial | 20 Crores | State agencies, academic and research institutes, central and sate pollution control boards, ministry of environment & forests, ministry of earth science, central and state department of science and technology, apps developer |
| 10 | Consider implementing innovative low carbon emission innovative livelihood programmes for coastal population and island dwellers at pilot scale | Partial | | State agencies, academic and research institutes, NGOs involved in capacity building, central and sate pollution control boards, ministry of environment & forests, ministry of earth science, central and state department of science and technology, ministry of agriculture, food processing units, state fisheries department |

**Note* '–' means financial investment could not be estimated tidal energy, hydro power, geothermal energy and biofuels. The emission from renewable sources is practically zero/negligible. For islands and mangrove forest dominated ecosystems like Sundarbans, tidal energy is one of the best options for generation of electric current, provided adequate precautionary measures are taken during the period of cyclonic depressions.

B. Blue carbon refers to the coastal vegetation which includes mangroves, sea grass, salt marsh grass etc. Blue carbon has specific status in combating the process of acidification as they scrub dissolved form of carbon from the ambient aquatic phase. It is no exaggeration to state that increased level of carbon dioxide has shifted the pH of ocean, seas, bays and estuaries from the normal value ranging between 8.1–8.3 and 8.0–7.8. This shifting has posed several adverse impacts on marine and estuarine systems specially on the molluscan community with calcareous structures. The survival and growth of mangroves in the in the intertidal zone can shift the pH value of the ambient aquatic phase to its equilibrium by retarding the process of acidification. Basically mangroves absorb carbon dioxide from the atmosphere and the water column and store it as carbon in their vegetative and reproductive parts. The principal carbon-fixing enzyme in plants is ribulose 1, 5- biphosphate carboxylase/oxygenase (rubisco). The magnitude of the mangrove vegetation pool thus exerts a regulatory role on the pH of the estuarine water as these halophytes absorb carbon dioxide for photosynthesis and shifts the equilibrium towards alkalinity. It is therefore the need of the hour to expand the area of blue carbon and other mangrove associate species. Expansion of blue carbon can be achieved by several low cost technologies like (i) massive mangrove afforestation programme considering the salinity of the region (salinity-specific mangrove afforestation programme), (ii) halophyte farming and (iii) triggering phytoplankton growth coupled with aquaculture through artificial method like iron fertilization.

C. Estuarine acidification (EA), being a major stressor poses adverse impacts on the biotic community as well as on the ambient environmental quality. Thus both biotic and abiotic components are exposed to negative impacts of acidification. There are methods by which stressors can be reduced by employing resilience management to assist the ecological systems and dependent stakeholders, industrials and communities in resisting and recovery from EA. Very scanty works have been conducted on the specific biological attributes that can offer resilience against EA in the lower Gangetic delta region. Till date, there is limited understanding of the major drivers that impart estuarine ecosystem and its services more resilience to estuarine acidification.

D. EA can be minimized to a great extent by following steps like imposing restrictions on recreational activities, shrimp farming at the expense of blue carbon, industrial discharges, land based activities and emissions from the surroundings. Waste discharges from fishing vessels/trawlers and fish landing stations need to be minimized or properly treated before releasing them to ambient aquatic phase. Regular monitoring needs to be undertaken to manage risks from invasive species, ballast discharge etc. Temporary closure may be done to safeguard the vulnerable spots.

# Annexure 8: Salinity—Specific Mangrove Distribution in Indian Sundarbans

A detailed survey was carried out during 2017 on the ground-zero level in Indian Sundarbans covering eighteen sampling stations (Table 8.11) to assess the biomass and stored carbon in five dominant species namely *Sonneratia apetala, Avicennia marina, Avicennia officinalis, Avicennia alba* and *Excoecaria agallocha.*

We observed two interesting points in the output datasets (Tables 8.12, 8.13, 8.14, 8.15 and 8.16) in terms of preference of the mangrove floral species to environment. Firstly, *Sonneratia apetala* has high preference for low saline regions as witnessed in the western and eastern Indian Sundarbans. *Avicennia* spp. grows well in the high saline zone of central Indian Sundarbans. *Excoecaria agallocha* shows a wide range of salinity tolerance. Secondly, *Excoecaria agallocha* although has wide range of tolerance to salinity, but because of the poor biomass of the species, its role in carbon sequestration is not admirable. Giving weightage to mangrove biomass in regulating EA, it is therefore suggested to plant *Sonneratia apetala* in western and eastern sectors of Indian Sundarbans and *Avicennia* spp. in central Indian Sundarbans.

On the basis of these report cards, we developed a bird's eye view (Table 8.17) for each of the selected species in terms of their adaptations to different salinity using carbon sequestration as proxy. A proxy variable is an easily measurable variable that

**Table 8.11** Sampling stations in the western, central and eastern sectors of Indian Sundarbans

| Sectors | | Sampling stations | Longitude | Latitude |
|---|---|---|---|---|
| Western sector | Stn. 1 | Chemaguri ($W_1$) | 88°08′53.55″ E | 21°38′25.86″ N |
| | Stn. 2 | Saptamukhi ($W_2$) | 88°23′27.18″ E | 21°40′02.33″ N |
| | Stn. 3 | Jambu Island ($W_3$) | 88°10′22.76″ E | 21°35′42.03″ N |
| | Stn. 4 | Lothian ($W_4$) | 88°20′29.32″ E | 21°38′21.20″ N |
| | Stn. 5 | Harin Bari ($W_5$) | 88°04′32.97″ E | 21°44′22.55″ N |
| | Stn. 6 | Prentice Island ($W_6$) | 88°17′55.05″ E | 21°42′47.88″ N |
| Central sector | Stn. 7 | Thakuran Char ($C_1$) | 88°31′25.57″ E | 21°49′53.17″ N |
| | Stn. 8 | Dhulibasani ($C_2$) | 88°33′48.20″ E | 21°47′06.62″ N |
| | Stn. 9 | Chulkathi ($C_3$) | 88°34′10.31″ E | 21°41′53.62″ N |
| | Stn. 10 | Goashaba ($C_4$) | 88°46′41.44″ E | 21°43′50.64″ N |
| | Stn. 11 | Matla ($C_5$) | 88°44′08.74″ E | 21°53′15.30″ N |
| | Stn. 12 | Pirkhali ($C_6$) | 88°51′06.04″ E | 22°06′00.97″ N |
| Eastern sector | Stn. 13 | Arbesi ($E_1$) | 89°01′09.04″ E | 22°11′43.14″ N |
| | Stn. 14 | Jhilla ($E_2$) | 88°57′57.07″ E | 22°09′51.53″ N |
| | Stn. 15 | Harinbhanga ($E_3$) | 88°59′33.24″ E | 21°57′17.85″ N |
| | Stn. 16 | Khatuajhuri ($E_4$) | 89°01′05.33″ E | 22°03′06.55″ N |
| | Stn. 17 | Chamta ($E_5$) | 88°57′11.40″ E | 21°53′18.56″ N |
| | Stn. 18 | Chandkhali ($E_6$) | 89°00′44.68″ E | 21°51′13.59″ N |

**Table 8.12** Carbon sequestration by *Sonneratia apetala* inhabiting Indian Sundarbans

| Species | Stations | AGB (tonnes/ha) | | AGC (tonnes/ha) | | Carbon sequestration = ΔAGC/Δt (tonnes/ha/yr) |
|---|---|---|---|---|---|---|
| | | 1 Year | 11 Year | 1 Year | 11 Year | |
| *Sonneratia apetala* | Muriganga | 0.280 | 121.65 | 0.147 (52.5%) | 56.202 (46.2%) | 5.606 |
| | Saptamukhi | 0.292 | 123.06 | 0.154 (52.74%) | 57.223 (46.5%) | 5.707 |
| | Jambu Island | 0.142 | 104.59 | 0.075 (52.82%) | 48.216 (46.1%) | 4.814 |
| | Lothian Island | 0.275 | 142.33 | 0.145 (52.73%) | 65.472 (46%) | 6.533 |
| | Sagar Island | 0.267 | 139.88 | 0.140 (52.43%) | 64.764 (46.3%) | 6.462 |
| | PrenticeIsland | 0.271 | 140.63 | 0.143 (52.77%) | 66.096 (47%) | 6.595 |
| | Thakuran | 0.105 | 43.62 | 0.053 (50.48%) | 19.978 (45.8%) | 1.993 |
| | Dhulibasani | 0.112 | 48.71 | 0.057 (50.89%) | 22.455 (46.1%) | 2.240 |
| | Chulkathi | 0.110 | 45.50 | 0.056 (50.91%) | 20.885 (45.9%) | 2.083 |
| | Goashaba | 0.102 | 41.39 | 0.052 (50.98%) | 18.791 (45.4%) | 1.874 |
| | Matla | 0.236 | 68.75 | 0.120 (50.85%) | 31.281 (45.5%) | 3.116 |
| | Pirkhali | 0.114 | 50.27 | 0.058 (50.88%) | 23.124 (46%) | 2.307 |
| | Arbesi | 0.452 | 166.85 | 0.239 (52.88%) | 78.586 (47.1%) | 7.835 |
| | Jhila | 0.300 | 158.43 | 0.159 (53%) | 75.254 (47.5%) | 7.510 |
| | Harinbhanga | 0.296 | 155.29 | 0.157 (53.04%) | 73.452 (47.3%) | 7.330 |
| | Khatuajhuri | 0.289 | 152.71 | 0.153 (52.94%) | 72.690 (47.6%) | 7.254 |
| | Chamta | 0.300 | 160.05 | 0.159 (53.00%) | 76.024 (47.5%) | 7.587 |
| | Chandkhali | 0.281 | 149.99 | 0.149 (53.02%) | 70.645 (47.1%) | 7.050 |

**Table 8.13** Carbon sequestration by *Avicennia marina* inhabiting Indian Sundarbans

| Species | Stations | AGB (tonnes/ha) | | AGC (tonnes/ha) | | Carbon sequestration = ΔAGC/Δt (tonnes/ha/yr) |
|---|---|---|---|---|---|---|
| | | 1 Year | 11 Year | 1 Year | 11 Year | |
| *Avicennia marina* | Muriganga | 0.151 | 39.863 | 0.077 (51.0%) | 18.018 (45.2%) | 1.794 |
| | Saptamukhi | 0.159 | 42.334 | 0.080 (50.3%) | 19.347 (45.7%) | 1.927 |
| | Jambu Island | 0.313 | 45.781 | 0.159 (50.8%) | 21.013 (45.9%) | 2.085 |
| | Lothian Island | 0.313 | 44.663 | 0.159 (50.8%) | 20.366 (45.6%) | 2.021 |
| | Sagar Island | 0.309 | 43.284 | 0.157 (50.8%) | 19.651 (45.4%) | 1.949 |
| | PrenticeIsland | 0.155 | 45.004 | 0.079 (51%) | 20.567 (45.7%) | 2.049 |
| | Thakuran | 0.533 | 66.281 | 0.279 (52.3%) | 31.749 (47.9%) | 3.147 |
| | Dhulibasani | 0.357 | 64.385 | 0.187 (52.4%) | 30.969 (48.1%) | 3.078 |
| | Chulkathi | 0.358 | 64.224 | 0.187 (52.2%) | 30.699 (47.8%) | 3.051 |
| | Goashaba | 0.366 | 66.883 | 0.191 (52.2%) | 32.104 (48%) | 3.191 |
| | Matla | 0.363 | 67.991 | 0.190 (52.3%) | 32.568 (47.9%) | 3.238 |
| | Pirkhali | 0.555 | 70.224 | 0.290 (52.3%) | 33.778 (48.1%) | 3.349 |
| | Arbesi | 0.157 | 39.112 | 0.078 (49.7%) | 17.718 (45.3%) | 1.764 |
| | Jhila | 0.152 | 35.668 | 0.076 (50%) | 16.122 (45.2%) | 1.605 |
| | Harinbhanga | 0.301 | 37.885 | 0.151 (50.2%) | 17.427 (46%) | 1.728 |
| | Khatuajhuri | 0.151 | 38.447 | 0.075 (49.7%) | 17.609 (45.8%) | 1.753 |
| | Chamta | 0.148 | 40.299 | 0.074 (50%) | 18.336 (45.5%) | 1.826 |
| | Chandkhali | 0.142 | 41.326 | 0.071 (50%) | 18.886 (45.7%) | 1.882 |

**Table 8.14** Carbon sequestration by *Avicennia alba* inhabiting Indian Sundarbans

| Species | Stations | AGB (tonnes/ha) | | AGC (tonnes/ha) | | Carbon sequestration = ΔAGC/Δt (tonnes/ha/yr) |
|---|---|---|---|---|---|---|
| | | 1 Year | 11 Year | 1 Year | 11 Year | |
| *Avicennia alba* | Muriganga | 0.415 | 41.224 | 0.212 (51.0%) | 18.716 (45.4%) | 1.850 |
| | Saptamukhi | 0.270 | 43.081 | 0.138 (51.1%) | 19.731 (45.8%) | 1.959 |
| | Jambu Island | 0.442 | 47.112 | 0.226 (51.1%) | 21.719 (46.1%) | 2.149 |
| | Lothian Island | 0.437 | 45.998 | 0.223 (51.0%) | 21.067 (45.8%) | 2.084 |
| | Sagar Island | 0.287 | 45.110 | 0.146 (50.8%) | 20.705 (45.9%) | 2.056 |
| | PrenticeIsland | 0.288 | 47.117 | 0.147 (51.0%) | 21.580 (45.8%) | 2.143 |
| | Thakuran | 0.326 | 68.130 | 0.170 (52.1%) | 32.771 (48.1%) | 3.260 |
| | Dhulibasani | 0.483 | 70.223 | 0.252 (52.1%) | 33.918 (48.3%) | 3.367 |
| | Chulkathi | 0.652 | 68.441 | 0.340 (52.1%) | 32.852 (48.0%) | 3.251 |
| | Goashaba | 0.653 | 67.965 | 0.341 (52.2%) | 33.099 (48.7%) | 3.276 |
| | Matla | 0.665 | 70.416 | 0.347 (52.2%) | 34.011 (48.3%) | 3.366 |
| | Pirkhali | 0.493 | 73.103 | 0.258 (52.3%) | 35.747 (48.9%) | 3.549 |
| | Arbesi | 0.264 | 43.269 | 0.132 (50.0%) | 19.817 (45.8%) | 1.969 |
| | Jhila | 0.263 | 39.808 | 0.132 (50.1%) | 18.192 (45.7%) | 1.806 |
| | Harinbhanga | 0.134 | 38.615 | 0.067 (50.0%) | 17.917 (46.4%) | 1.785 |
| | Khatuajhuri | 0.397 | 40.109 | 0.199 (50.1%) | 18.450 (46.0%) | 1.825 |
| | Chamta | 0.136 | 41.967 | 0.068 (50.0%) | 19.263 (45.9%) | 1.920 |
| | Chandkhali | 0.277 | 41.326 | 0.139 (50.2%) | 18.969 (45.9%) | 1.883 |

**Table 8.15** Carbon sequestration by *Avicennia officinalis* inhabiting Indian Sundarbans

| Species | Stations | AGB (tonnes/ha) | | AGC (tonnes/ha) | | Carbon sequestration = ΔAGC/Δt (tonnes/ha/yr) |
|---|---|---|---|---|---|---|
| | | 1 Year | 11 Year | 1 Year | 11 Year | |
| *Avicennia officinalis* | Muriganga | 0.428 | 42.008 | 0.220 (51.4%) | 19.786 (47.1%) | 1.957 |
| | Saptamukhi | 0.574 | 44.116 | 0.295 (51.3%) | 20.690 (46.8%) | 2.040 |
| | Jambu Island | 0.420 | 49.004 | 0.215 (51.2%) | 23.032 (47.0%) | 2.282 |
| | Lothian Island | 0.447 | 46.137 | 0.230 (51.4%) | 21.177 (45.9%) | 2.095 |
| | Sagar Island | 0.451 | 46.220 | 0.231 (51.2%) | 21.631 (46.8%) | 2.140 |
| | PrenticeIsland | 0.415 | 48.217 | 0.213 (51.3%) | 22.517 (46.7%) | 2.230 |
| | Thakuran | 0.970 | 68.995 | 0.509 (52.4%) | 31.600 (45.8%) | 3.109 |
| | Dhulibasani | 0.649 | 75.116 | 0.340 (52.4%) | 36.131 (48.1%) | 3.579 |
| | Chulkathi | 0.505 | 72.553 | 0.265 (52.5%) | 34.680 (47.8%) | 3.442 |
| | Goashaba | 0.672 | 70.134 | 0.352 (52.3%) | 33.664 (48.0%) | 3.331 |
| | Matla | 0.859 | 73.228 | 0.450 (52.4%) | 34.857 (47.6%) | 3.441 |
| | Pirkhali | 0.683 | 75.004 | 0.358 (52.4%) | 35.102 (46.8%) | 3.474 |
| | Arbesi | 0.272 | 42.819 | 0.137 (50.4%) | 20.467 (47.8%) | 2.033 |
| | Jhila | 0.283 | 36.208 | 0.143 (50.5%) | 17.380 (48.0%) | 1.724 |
| | Harinbhanga | 0.143 | 37.191 | 0.072 (50.3%) | 17.666 (47.5%) | 1.759 |
| | Khatuajhuri | 0.297 | 40.120 | 0.150 (50.5%) | 18.816 (46.9%) | 1.867 |
| | Chamta | 0.140 | 38.178 | 0.070 (50.0%) | 17.791 (46.6%) | 1.772 |
| | Chandkhali | 0.283 | 41.227 | 0.142 (50.2%) | 19.418 (47.1%) | 1.928 |

**Table 8.16** Carbon sequestration by *Excoecaria agallocha* inhabiting Indian Sundarbans

| Species | Stations | AGB (tonnes/ha) | | AGC (tonnes/ha) | | Carbon sequestration = ΔAGC/Δt (tonnes/ha/yr) |
|---|---|---|---|---|---|---|
| | | 1 Year | 11 Year | 1 Year | 11 Year | |
| *Excoecaria agallocha* | Muriganga | 0.256 | 38.112 | 0.132 (51.6%) | 18.103 (47.5%) | 1.797 |
| | Saptamukhi | 0.131 | 40.667 | 0.068 (51.9%) | 19.154 (47.1%) | 1.909 |
| | Jambu Island | 0.252 | 42.814 | 0.130 (51.5%) | 19.823 (46.3%) | 1.969 |
| | Lothian Island | 0.261 | 43.559 | 0.135 (51.7%) | 19.950 (45.8%) | 1.982 |
| | Sagar Island | 0.129 | 40.664 | 0.066 (51.1%) | 18.421 (45.3%) | 1.836 |
| | PrenticeIsland | 0.127 | 42.917 | 0.066 (51.9%) | 19.742 (46.0%) | 1.968 |
| | Thakuran | 0.273 | 62.35 | 0.142 (52.0%) | 28.307 (45.4%) | 2.817 |
| | Dhulibasani | 0.260 | 59.581 | 0.135 (51.9%) | 28.003 (47.0%) | 2.787 |
| | Chulkathi | 0.123 | 58.017 | 0.064 (52.0%) | 27.442 (47.3%) | 2.738 |
| | Goashaba | 0.256 | 62.206 | 0.133 (51.9%) | 29.610 (47.6%) | 2.948 |
| | Matla | 0.122 | 63.459 | 0.063 (51.6%) | 29.826 (47.0%) | 2.976 |
| | Pirkhali | 0.278 | 65.883 | 0.145 (52.1%) | 30.636 (46.5%) | 3.049 |
| | Arbesi | 0.131 | 36.497 | 0.066 (50.3%) | 17.081 (46.8%) | 1.702 |
| | Jhila | 0.123 | 33.22 | 0.063 (51.2%) | 15.713 (47.3%) | 1.565 |
| | Harinbhanga | 0.130 | 34.426 | 0.066 (50.7%) | 16.215 (47.1%) | 1.615 |
| | Khatuajhuri | 0.125 | 36.108 | 0.063 (54.4%) | 16.790 (46.4%) | 1.673 |
| | Chamta | 0.126 | 37.515 | 0.064 (50.7%) | 17.294 (46.0%) | 1.723 |
| | Chandkhali | 0.259 | 38.777 | 0.131 (50.5%) | 18.186 (46.8%) | 1.806 |

**Table 8.17** Species-wise carbon sequestration potential of mangroves inhabiting three different sectors of Indian Sundarbans with variable salinity

| Species | Western Indian Sundarbans | Central Indian Sundarbans | Eastern Indian Sundarbans |
|---|---|---|---|
| *Sonneratia apetala* | [dark red] | [yellow] | [red] |
| *Avicennia marina* | [yellow] | [orange] | [yellow] |
| *Avicennia alba* | [yellow] | [orange] | [yellow] |
| *Avicennia officinalis* | [yellow] | [orange] | [yellow] |
| *Excoecaria agallocha* | [yellow] | [yellow] | [yellow] |

**Index (based on carbon sequestration values stated in tables 8.11 to 8.16)**

**1.000- 2.900-** [yellow] **Not recommended for plantation**

**3.000- 4.900-** [orange] **Recommended for plantation**

**5.000- 6.900-** [dark red] **Strongly recommended for plantation**

**7.000- 8.900-** [red] **Most strongly recommended for plantation**

is used in place of a variable that cannot be measured or is difficult to measure. The proxy variable can be something that is not of any great interest itself, but has a close correlation with the variable of interest (here the adaptation of mangroves to different salinity gradients). Our approach to expand mangrove vegetation in different sectors of lower Gangetic delta as a function of salinity (Table 8.17) may be a viable road map for mitigation of carbon dioxide rise at local level.

## References


Adams JM, Faure H, Faure-Denard L, McGlade JM, Woodward FI (1990) Increases in terrestrial carbon storage from the last glacial maximum to the present. Nature 348:711–714

Ajtay GL, Ketner P, Duvigneaud P (1979) Terrestrial primary production and phytomass. In: In Bolin B, Degens ET, pp 129–182

Alexander HD, Dunton KH (2006) Treated wastewater effluent as an alternative fresh water source in a hyper saline salt marsh: impacts on salinity, inorganic nitrogen, and emergent vegetation. J Coast Res 22:377–392

Bazilevich NI, Rodin LY, Rozov NN (1971) Geographical aspects of biological productivity. Sov Geogr Rev Transl 12:293–317; and Pap. Vth Congr. USSR Geographical Society, Leningrad, 1970

Bazilevich NI (1974) Energy flow and biological regularities of the world ecosystems. In: Cavé

Bohn HL (1976) Estimate of organic carbon in world soils. Soil Sci Soc Am J 40:468–470

Bohn HL (1978) Organic soil carbon and $CO_2$. Tellus 30:472–475

Bohn HL (1982) Organic carbon in world soils. Soil Sci of Am 46:1118–1119

Bolin B, Degens ET, Duvigneaud P, Kempe S (1979) The global biogeochemical carbon cycle. In: Bolin B, Degens ET, Kempe S, Kenter P (eds) The global carbon cycle, 1st edn. Wiley, Chichester, pp 1–56

Buringh P (1983) Organic carbon in soils of the world

Conkle JL, White JR, Metcalfe C (2008) Removal of pharmaceutically active compounds by a lagoon- wetland waste water treatment system in Southeast Louisiana. Chemosphere 73:1741–1748

Day JW, Ko JK, Rybczyk J, Sabins D, Bean R, Berthelot G, Brantley C, Cardoch L, Conner W, Day JN, Englande AJ, Feagley S, Hyfield E, Lane R, Lindsey J, Mitsch J, Reyes E, Twilley R (2004) The use of wetlands in the Mississippi delta for waste water assimilation: a review. Ocean Coast Manage 47:671–697

Dixon RK, Krankina ON (1993) Forest fires in Russia: carbon dioxide emissions to the atmosphere. Can J Forest Res 23:700–705

Eswaran H, Den Berg EV, Reich P (1993) Organic carbon in soils of the world. Soil Sci Soc Am J 57:192–194

Gorham E (1990) Biotic impoverishment in northern peatlands. In: Woodwell GM (ed) The earth in transition: patterns and processes of biotic impoverishment. Cambridge University, New York, NY, pp 65–98

Gorham E, Janssens JA (1992) Concepts of fen and bog reexamined in relation to bryophyte cover and the acidity of surface waters. Acta Societalis Botanicorum Poloniae 61:7–20

Harmon ME, Ferrell WK, Franklin JF (1990) Effects on carbon storage conversion of old-growth forests to young forests. Sci 247:699–702

IPCC-Working Group II (1990) Chapter 5: world ocean and coastal zones

Kadlec RH, Wallace SD (2009) Treatment wetlands, 2nd edn. CRC Press, Boca Raton, FL, p 106

Markov VD, Olunin AS, Ospennikova LA, Skobeeva EI, Khoroshev PI (1988) World peat resources. Moscow 'Nedra': 383 (In Russian)

Mitra A (2013) In: Sensitivity of Mangrove ecosystem to changing climate by Dr. Abhijit Mitra. Publisher Springer New Delhi Heidelberg New York Dordrecht London, edition (August 31, 2013); ISBN-10:8132215087; ISBN-13:978-8132215080., copyright Springer, India 2013; ISBN 978-81-322-1509-7 (eBook)

Mitra A, Zaman S (2016) Basics of marine and estuarine ecology; publisher, Springer India, Book, ISBN 978-81-322-2707-6; pp 1–481

Mitra A, Zaman S (2020) Environmental science—a ground zero observation on the Indian subcontinent, Published by Springer Nature Switzerland AG ISBN 978-3-030-49130-7 ISBN 978-3-030-49131-4. eBook https://doi.org/10.1007/978-3-030-49131-4

Olson JS, Watts JA, Allison LJ (1983) Carbon in live vegetation of major world ecosystems. Oak ridge national laboratory, Oak Ridge, ORNL-5862

Post WM, Emanuel WR, Zinke PJ, Stangenberger AG (1982) Soil carbon pools and world life zones. Nature 298:156–159

Schlesinger WH (1984) Soil organic matter: a source of atmospheric $CO_2$. In: Woodwell GM (ed) The role of terrestrial vegetation in the global carbon cycle: measurement by remote sensing. Wiley, New York, pp 111–127 (e-publishing)

Sjors H (1980) Peat on earth: multiple uses or conservation? Ambio 9:303–308

Vymazal J (2010) Constructed wetlands for wastewater treatment. Water 2:530–549

Whittaker RH, Likens GE (1975) The biosphere and man; In: Lieth H, Whittaker RH (eds) Primary productivity of the biosphere, vol 14. Springer, Berlin. E. col Study, pp 305–328

Zhang X (1995) Use of a natural swamp for wastewater treatment. Ph.D. dissertation, department of agronomy, Louisiana State University, Baton Rouge, LA

## *Internet References*

https://www.census.gov/newsroom/stories/world-population-day.html
https://www.projectstoday.com/news/west-bengal-to-set-up-wind-power-station
https://www.whoi.edu/know-your-ocean/ocean-topics/ocean-chemistry/ocean-acidification/